Springer Series in
Surface Sciences

9

Editor: Robert Gomer

Springer Series in **Surface Sciences**

Editors: G. Ertl and R. Gomer   Managing Editor: H. K. V. Lotsch

V. F. Kiselev   O. V. Krylov

Adsorption and Catalysis on Transition Metals and Their Oxides

With 105 Figures

Springer-Verlag Berlin Heidelberg New York
London Paris Tokyo

Professor Dr. Vsevolod F. Kiselev

Physics Department, Lomonosov Moscow State University,
Moscow 117234, USSR

Professor Dr. Oleg V. Krylov

Institute of Chemical Physics, Academy of Sciences,
Kosygina St. 4, Moscow 117334, USSR

Series Editors

Professor Dr. Gerhard Ertl

Fritz-Haber-Institut der Max-Planck-Gesellschaft, Faradayweg 4–6
D-1000 Berlin 33

Professor Robert Gomer, Ph. D.

The James Franck Institute, The University of Chicago, 5640 Ellis Avenue,
Chicago, IL 60637, USA

Managing Editor: Dr. Helmut K. V. Lotsch

Springer-Verlag, Tiergartenstrasse 17
D-6900 Heidelberg, Fed. Rep. of Germany

ISBN-13:978-3-642-73889-0   e-ISBN-13:978-3-642-73887-6
DOI: 10.1007/978-3-642-73887-6

Library of Congress Cataloging-in-Publication Data. Kiselev, V.F. (Vsevolod Fedorovich) Adsorption an catalysis on transition metals and their oxides. (Springer series in surface science ; 9) Completely rev. ed. of: Adsorbtsiia i kataliz na perekhodnykh metallakh i ikh oksidakh / O.V. Krylov, V.F. Kiselev. 1981. 1. Adsorption. 2. Catalysis. 3. Transition metal catalysis. 4. Metallic oxides. I. Krylov, O.V. (Oleg Valentinovich) II. Krylov, O.V. (Oleg Valentinovich). Adsorbtsiia i kataliz na perekhodnykh metallakh i ikh oksidakh. III. Title. IV. Series. QD547.K518 1988 546'.6 88-24858

The text was word-processed using PS™ software and was printed with a Toshiba P321

2154/3150-543210 – Printed on acid-free paper

Preface

This book deals with adsorption and catalysis on the surface of transition elements and their compounds, many of which are interesting because of their particular electronic structure. The authors have worked through a vast body of experimental evidence on the structure and properties of surfaces of transition metals and relevant oxides. Consideration is given mostly to simple (as opposed to mixed) oxides of transition elements, to common metals and to the adsorption of simple gases. A great deal of attention is paid to the nature of active surface sites responsible for chemisorption and catalytic transformations. The description relies mainly on the simplified ligand—field theory, which, however, proves quite satisfactory for predicting the adsorptive and catalytic activity of species. In many cases simple systems were explored with the aid of novel techniques, and it is only for such systems that the mechanism of the elementary act of adsorption and catalysis can be given adequate treatment.

The present monograph has emerged from our earlier work in Russian, which appeared in the Khimiya Publishing House (Moscow) in 1981. This English edition has, however, been revised completely to broaden its scope and to include more recent achievements.

For fruitful discussions the authors are grateful to A.A. Kadushin, Z.L. Krylova, L.Ya. Margolis, O.V. Nikitina, K.N. Spiridonov, I.I. Tretyakov, and G.M. Zhidomirov. They acknowledge the cooperation of Mr. A.S. Dobroslavski in the English translation of the manuscript, and appreciate the assistance of the copy editor Dr. T.P. Culviner.

Moscow, June 1987   O.V.Krylov   V.F. Kiselev

Contents

1. Introduction

The majority of industrial chemical products are produced with the aid of catalysis. Despite the development of novel means of activation (radiochemistry, plasma chemistry, laser chemistry, etc.), the importance of catalysis in its industrial applications steadily grows on not only an absolute production scale, but in comparison with alternative techniques as well. According to some estimates, in monetary terms industrial catalysis products are worth 18% of the world's total industrial output, second only to the machine-building industry.

Out of the great variety of industrial catalysts the prominent place belongs to transition elements. Whether used in the form of homogeneous complexes, or isolated ions in a solid matrix, or in the form of metals (where all the atoms of the transition element are bound into a single chain), the transition elements usually outdo the nontransition ones in their adsorptive and catalytic properties. The peculiar features of transition elements (in particular, the existence of d electrons), ought to be taken into account by any sensible theory of adsorption and catalysis.

So far there exists, alas, no reliable rule for choosing an appropriate catalyst, so this task must be approached in a trial-and-error way. It must be said, however, that the situation with the theory of catalysis is only little worse than with the theory of reactivity in general: the latter also fails to pave the way for quantitative assessment of a given compound in advance, relying instead (as does the theory of catalysis) on

1

broadly understood analogies and correlations. At the same time considerable progress has been achieved in our understanding of adsorption and catalysis. Advances in quantum chemistry made is possible to compute the configuration of orbitals, to calculate the relevant energy levels, and to analyze the paths of energy exchange in the catalytic act. The understanding of these phenomena benefited greatly from rapid development of novel research techniques in surface sciences. Many important break-throughs were made in the 1970s, when UV and IR optical spectroscopies, as well as EPR, NMR and Mössbauer spectroscopy found their way into all leading centers of research. Great progress was made in X-ray and UV photoelectron spectroscopy, and in the low-energy electron diffraction techniques. Recent innovations are so numerous that it would be futile even to attempt to list them all. Accordingly, we shall restrict ourselves only to those of them which led to the most fundamental discoveries of the past decade, as our primary purpose is to discuss the phenomena of adsorption and catalysis from the viewpoint of the latest developments in this field.

The past decade has also seen rapid development of the quantum chemistry of surface phenomena. Detailed structures of molecular orbitals have been calculated for simpler complexes of transition elements, formed on the surface of metals and oxides. Even the simplified crystal-field and ligand-field theories have been fruitfully employed for interpreting the chemical bonding in surface compounds. In many cases these simple and clear-cut models have made it possible to explain and predict the changes in adsorptive and catalytic activity in terms of the number of d electrons - i.e., as a function of the element's location in the periodic table. Such models are extensively used in the present monograph.

In a sense this book is a sequel to our previous monographs [1.1,2], where we analyzed the most recent experimental data and offered theoretical explanations for some fundamental surface phenomena in adsorption and catalysis with examples of simple adsorbents: elementary semiconductors (silicon and germanium), aluminum and silicon oxides, and some others. Incidentally, such systems are sometimes of greater importance for microelectronics than they are for catalysis.

Here we consider the electronic properties of transition elements and their oxides, which often are good as catalysts. Recent techniques have mostly been employed in studying the simplest systems - eighth-group transition metals and their simple (nonmixed) oxides. This explains our emphasis on simple systems, while the more complex systems are dealt with only occasionally. As a rule, the results for the latter arise from the authors' research.

In Chap.2 we introduce the ligand-field theory which underlies the current conceptions of the structure of transition-element compounds. We also give a brief account of the modern techniques for investigating the events taking place on the surfaces of catalysts and adsorbents.

Chapter 3 deals with the electronic structure of transition atoms on oxide surfaces, which are viewed as diluted systems where the interaction between transition metal atoms is weakened. We also discuss the surface structure of the principal types of oxide systems, paying special attention to those which are important catalysts in industry.

In Chap.4 we demonstrate how the crystal-field and ligand-field theories can be applied to the processes of adsorption and catalysis. Some promising directions of future research are identified, e.g., the application of orbital symmetry conservation rules to adsorption and catalysis. We quote from experimental data obtained with modern techniques for adsorption of simple gases on oxides. The emphasis here is on systems studied by the authors and their immediate colleagues. This chapter ends with a brief discussion of catalysis in simple systems.

Surface structure of transition metals is dealt with in Chap.5. We discuss the effects arising from the overlapping of d orbitals of adjacent atoms and proceed to consider the surface structure of pure metals, as well as that of supported metals and alloys, on the basis of most recent experimental findings.

Chapter 6 is dedicated to adsorption and catalysis by transition metals. The discussion is based on recent results about the geometric and electronic structures of adsorbed layers of simple gases on the surfaces of metallic single crystals. We also consider some effects due to the use of deposited and alloyed transition metals.

Note that especially rapid progress has lately been associated with adsorption of simple gases on the surfaces of metallic single crystals. It is just impossible to embrace all these results, and so we had to restrict ourselves to considering only some of them, related directly to chemical bonding of adsorbates on the surface.

The conclusion summarizes our knowledge on the high adsorptive and catalytic activity of transition metals. Since the most sound conclusions relate more to the surface structure and adsorption than to catalysis, here we outline the main directions of future catalytic research. Full understanding of catalytic mechanisms requires information on the structure and the energy levels of surface compounds, obtainable now directly in the course of a catalyzing reaction. In particular, time-domain techniques can and must be used for measuring the elementary reaction rate constants. Considerable progress in this field ought to be expected during the next few years.

2. The Electronic Structure of Transition-Metal Atoms

The active centers for adsorption and catalysis on transition elements and their compounds are coordination–unsaturated surface atoms with underpopulated d orbitals. The individual properties of an active center turn out to be more important for adsorption and catalysis than are the collective electronic properties of the solid lattice. Therefore, the main research on adsorption and catalysis has shifted to a study of the chemistry of complex compounds.

The chemistry of complex compounds owes its success to the development of crystal-field theory and ligand-field theory. They reveal the structure of complex compounds, the configuration of energy levels (molecular orbitals), and the way in which the latter are occupied by electrons. The establishment of a chemical bond between transition-metal atoms and other atoms both on the surface and in the bulk of a solid exhibits the same regularities observed in complex compounds. Consequently, crystal-field theory and ligand-field theory were successfully used for explaining the surface properties of adsorbents and catalysts, as well as adsorption and catalysis in general.

2.1 The Principles of the Crystal-Field and Ligand-Field Theories

The outermost occupied electron levels of transition–metal atoms are the d levels (d orbitals). The specific properties of

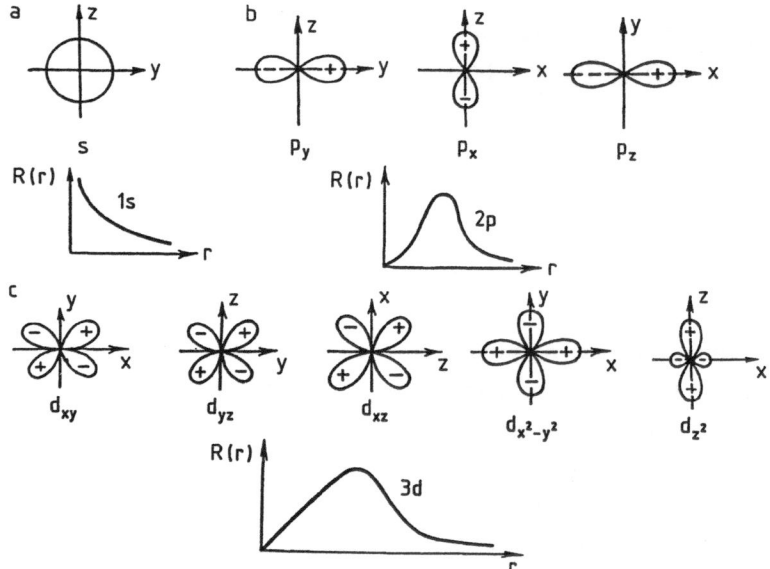

Fig.2.1. Angular (s, p_x, p_y, p_z, d_{xy}, d_{yz}, d_{xz}, $d_{x^2-y^2}$, d_{z^2}) and radial (R(r)) distribution of atomic wave functions: (a) s electrons; (b) p electrons; (c) d electrons

the angular and the radial distribution of the wave functions of d orbitals, which distinguish them from s and p orbitals (Fig.2.1), account for the peculiar chemical properties of transition-metal atoms both on the surface and in the bulk.

The field around the isolated transition-metal atom is spherically symmetric. It incorporates five d orbitals with equal energy (quantum numbers $\ell=2$; $m=0,\pm1,\pm2$). For an isolated ion their angular distribution can be depicted in a rather arbitrary way, since it shows up only after degeneracy has been removed by an external agent. In view of this, linear combinations of wave functions of the five d orbitals are usually chosen as follows (Fig.2.1). Should the orbitals d_{z^2} and $d_{x^2-y^2}$ exhibit the maximum electron density (i.e., squared modulus of the wave function) along the coordinate axes x, y, and z, aligned with the octahedron axes, then the orbitals d_{xy}, d_{yz}, and d_{xz} will be directed along the bisectors of angles between the coordinates axes. The orbitals d_{xy}, d_{yz}, and d_{xz} are sometimes called the t_{2g} orbitals (or d_ϵ orbitals), while d_{z^2} and $d_{x^2-y^2}$ are known as e_g orbitals (or d_γ orbitals).[1]

The radial distribution $R(r)$ of the transition-metal wave function (so far as d orbitals are concerned), as seen from Fig.2.1, differs from the radial distribution of s and p orbitals in that it is more diffuse and spatially delocalized as compared with compact s and p orbitals. This is true at least for 3d orbitals and 3p and 3s orbitals.

An external field may remove the degeneracy of the d orbitals. The effects of symmetry of the external field and its strength on the degeneracy of d orbitals are usually considered in the approximation of crystal-field theory, originally proposed by BETHE [2.3]. This theory considers the Stark splitting of d or f electron levels of the central atoms in the ligand fields. A ligand is a particle (either an atom, or an anion, or a molecule), found in the first coordination sphere that is in the immediate proximity to the central metallic ion and interacting with the latter. The central atom and the ligand form a complex. Ligands are considered as point sources of the external electrostatic field, disregarding the structure of their electron orbitals. Possible formation of a covalent bond is completely neglected. Further development of the crystal-field theory [2.4-9] explained the chemical and magnetic properties of ions in crystals, and optical and EPR spectra, as well as some other features of transition-metal compounds.

The nature of splitting of the initial d level and the occupancy of the d orbitals by electrons depend on the symmetry of the complex and the properties of the ligands. If the transition-metal atom falls under the action of a truly spherical field created by a negative charge, its d level is raised by the value E_0 with respect to that of an isolated ion (Fig.2.2) owing to the action of repulsive forces. The field in a real complex or in a solid is always less than spherical. The splitting of d

---

[1] The meaning of this literal notation, borrowed from group theory is explained in [2.3-6].

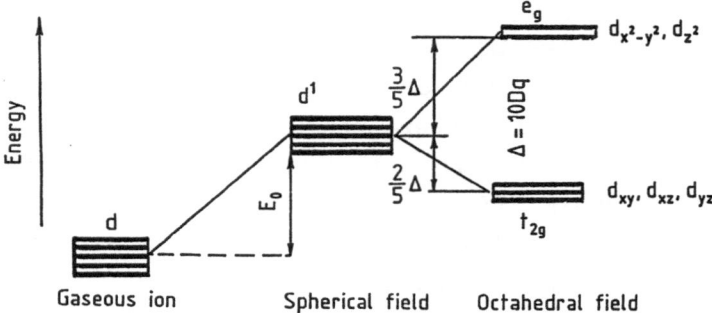

Fig.2.2. Splitting of the d level of isolated ion in octahedral field

orbitals occurs in such a way that their mean energy is still equal to E_0 of the spherical field. If the central ion is surrounded by six ligands along the octahedron's axes, their electrons are repelled from the electrons of the e_g orbitals (d_{z^2} and $d_{x^2-y^2}$). The action of repulsive forces elevates the energy of these orbitals with respect to the energy of the initial fivefold degenerate level; at the same time the t_{2g} orbitals (d_{xy}, d_{xz}, d_{yz}) become more energetically advantageous. The initial d level is thus split into two groups of sublevels (Fig.2.2): the lower group consisting of triply degenerate t_{2g} orbitals, and the upper of doubly degenerate e_g orbitals. The gap between the t_{2g} and e_g orbitals is denoted by Δ or 10Dq and is called the parameter of crystal-field splitting. The mean energy E_0 of the d electrons is conserved, so the upper two orbitals lie $(3/5)\Delta$ =6Dq above the initial (unsplit) d level, while the lower three orbitals lie $(2/5)\Delta$ =4Dq below the initial d level E_0.

If the central ion is located in a tetrahedral surrounding, its t_{2g} orbitals come closer to the ligands than do its e_g orbitals. Consequently, the three t_2 orbitals lie $(2/5)\Delta$ above the initial d level (corresponding to the spherical field), and the two e levels are $(3/5)\Delta$ below the initial one (Fig.2.3). The gap Δ_t now is smaller than the gap Δ_0 in the case of octahedral surrounding by the same ligands:

8

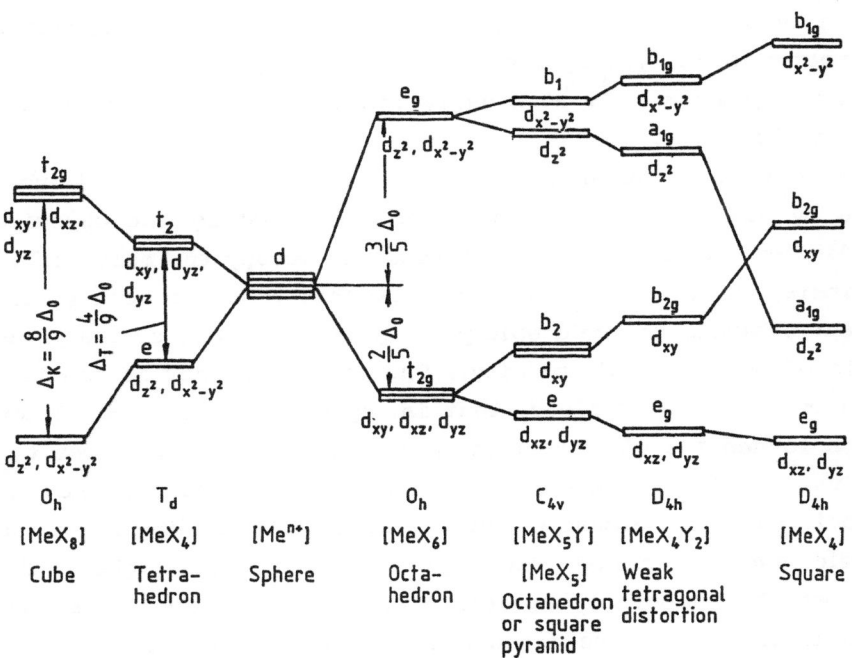

Fig.2.3. Splitting of d level in fields of different symmetry

$$\Delta_t = (4/9)\Delta_0 \ . \tag{2.1}$$

The energy E_0 of the initial d' level (Fig.2.2) is also lower than in the case of an octahedral surrounding. For the case of a cubic surrounding it can be shown [2.5,9] that

$$\Delta_c = (8/9)\Delta_0 \ . \tag{2.2}$$

Figure 2.3 shows the changing pattern due to splitting as the symmetry of the complex is further reduced, including complexes with ligands of different kinds. For instance, in a square pyramid (C_{4v} symmetry group) the energies of the two e_g orbitals will no longer be equal: the electron in the $d_{x^2-y^2}$ orbital is repelled by the ligands stronger than the electron in the d_{z^2} orbital, so the energy of the $d_{x^2-y^2}$ orbital is somewhat increased, whereas the energy of the d_{z^2} orbital is lowered. The t_{2g} level is also split into two: b_2 orbital (d_{xy})

9

and e orbital (d_{xz}, d_{yz}). In tetragonal and square complexes (symmetry group D_{4h}) splitting goes further, and further reduction of symmetry will eventually remove the degeneracy of all d orbitals.

Consider now the effects of sequential occupancy of d orbitals by electrons. This process is dominated by two opposite tendencies: on the one hand, it tends to occupy the more stable orbitals, on the other hand, the electron tends to occupy different orbitals to avoid spin pairing (Hund's rule, or the principle of maximum multiplicity). The first tendency is conditioned by the crystal field, that is, by electrostatic repulsion which forces the electrons out (for instance, to the lower t_{2g} orbital in the octrahedral complex). The second tendency is associated with exchange interactions: the stability of an atomic system against exchange interaction is higher, the greater the number of unpaired electrons with parallel spins, each taking one of the degenerate orbitals. In a strong ligand field (large values of Δ) the first tendency (creation of low-spin complexes) prevails; in a weak field the second tendency takes over (creation of high-spin complexes). To every given metal ion corresponds a certain critical value of Δ, above which the low-spin complexes are created, and below which the high-spin ones are found. The low-spin configuration is realized when the pairing energy E_p (the difference in energies of interelectron interaction in the low-spin and the high-spin configurations, divided by the number of pairing electrons) is smaller than Δ. At $E_p > \Delta$ the high-spin configuration is realized.

If the ligands are of the same kind, then $E_p(d^6) < E_p(d^7) < E_p(d^4) < E_p(d^5)$, and the strong-field complexes are formed most easily with the configuration d^6(Co^{3+}, Fe^{2+} ions).

The magnitude of Δ is usually not computed out, but rather is found experimentally from the analysis of molecular spectra. The predominantly ionic complexes produce weak fields; the covalent complexes give strong fields. The so-called spectrochemical row of ligands was constructed in an empirical way, in which they stand in ascending order of Δ [2.7]:

$I^- < Br^- < Cl^-$, $\underline{SCN}^- <$ diethyl thiophosphate, $NO_3^- <$ carbamide, OH^-, $\underline{O}NO^- < HCOO^-$, oxalate (O) $< H_2O$, O_2^-, malonate (O) $< SC\underline{N}^-$, ethylenediaminetetraacetate (N,O), aminoacetate $<$ pyridine, NH_3, ethylenediamine (N), diethylenetriamine (N), triethylene-tetraamine (4N) $< S^{2-}$, $\underline{S}O_3^{2-} < \underline{N}O^-$, phenantroline $< \underline{C}N^-$.

Here the ligand atoms which coordinate with metallic atoms are underlined or parenthesized.

For oxides and hydroxides of bivalent metals M^{2+} (as well as for complexes of M^{2+} with oxygen-containing ligands in solution) the parameter Δ amounts to 7500-12500 cm^{-1} (0.9-1.6 eV or 80-140 kJ/mol); for the corresponding complexes of trivalent metals it equals 13500-21000 cm^{-1}, while for the tetravalent metals (Pt^{4+}) the parameter Δ is about 30.000 cm^{-1}. In the oxygen compounds of first-row transition elements (3d elements) the strong-field complexes are formed only for the d^6 systems. Nitrogen-containing ligands may form strong-field complexes with transition metals having a number of d electrons other than six. The critical value of Δ (the transition from the low-field to the strong-field complexes) in the spectrochemical series lies somewhere between H_2O and NH_3. In the second and third long rows the low-spin complexes are more common, with Δ 30 to 70% higher.

When the first d electrons take the lower t_{2g} orbitals, the octahedral configuration becomes energetically more advantageous than the spherical one. This is the so-called crystal-field stabilization energy (CFSE). If the configuration of d orbitals of a transition ion in an octahedral ligand field is $(t_{2g})^m (e_g)^n$, m and n being the numbers of electrons on t_{2g} and e_g orbitals, respectively, then

$$CFSE = \Delta_0 (4m - 6n)/10 = (4m - 6n)Dq \ . \tag{2.3}$$

For tetrahedral configuration $(e)^n (t_2)^m$ we have

$$CFSE = \Delta_t (6n - 4m)/10 = (6n - 4m)Dq \ . \tag{2.4}$$

Table 2.1. The distribution of 3d electrons in octahedral complexes

No. of 3d electrons	Ion	High spin (weak field)			Low spin (strong field)		
		t_{2g}	e_g	CFSE	tg_{2g}	e_g	CFSE
0	Ti^{4+}, Sc^{3+}, Ca^{2+}			0.0			0
1	Ti^{3+}, V^{4+}	↑		0.4	↑		0.4
2	Ti^{2+}, V^{3+}	↑ ↑		0.8	↑ ↑		0.8
3	V^{2+}, Cr^{3+}, Mn^{4+}	↑ ↑ ↑		1.2	↑ ↑ ↑		1.2
4	Cr^{2+}, Mn^{3+}, Fe^{4+}	↑ ↑ ↑	↑	0.6	↑↓ ↑ ↑		1.6
5	Mn^{2+}, Fe^{3+}	↑ ↑ ↑	↑ ↑	0	↑↓ ↑↓ ↑		2.0
6	Fe^{2+}, Co^{3+}	↑↓ ↑ ↑	↑ ↑	0.4	↑↓ ↑↓ ↑↓		2.4
7	Co^{2+}, Ni^{3+}	↑↓ ↑↓ ↑	↑ ↑	0.8	↑↓ ↑↓ ↑↓	↑	1.8
8	Ni^{2+}	↑↓ ↑↓ ↑↓	↑ ↑	1.2	↑↓ ↑↓ ↑↓	↑ ↑	1.2
9	Cu^{2+}	↑↓ ↑↓ ↑↓	↑↓ ↑	0.6	↑↓ ↑↓ ↑↓	↑↓ ↑	0.6
10	Zn^{2+}, Cu^+	↑↓ ↑↓ ↑↓	↑↓ ↑↓	0	↑↓ ↑↓ ↑↓	↑↓ ↑↓	0

Table 2.1 shows the occupation by d electrons of orbitals in octahedral complexes, and the values of CFSE in parts of Δ (disregarding interaction between electrons). In a weak octahedral field the CFSE is found to be the largest for the configurations d^3 and d^7, and the smallest for d^5, whereas in a strong field it is the largest for d^6. Table 2.2 lists values of CFSE for different configurations in the weak-field approximation [2.5,9]. From Tables 2.1,2 it is clear that crystal-field stabilization of d electrons in any weak ligand field takes place with all transition-metal ions with the exception of those having the configurations d^0 (Sc^{3+}, Ti^{4+}, V^{5+}), d^5 (Mn^{2+}, Fe^{3+}), and d^{10} (Cu^+, Zn^{2+}). Lack of CFSE is a common feature of these ions and of nontransition-metal ions of the same charge and radius. The ions with the configurations d^0, d^5, and d^{10} have a more symmetric surrounding. In a strong field the same properties are exhibited by d^0 and d^{10} ions, while the d^5 ion is stabilized by the crystal field.

DUNITZ and ORGEL [2.10] applied crystal-field theory to explaining the structure of solid-state ionic compounds of transition metals (particularly the oxides). Table 2.2 indicates that the octahedral configuration corresponds to larger CFSE

Table 2.2. The crystal-field stabilization energy of 3d electrons in the weak-field approximation (in Dq units)

Electron configuration	Triangle D_{3h}	Tetrahedron T_d	Square D_{4h}	Square pyramid C_{4v}	Octahedron O_h
d^0	0	0	0	0	0
d^1	3.86	2.67	5.14	4.57	4.0
d^2	7.72	5.34	10.28	9.14	8.0
d^3	10.92	3.56	14.56	10.0	12.0
d^4	5.46	1.78	12.28	9.14	6.0
d^5	0	0	0	0	0
d^6	3.86	2.67	5.14	4.57	4.0
d^7	7.72	5.34	10.28	9.14	8.0
d^8	10.92	3.56	14.56	10.0	12.0
d^9	5.46	1.78	12.28	9.14	6.0
d^{10}	0	0	0	0	0

values than the tetrahedral one. This fact might be important for catalysis as well. In the most common transition-metal compound catalysts, the cation is found in the center of a more or less regular octahedron. Such are the structures of NaCl type (TiO, VO, MnO, MnS, CoO, FeO, NiO), NiAs type (FeS, CoS, NiS), pyrite (MnS_2, FeS_2, CoS_2, NiS_2), rutile ($NiCl_2$), corundum (V_2O_3, Cr_2O_3, Fe_2O_3, Co_2O_3), Mn_2O_3 and $CrCl_3$ types, and spinel (Mn_3O_4, Fe_3O_4, Co_3O_4). The largest difference in CFSE for the octahedral and the tetrahedral configurations is found with d^3(Cr^{3+}) and d^8(Ni^{2+}) ions. Consequently, these elements almost never occur in a tetrahedral configuration in the solid form. Tetrahedral structures – sphalerite (ZnS, $CuCl_2$), wurtzite (ZnS, ZnO), and fluorite (VO_2) – and tetrahedral centers in spinels are observed only with those systems where the difference in CFSE between the octahedral and the tetrahedral configuration is small (d^1 and d^6) or where CFSE is nonexistent as such (d^0, d^5, d^{10}). For instance, in the sulfide row, MnS (d^5) forms a tetrahedral sphalerite lattice, whereas its neighbors VS (d^3), CrS (d^4), and NiS (d^8) crystallize to a hexagonal lattice of the NiAs type.

The splitting of levels (lifting of degeneracy) in fields of symmetry lower than octahedral or tetrahedral (Fig.2.3) is described by the Jahn-Teller theorem (also called the Jahn-Teller effect). As regards complexes, this can be expressed as follows: if the ground orbital state of an ion in a complex is electronically degenerate, the vibrational motion will change the nuclear configuration to bring atoms to an arrangement which corresponds to the lowest possible symmetry and the lowest possible energy, thus eliminating the existing degenercy. A more general and strict proof of the Jahn-Teller theorem can be found in [2.9] by BERSUKER.

The Jahn-Teller effect is responsible for the fact that the transition-metal compounds often crystallize into structures with reduced symmetry [2.4]. For instance, the Cu^{2+} ion in an octahedral coordination has the electron configuration $d^9 = t_{2g}^6 e_g^3$ (Table 2.1, Fig.2.2). The d^{10} ion exhibits spherical symmetry whatever the field. The d^9 ion differs in that it has a hole in the d shell, which may take either of the e_g orbitals ($d_{x^2-y^2}$ or d_{z^2}). In the first case, the charge is less shielded in the x and y directions, and the ligands (anions) are attracted more strongly, so the bond lengths along x and y are shorter than those along the z axis (Fig.2.4). On the other hand, in the second case the bond length along the z axis will be the shorter one. Experiment reveals that the first case is realized with $CuCl_2$, $CuBr_2$, and CuI_2: a distorted octahedral (that is, D_{4h} tetragonal) structure is formed with two long and

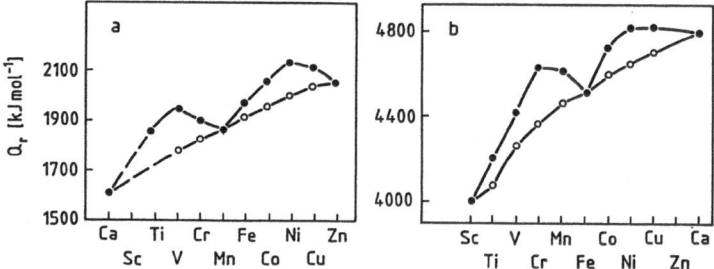

Fig.2.4. The heat of hydration Q_h of aquacomplexes of bivalent (a) and trivalent (b) ions of forth-row metals [2.4, 11]: x – experimental points; o – Q_h minus CFSE

four short bonds. For example, in $CuCl_2$ there are four Cu–Cl bonds each 0.230 nm long and two Cu–Cl bonds of 0.295 nm each. Such a lattice exhibits splitting of the t_{2g} and e_g levels and extra stabilization by the crystal field (Fig.2.3). In the extreme case the two distant ligands will be removed altogether thus reducing the structure to a square. This is what really occurs with CuO, which has a monoclinic tenorite structure. The structures having two long and four short bonds, similar to d^9, occur in high–spin complexes with d^4 configurations (Cr^{2+}, Mn^{3+}) and in low–spin d^7 complexes (Co^{2+}, Ni^{3+}).

In the high–spin octahedral complexes d^1, d^2, d^6, and d^7 and in the low–spin d^4 and d^5 complexes, the Jahn–Teller deviations of the crystal structure, instead of being purely octahedral, are much smaller. This is because the axes of symmetry of the t_{2g} orbitals (d_{xy}, d_{xz}, d_{yz}) are not directed toward the ligands. Consequently, the asymmetric occupation of t_{2g} orbitals by electrons has a much smaller effect on the bond lengths than does the occupation of e_g orbitals ($d_{x^2-y^2}$ and d_{z^2}). No deviations will be observed with structures with a full (or half–full) t_{2g} shell and empty e_g shell, as well as with high–spin d^3, d^5, and d^8 complexes and low–spin d^6 complexes [2.8,10–12].

Elimination of degeneracy in tetrahedral complexes leads to trigonal deformations: compression or elongation of the tetrahedron along the third–order axis. For instance, $TiCl_4$ (d^0 configuration and zero Jahn–Teller effect) has the structure of an almost regular tetrahedron, while VCl_4 (d^1) has the structure of a compressed tetrahedron.

Note that crystal–field stabilization is just one of a number of factors contributing to stability of transition–metal complexes. Figure 2.5 shows experimental data on the heat of hydration Q_h of water complexes of bivalent and trivalent ions of 3d elements [2.8,13]. The experimental points (crosses) describe a smooth curve (indicated by open circles), which rises steadily toward the end of the row. This (main) contribution to

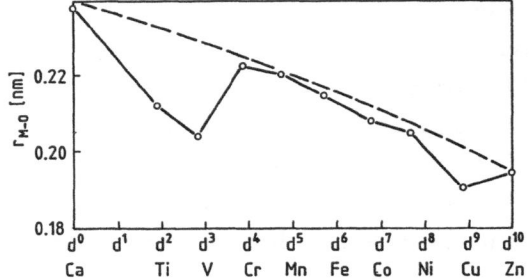

Fig.2.5. The separation between metal and oxygen r_{M-O} in the oxides of fourth-row metals

stability of the complex depends on the attraction of ligands to the skeleton of the spherosymmetric metal ion. The stability of complexes of the first-row bivalent metal ions with almost any ligand increases in the series $Mn^{2+} < Fe^{2+} < Co^{2+} < Ni^{2+} < Cu^{2+} > Zn^{2+}$. This apparently applies to the oxygen complexes as well, including the bonding between metallic ions and oxygen ions in the solid.

The two-peak law of variation of different properties in the row of transition-metal compounds seems to be quite common. Figure 2.6 shows variation in the distance between oxygen O and metal M atoms in the row of simplest Mo oxides of fourth-period metals. The two-peak dependence from the beginning through the end of the period is also observed here – against the overall compression of the lattice (dashed line).

Initiated by VAN VLECK [2.4], a more sophisticated theory, based on the method of molecular orbitals, has evolved for describing the structure and properties of transition-metal compounds. It became known as the ligand-field theory. Unlike the

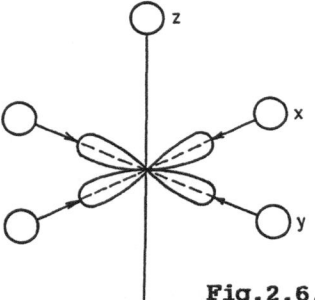

Fig.2.6. The bonding in a complex of d^9 ion (Cu^{2+})

crystal-field theory, it takes into account the electron structure of not only the central ion, but of ligands as well. This
theory is detailed in a number of monographs [2.5–9]. In
general, the method differs little from the LCAO–MO technique
(linear combination of atomic orbitals – molecular orbitals)
commonly used in quantum chemistry.

The molecular orbitals of the complex of the transition-
metal ion with ligands are formed by way of combination of
central ion orbitals with ligand orbitals, having similar symmetry properties with respect to one and the same coordinate
system. Before composing such a combination, one has to find
out the combinations of ligand orbitals with one another
according to the rules of group theory – the so-called group
orbitals. When combining the orbitals of metal and ligands to
find the molecular orbitals of the complex the criteria of symmetry and overlapping must be used.

Figure 2.7 displays the scheme for the formation of energy
levels (molecular orbitals) in an octahedral complex of 3d, 4s,
and 4p atomic–metal orbitals and group–ligand orbitals (disregarding π bonding). The metal-ion orbitals are on the left, the
ligand orbitals on the right, and the molecular orbitals of the
ligand are in the center of the diagram. The 4s metal orbital
combines with the a_{1g} group–ligand orbital $\sigma_1+\sigma_2+\sigma_3+\sigma_4+\sigma_5+\sigma_6$
(σ_i being the individual ligand orbitals) to form the bonding
a_{1g} orbital and the antibonding a_{1g}^* orbital. The three 4p orbitals of the metal ion combine with three t_{1u} orbitals of the
ligands to form the bonding t_{1u} orbital and the antibonding
t_{1u}^* orbital. The two e_g orbitals (d_{z^2} and $d_{x^2-y^2}$) combine with
the e_g orbitals of the ligands to form the e_g bonding and the
e_g^* antibonding orbitals. In this scheme the t_{2g} orbitals of
metal, according to symmetry conditions, have no counterparts
among the ligand orbitals and so remain as nonbonding atomic
orbitals. In Fig.2.7 the energy of these orbitals corresponds to
the energy of the 3d orbitals of the initial metal atom. The incipient molecular orbitals must take on the twelve electrons

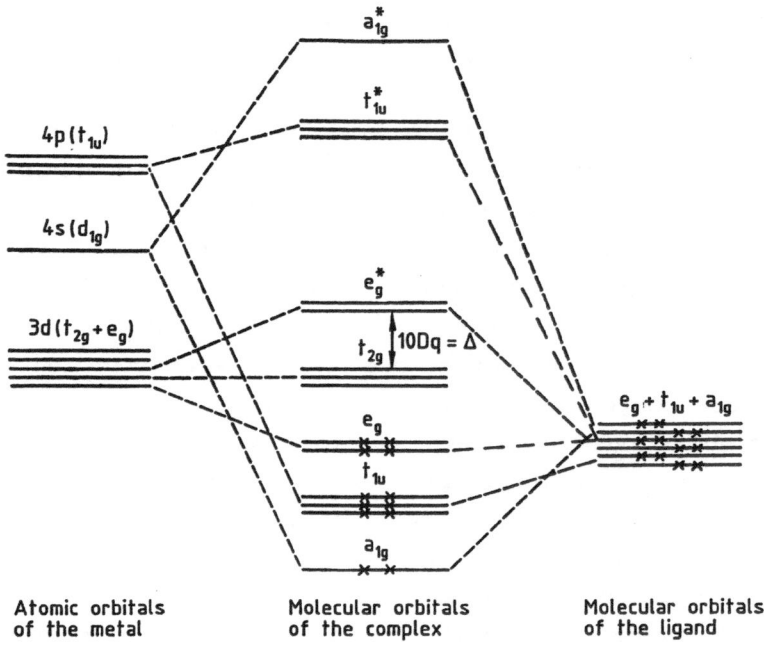

| Atomic orbitals | Molecular orbitals | Molecular orbitals |
| of the metal | of the complex | of the ligand |

Fig.2.7. The formation of orbitals in the octahedral complex of a fourth-row transiton metal ions (disregarding π bonding). Crosses indicate the arrangement of 12 electrons of ligands on the molecular orbitals of the complex

of the ligands (two on each of the six ligand σ orbitals) and n electrons of the central $3d^n$ metal ion. The former are arranged on the bonding σ orbitals (a_{1g}, t_{1u}, and e_g), and the remaining n electrons are distributed between the nonbonding t_{2g} orbital of the metal atom and the antibonding molecular e_g^* orbital, whose capacities are 6 and 4 electrons, respectively.

The gap between the last two orbitals $\Delta = 10Dq$ is found from spectroscopic data and has exactly the same meaning as the corresponding quantity in crystal-field theory. However, the origin of levels is now different: the t_{2g} orbital is a nonbonding one and has the same energy as the initial degenerate d level, whereas the antibonding orbital e_g^* lies above the t_{2g} orbital. In this scheme the occupation of the t_{2g} orbital does not lead to stabilization (no energy gain), while the occupation of e_g^* orbital is energetically disadvantageous. The ligand-field stabilization energy can be defined as the change in energy upon transition from a structure having entirely uniform distribution of electrons over all nonbonding and antibonding orbitals to a structure having the actual electron distribution [2.14]. Nevertheless, the relative energy of these orbitals and the order of their occupation by electrons in strong and weak ligand fields remain the same as in crystal-field theory (Table 2.1).

The characteristic two-peak pattern of various properties of complexes of the 3d transition elements against the number of d electrons can be qualitatively explained within the framework of ligand-field theory as well. There also exist alternative explanations of the two-peak pattern (e.g., the angular overlap MO model [2.15]).

The advantages of ligand-field theory show up when one comes to consider the complexes of ligands capable of π bonding, and such are the majority of ligands: oxygen, nitrogen, olefins, etc. In an octahedral complex the twelve group π orbitals of ligands with t_{2g}, t_{2u}, t_{1g}, and t_{1u} symmetry may combine with the orbitals of metal: e.g., the t_{1u} ligand orbitals with 4p metal orbitals, and t_{2g} ligand orbitals with t_{2g} metal orbitals (d_{xy}, d_{xz}, d_{yz}). In this case the latter will no longer be the nonbonding ones. However, the overlapping of the t_{2g} metal orbitals with the ligand orbitals (other conditions being equal) in an octahedron is much lower than the overlapping of e_g orbitals (as well as that of 4s and 4p orbitals). The splitting of π orbitals in two - the bonding t_{2g} orbital and the antibonding t_{2g}^* orbital - is therefore small. Both these orbitals remain close to the initial 3d level of the ion, the t_{2g} orbital being a little lower, and the t_{2g} a little higher.

Figure 2.8 indicates that the parameter Δ of "spectrochemical" splitting between the t_{2g} and e_g levels upon establishment of π bonds may either decrease (if π orbitals are occupied) or increase (if π orbitals are free). The first case ("weak field"

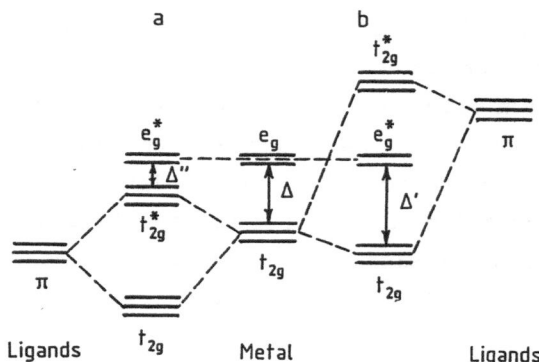

Fig.2.8 The effects of π orbitals of ligands on the splitting Δ between t?$_{2g}$ and e$_g$ levels. (a) occupied π orbitals; (b) vacant π orbitals

in the spectrochemical row) is observed, for instance, with halogens and H_2O, the second case ("strong field") with CN^- and CO [2.8].

The method of molecular orbitals clearly indicates the way by which the ligand electrons and the d, s, and p metal electrons participate in the formation of a chemical bond. The schemes quoted here point to the fact that the actual cases may range from covalent bonding to polar high-ionicity bonding. If the bonding is formed via an unshared pair of electrons of the ligand and the vacant acceptor (metal) orbital, it is called the donor-acceptor bond (or sometimes the coordination bond; however, the latter designation is applied to almost any type of bonding in complex compounds of metals). We discussed the concept of donor-acceptor bond extensively in [2.1].

If the electrons are displaced the other way: from occupied metal orbitals to vacant ligand orbitals, the metal atom becomes the donor and the ligand the acceptor. This kind of bonding is often called the dative or back-donation bond. For instance, the formation of π bonds may be accompanied by displacement of electron density from the (d_{xy}, d_{xz}, d_{yz}) orbitals of the metal atom (in octahedral coordination) towards the t$_{2g}$ ligand orbitals. The tendency toward establishing back-donation bonds is therefore exhibited by metals with fully or partly occupied d orbitals, including ions with d^{10} configuration (Hg^{2+}, Cd^{2+}, Cu^+, Ag^+, Pt^0).

20

Theoretical calculations of the energy of molecular orbitals in transition-metal compounds are based on a variety of approximate techniques: EHM (extended Hückel method), MWH (Mulliken-Wolfsberg-Helmholtz method), CNDO (complete neglect of differential overlapping) and others. These techniques have been reviewed in [2.9,16,17]. Calculations of this kind can at best indicate the correct order of the levels. Considerable progress in calculations of the energetic structure of complex compounds was made in the 1970s thanks to the development of a new nonempirical method of molecular orbitals, not involving the LCAO approximation. This technique was proposed by SLATER and JOHNSON and is called SCF-X_α-SW (self-consistent field-X_α-scattered waves) [2.18-20]. It allowed computation of the energetic structure of a number of complexes, at the same time saving a considerable amount of machine time in comparison with other nonempirical methods providing the same extent of adequacy in reproduction of one-electron levels. The X_α technique was successfully employed for the calculation of energy spectra of solids both on the surface and in the bulk. The results fit in well with experimental data on photoelectron spectra.

Despite the development and use of new calculation methods, the old crystal theory provides the simplest and best means for direct study of the chemistry of coordination compounds. It is particularly useful in considering the properties related to electron structure of the central atom.

2.2 Experimental Techniques for Investigating Surface Compounds and Transition-Metal Atoms in Solids

Physical techniques (conductivity, work function measurements, etc.) are widely employed for studying the structure of metals. Unfortunately, they are poorly adapted for the investigation of wide-band semiconductors and insulators, especially powdered oxide adsorbents and catalysts [2.2]. More reliable information is obtained from spectroscopic measurements which have been substantially improved in the recent years. Since the transmission of transition-metal oxides is relatively low at those wavelengths that are usually employed in the measurements, they are especially valuable for studying the structure of sub-surface layers. In most cases these techniques are sensitive enough for the investigation of transition-metal ions directly on the surface, as well as adsorbed molecules.

2.2.1 Optical Spectra

The energy structure of upper levels and the coordination of transition-metal atoms are usually assessed with the aid of UV and visible light spectroscopy, since the hardware required is relatively simple and inexpensive. The transition energies between these levels lie in the range of 2–6 eV.

Powder-like materials, used as catalysts and adsorbents, are studied by their diffuse reflection spectra. In this case the reflected radiation, scattered from the surface, consists of two components: the direct one which is bounced off the surface without penetrating to any depth into the crystal, and the diffuse component which is absorbed and thereafter reradiated by the solid.

The processing of diffuse-reflection spectra from powders is based on various approximate expressions [2.21,22]. The maximum of the band, found from diffuse reflection spectra, usually coincides with the maximum in the spectrum of the same sub-stance taken in the form of a single crystal or solution. This allows assessment of the energy of electron transitions using the maxima in the spectrum of diffuse reflection of a powder.

The concentration of the transition-metal ions is harder to determine. In the case of homogeneous systems (e.g., liquids) the intensity of spectral bands is higher, the larger the concentration of chromophore (i.e., transition metal). The required relationship of powders is not always established so easily, since the optical spectra of oxides give information on the electron structure of the layer not more than 50 nm thick, the radiation being unable to penetrate to a greater depth [2.23].

Until recently, optical spectroscopy had been limited in exploring the surface structure of single crystals because of its insufficient sensitivity. The situation has changed lately due to the development of novel techniques, taking advantage of multiple internal reflections and making it possible to study the "surface" levels of monocrystals.

The crystal-field theory, discussed in Sect.2.1, describes the transitions between d orbitals that usually show up in the visible spectra (transition energy 1.5-2.5 eV). The intensity of these spectra is relatively low. They are usually characterized by the coefficient of extinction (molar absorption coefficient at a given wavelength and temperature)

$$\epsilon = A/(cb) = \log(I_0/I) , \qquad (2.5)$$

where A is total absorption, c is the concentration of chromophore, b is the thickness of absorbing layer, and I_0 and I are the intensities of the incoming and the outgoing light, respectively.

For octahedral complexes $\epsilon \simeq 10$. The d-d transitions here are forbidden by the so-called Laporte selection rule (which outlaws internal electron transitions in centric molecules). This restriction is made less strict by the existence of normal vibrations of the octahedral complex and consequent distortion of symmetry. In complexes with an Mn^{2+} ion (configuration d^5, spin 5/2) the d-d transitions change the total spin of the complex. Such transitions are spin forbidden, and the complexes would be achromatic were it not for the existence of electron-vibrational interactions. The spectra of tetrahedral complexes and other complexes with lower symmetry are stronger ($\epsilon \simeq 10^2-10^3$).

The charge transfer spectra in transition-metal complexes are much stronger ($\epsilon \simeq 10^4$-10^5) and are characterized by bands in the UV (and sometimes in the visible) range, which corresponds to a transition energy of 3 to 6 eV. The charge transfer spectra relate to the transition of an electron between orbitals pertaining to different atoms (or groups of atoms). The theory of charge-transfer complexes associated with the formation of molecular donor–acceptor compounds was first proposed by MULLIKEN [2.24]. In the LCAO-MO scheme the wave function of such a complex is described by a sum of wave functions for the bonding and the antibonding orbitals, which are usually localized at different atoms of the system: for instance, the bonding MO on ligands L and the antibonding MO belongs to metal M. The electron in this case goes over from ligand to metal: ML → M⁻L⁺. Charge transfer spectra contain, in particular, the transition from the π orbitals of the ligand to the d orbitals of the metal (donor–acceptor bond), and the transitions from d orbitals to antibonding ligand π* orbitals (back–donation bond).

The charge transfer spectra allow assessment of the location of d levels of transition metal in the band structure of an oxide. This point will be discussed in greater detail in Sect.3.5.

Optical spectroscopy is little used for studying metallic surfaces because of their high reflectance and conductivity. Absorption spectra of metals in the visible and UV spectral ranges can be employed for studying the so-called plasmons, produced by collective vibrations of a metal's free-electron plasma. Plasma vibrations are responsible for the hues of transition metals, as well as those of copper, silver, and gold [2.25].

2.2.2 Photoelectron and Auger Spectra

The ambiguity in the interpretation of charge transfer bands in visible and UV spectra was somewhat reduced thanks to the development of photoelectron spectroscopy (also called ESCA:

electron spectroscopy for chemical analysis). Monochromatic X-ray or UV radiation is used to knock electrons out of the sample, whose energy is then registered with an appropriate detector. By the law of energy conservation

$$h\nu = E_b + E_{kin} + \phi_{sp} \ , \tag{2.6}$$

where E_b is the electron's binding energy, E_{kin} is the kinetic energy of the electron after photoionization, and ϕ_{sp} is the work function of the sensitive element.

Knowing $h\nu$ and ϕ_{sp} and measuring E_{kin} one can find E_b – the energy of electron levels in the solid.

The ESCA approach includes two major techniques: X-ray photoelectron spectroscopy (XPS) and ultraviolet photoelectron spectroscopy (UPS). The source of exciting radiation in X-ray spectroscopy is usually represented by K_α lines of aluminium ($h\nu$=1487 eV) and magnesium ($h\nu$=1254 eV), serving as X-ray tube anode materials. Such excitation is strong enough to knock out photoelectrons corresponding to both outer and deep inner levels. Relating variations in the location of levels to changes in the chemical composition of the sample, one can measure the so-called chemical shifts, which reflect changes in the chemical status of the atom. The magnitude of the chemical shift may amount to several electron volts. The XPS and UPS techniques have been reviewed in [2.26-30].

The resolution achieved in X-ray photoelectron spectrometers in energy terms usually constitutes about 10^{-4} of the $h\nu$ value. The accuracy in measuring E_b can hardly be made better than 0.1 eV when K_α lines of Mg or Al are used for the source of excitation. This prevents one from studing the fine structure of solids using XPS techniques. Likewise, the XPS high-energy spectra allow one to distinguish atoms in different chemical states, but the knowledge of E_b is by itself not sufficient for calculating the atom's coordination and surrounding.

The UPS techniques are based on low-energy excitation sources: helium lamps emitting resonant radiation at 21.2 and 40.8 eV and other sources of vacuum ultraviolet radiation. These techniques allow one to assess the energy of outer valence-band electrons, participating in chemical bonds, as well as the energy of occupied levels (bands) of the solid within 21 eV (or 40 eV) from the Fermi level, such as 2p and 2s levels of oxygen and 3d levels of transition elements.

Still more advantageous for the study of the electron structure of solids is the synchrotron radiation (magnetic bremsstrahlung): emission of photons by charged relativistic particles traveling in a magnetic field in circular and helical paths. Such radiation is produced by electrons in electron synchrotron accelerators, whence it got its name. The high strength of synchrotron radiation sources and their capability of producing a nearly continuous-spectrum of exciting radiation allows exploration of the electron structure in a wide range of energies, stretching from UV to the deepest atomic levels, and thus the extraction of information about both the valence bands and the inner electron shells. Synchrotron radiation is now widely used to study surface phenomena [2.31,32].

An important feature of synchrotron radiation is its linear polarization, which makes it possible to employ it in the so-called angle-resolved ultraviolet photoelectron spectroscopy (ARUPS). By studying the intensity of photoemission as a function of polarization of the exciting radiation in a predetermined solid angle, one can examine the orientation of surface orbitals [2.33].

Synchrotron radiation is very useful in the study of the band structure of solids because the ionization cross section of electrons on different levels varies differently with excitation quanta energy $h\nu$. For instance, for an oxide of transition metal the intensity of the photoelectron spectrum from oxygen 2p levels drops with respect to 3d levels of metal, and $h\nu$ increases. This circumstance facilitates interpretation of photo-electron spectra.

Highly important is the question of how to distinguish between surface and bulk atoms. The depth from which the electrons are expelled depends on the material, on the one hand, and on the kinetic energy of the electron, on the other. Generally, the depth is smaller, the larger the atomic number Z. As the kinetic energy increases, the value of electron–ejection depth goes down, passes a minimum, and goes up again. The minimum corresponds to E_{kin} of 50 to 100 eV. For instance, in the case of silver at E_{kin} = 72 eV the average depth from which the electrons come is just about 0.4 nm; in other words, the electrons come out from a thin surface layer. The binding energy of electrons pertaining to the bulk and the surface atoms also might be expected to differ.

This problem can conveniently be approached with the help of synchrotron radiation. By varying the energy of the exciting quanta hν one can adjust E_{kin} to achieve selective sensitivity to either the bulk energy levels or the surface ones.

Auger electron spectroscopy (AES) is based on exciting the electrons of the inner shells of atoms in the subsurface region by electrons or hard X-ray quanta. In an Auger transition, a vacancy formed by an external ionizing agent in one of the inner shells (e.g., the K shell) is filled by an electron from a higher level (e.g., the L shell). The energy released in the transition is transferred to some other inner electron, which is expelled from the solid into the surrounding vacuum [Ref.2.1, Fig.3.7]. The energy distribution of Auger electrons characterizes the surface-building elements.

The reviews on Auger spectroscopy [2.34–37] quote theoretical values of the ionization cross section for a given atomic level, which define the probability of forming a vacancy in the corresponding shell. The ionization cross section for K and L shells having the same binding energy are nearly equal, whereas that for the M shell is much smaller. The value of ionization cross section drops rapidly with increasing atomic number Z. What is more, the heavier atoms exhibit extremely complicated Auger spectra, which are very hard to interpret. For this reason the AES technique has been successful in the study of light atoms, especially in the investigation of the chemical composition of a surface (constituents expressed as the percentage of a monolayer).

2.2.3 Low-Energy Electron Diffraction (LEED)

Modern equipment is designed to measure, in one and the same sample cell, the Auger spectra and the LEED spectra of the single-crystal surfaces [Ref.2.1, Chap.3]. In this way it becomes possible to study both the chemical compositon (AES) and the surface structure (LEED). In LEED a beam of electrons with an energy of 20–200 eV is directed onto the surface. The diffraction pattern created by elastically scattered electrons is made visible on a fluorescent screen. As indicated above, the electrons having energy of about 100 eV come mostly from the surface layer, and so the diffraction pattern reflects the structure of the crystalline surface. Studying the diffraction pattern as a function of accelerating voltage, one can derive the location of surface atoms [2.35,38–43].

Analyzing the intensity of the diffracted beam as a function of energy of the incident electrons $I(E)$ and of the angle of incidence $I(\phi)$, one can deduce the configuration of surface atoms with respect to the bulk lattice, as well as make estimates regarding the amplitude of surface atoms' vibrations (x_s) [2.43–45]. A theoretical analysis indicates that the mean-square amplitude $\langle x_s^2 \rangle$ is greater than that for the bulk atoms, $\langle x_v^2 \rangle$. Moreover, although the vibrations of the bulk atoms may be isotropic, the surface atoms can exhibit vibrations of an anisotropic kind. Some qualitative correlations were established between the nature of chemical bonds and the amplitude $\langle x_s^2 \rangle$. WALLIS [2.46] pointed out that the bonding constants for the surface atoms may be 25–30% lower than those for the bulk atoms. The calculations mostly employed highly idealized models of surfaces, assuming conserved symmetry of surface atoms and invariable dynamic bond constants. So far, we can only speak of qualitative agreement between theory and experiment.

2.2.4 Vacuum Tunneling Spectroscopy

The XPS, UPS, AES techniques described in Sect.2.2.2 are used for investigating the energy levels of atoms in the subsurface region of the solid. Recently new techniques of electron spectroscopy have been developed in which the emission of electrons is determined by the exponential parts of the wave functions of surface atoms, from which the electrons tunnel into the adjacent vacuum. For this reason all these techniques are grouped together under the term "vacuum tunneling spectroscopy" [2.47].

Elsewhere [Ref.2.1,Sect.3.1.1] we have already described the field-emission electron microscope (FEM) and field-emission ion microscope (FIM), which enable one to measure the work function for microfaces of sharp metallic tips, and to even resolve in some cases the individual metal atoms. Ions or electrons are emitted from the tips under the action of a strong electric field (about 10^8 V/cm). Further development of FEM and FIM devices resulted in putting them to use for spectroscopic purposes.

The theory of autoelectronic emission was developed by YOUNG [2.47], who found the energy distribution of electrons at a given point to be described by

$$j_0{}'(E) = j\ e^{\epsilon/b}\ f(E)/b\ ,\qquad\qquad (2.7)$$

where j is the density of the electron current from the tip, E is the energy of electrons in the metal with respect to the Fermi level, f(E) is the Fermi-Dirac distribution function, and b depends on both work function ϕ and field strength ϵ.

At low temperatures the dependence of log(j) on E is depicted by a straight line with 1/b slope almost reaching the

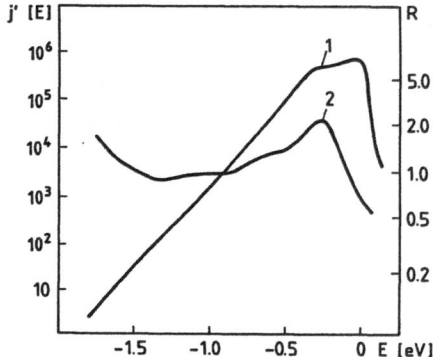

Fig.2.9. FEM spectrum of W surface [2.47,53]. (1) the energy distribution of electrons j'(E), emitted from the (100) face of tungsten tip at -196°C and ϵ =3.52·10^{7}V/cm; (2) R(E) = $j(E)/j'_0(E)$

Fermi level, followed by a drop (Fig.2.9, Curve 1). Such a distribution of emission over energy indicates that the height of the barrier which the electron has to surmount increases with the decrease in energy.

Equation (2.7) was derived assuming a free-electron gas in a bandless metal, and has nothing to say about the band structure. Further theory [2.48-51] has demonstrated that the fine structure of the electron energy distribution contains information about the surface states whose detection, however, requires high accuracy in the determination of the electrons' energy. Improvements made on the electron detectors (which may be positioned behind an opening in the screen), has brought the resolving power to 15-75 meV [2.52-54]. This has enabled one to verify the predicted tendency, and to convert a field-emission microscope into a spectroscopic instrument (field-emission electron spectroscopy: FES).

Figure 2.9 shows graphs which contain information about the energy spectrum of the surface, derived from the measured distribution of electrons over energy j'(E) for the (100) face of a tungsten tip at 77 K [2.50,53]. Curve 2 gives R(E) = $j(E)/j'_0(E)$ and is normalized to unity at E = -1.0 eV [$j'_0(E)$ being the theoretical value according to (2.7)]. The curve R(E) is observed to exhibit noticeable structure, characteristic of each of the tungsten's crystal faces. The peak at E = −0.35 eV relates to the surface and reflects the high density of surface d states. It exhibits extreme sensitivity to contamination and adsorption.

Field-emission ion microscopes also formed the basis for a spectroscopic technique: field-ion spectroscopy (FIS). It differs from FES in that it is sensitive to the vacant electron levels and energy bands of the surface, whereas FES describes the occupied levels. The distribution of ions over energy directly reflects the energy spectrum of vacant levels, since the probability of ionization of a helium atom at a given distance from the surface is proportional to the number of states accessible for a tunneling electron. We see that the FES and FIS techniques complement each other and can supply a complete picture of the surface states near (both above and below) the Fermi level. The FIS technique allowed observation of empty localized states on a tungsten surface up to 4 eV above the Fermi level E_F [2.55]

The ion neutralization spectroscopic technique (INS) was proposed by HAGSTRUM [2.56]. It is based on a two-electron Auger process. A beam of noble gas ions (He^+ or Ar^+) is directed onto the surface (Fig.2.10) the ion is indicated some distance away from the surface, it creates a potential pit whose depth equals the ionization potential of the He or Ar atom. An electron from the surface state or from the occupied valence band tunnels to the ion's orbital to form a neutral atom. The energy of this transition is spent on exciting a second electron in the occupied band or local level, which may then surmount the surface

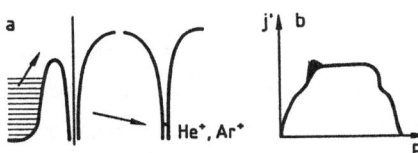

Fig.2.10 The schemes of ion neutralization spectroscopy (a) and electron energy distribution (b). Filled-in portion denotes additional emission at the expenses of surface states

potential barrier. The spectrum of these electrons is registered by a detector and manifests to the structure of surface levels.

Since the Auger process employed in the INS technique is a profoundly two-electron process, the resulting spectra are interpreted with less ease then the UPS or FES spectra. If both electrons come from one and the same local level, i.e., from one and the same atom of, say, helium, the energy of the electron ejected will reflect the difference in E_{He} and the sum of the atom's first and second ionization potentials. The INS technique is more sensitive towards surface states as compared with other techniques, since the spectrum here results from overlapping of He or Ar wave functions with projecting orbitals of surface atoms. The d_{z^2} and p_z orbitals must apparently display a greater extent of overlapping with the ψ_{He} or ψ_{Ar} wave functions than do other d or p orbitals. The INS technique is more sensitive to absorbed atoms projecting above the surface than to those immersed into the subsurface layer [2.34,57,58].

A relatively small number of investigations using the INS technique were concerned with the study of orbitals on the Ni(100) face, as well as on the surfaces of Hg and Te. The INS spectrum from an atomically clean Ni(100) face shows a peak at −0.8 eV (which may be attributed to the e_g orbital); the UPS spectra, apart from this peak, have a number of maxima between 4 and 14 eV, which reflect the structure of the surface layer [2.55].

Yet another technique employs deexcitation of metastable (excited) helium atoms on the surface, the surface electron tunneling to the lowest free level of a helium atom, while the helium's excited electron is emitted and registered. This technique is called surface Penning ionization (SPI), it is akin to INS but allows interpretation more readily [2.59,60a].

A new and remarkable method for surface studies has appeared in recent years, namely tunneling electron microscopy (TEM). It permits observation not only of individual atoms on a smooth metal surface, but also of the direction of outward surface orbitals [2.60b].

2.2.5 Magnetic Techniques

Earlier methods for studying the transition metals and their compounds include magnetic susceptibility measurements. Magnetic techniques are reviewed in [2.61-64].

The transition-element atoms in a crystalline structure may be engaged in collective (cooperative) interactions, which distinguish them from the same atoms in isolated complexes and from nontransition-element atoms. The interelectron exchange interaction governs the orientation of adjacent atomic magnetic moments, due to unpaired spins of d electrons. Crystalline solids are called ferromagnets, if in the absence of external magnetic field they possess regions (domains) with parallel spins. Ferromagnets include such metals as Co, Ni, and Fe. Solids with the spins in domains oriented antiparallel are called antiferromagnets, these include, for instance, MnO, NiO, CoO, and FeO oxides with face-centered-cubic lattices. The crystallographic sites, occupied in these oxides by the transition-metal atoms, are equivalent to one another. In an antiferromagnetic, one can single out two equivalent magnetic sublattices with the spins in the one being antiparallel to those in the other. A third group of materials, according to their magnetic properties, comprises ferrimagnets, these also have antiparallel spins, but whose net magnetic moment is nonzero ($m_0 \neq 0$). This lack of compensation is due either to dissymmetry of crystallographic sites of transition atoms in the two sublattices, or to differences in their electronic surrounding. Ferrimagnetic include some commonly used catalysts such as spinel-structure ferrites, manganites, chromites, and other oxide systems.

By measuring the magnetic susceptibility of species with isolated paramagnetic ions, it is possible to assess the degree of oxidation of a paramagnetic ion. Measurements of magnetic susceptibility in oxide systems are used for finding the magnitude of interaction between paramagnetic ions.

The "magnetic order" in antiferromagnets and ferromagnets is observed at not-too-high temperatures. At a certain critical temperature – the so-called Curie point, or, in the case if antiferromagnets, the antiferromagnetic Curie point or Néel point – spontaneous magnetization is destroyed by thermal motion, and a ferromagnet (antiferromagnet) becomes a paramagnetic. In the paramagetic region at temperatures well above the Curie point the magnetic susceptibility κ decreases with the rise in temperature T according to the Curie-Weiss law

$$\kappa = C/(T - \theta) , \qquad\qquad (2.8)$$

C and θ being material constants; here θ is called the Weiss constant. For ferromagnets $\theta > 0$ and is usually somewhat higher than the Curie point; for antiferromagnets $\theta < 0$.

Magnetic properties of oxide catalysts as functions of dispersity were studied in detail by SELWOOD [2.62,65,66]. He concludes that grinding of an oxide may lead to a point when the number of ions in regular lattice sites (in the bulk) will be comparable with the number of ions in odd sites (close to the surface), thus destroying antiferromagnetic interactions. SELWOOD [2.66] computed the average coordination number Z for a cation in a corundum-type oxide lattice as the function of the number of atomic layers and found Z = 3 for one atomic layer, Z = 6 for two layers, and Z = 7 for three layers. For a paramagnetic cation well inside the lattice, Z = 9. In other words, there is no need to take into account the surface effects on magnetic susceptibility if the number of layers is large enough. It is easy to calculate that particles which are three atoms thick correspond to a specific surface value of about 10^4 m^2/g, and that particles 16 atoms thick correspond to a specific surface value of 20 m^2/g (Z = 9). The magnetic properties may be assumed to vary accordingly with dispersity.

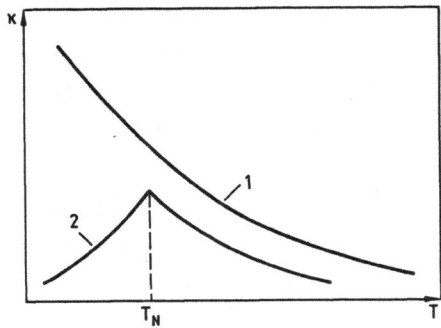

Fig.2.11 Magnetic susceptibility κ versus temperature T
for high–dispersion (1) and
crystalline (2) transition
metal oxide. T_N is the Néel
point

The magnetic susceptibility of high–dispersion amorphous
chromium oxide gel at –170°C is 5 times as great as that in the
crystalline state (κ equals 170 and 33, respectively). This
points to an almost complete lack of antiferromagnetism in the
gel. The Curie–Weiss law holds for Cr_2O_3 gel in the entire temperature range with θ =104°C [2.67]. Temperature dependencies
of κ for the high–dispersion and the crystalline oxides are
plotted in Fig.2.11.

The measurements of magnetic susceptibility allow one to
assess the extent of dilution of transition–metal oxides by
nonparamagnetic oxides (e.g., F_2O_3 in Al_2O_3). Dilution leads to
an increase in κ per one transition–metal ion. This rise is
especially noticeable at low concentrations (<1%), when isolated transition ions are available in sufficient numbers. At
higher transition–metal concentration the cation–cation interaction grows stronger, κ decreases, the oxide begins to display
ferromagnetic or antiferromagnetic properties, and κ as a function of temperature begins to follow the Curie–Weiss law (2.8).
The difference in the behavior of κ with variations in temperature is also roughly depicted by curves 1 and 2 in Fig.2.11 for
diluted and concentrated oxide systems, respectively.

Resonant techniques turned out to be very promising for the
study of bulk properties of magnetic materials [2.68]. Three
main types of resonant centers can be singled out in magnetic
materials: (1) Electron resonances – spin–ferromagnetic, anti-

ferromagnetic, ferrimagnetic, and paramagnetic (because of the outstanding importance of the EPR technique in the study of transition atoms, this method will be discussed specially in Sect.2.2.6); (2) domain boundaries ("domain resonance"); and (3) nuclear spins (nuclear magnetic resonance: NMR). Thanks to the existence of strong internal magnetic fields in many magnets, the resonant phenomena can often be observed in the absence of an external field; this is the so-called natural magnetic resonance. The width and shape of resonant lines carry vast information about the mechanisms of relaxation of spin and domain systems in magnets. High-dispersion samples have to be used for obtaining information regarding the surface phase of magnetic materials.

Magneto-optical techniques proved to be especially useful in the study of surface phases of magnets. Owing to the small penetration depth of light into magnets, investigating the dispersion of magneto-optical characteristics in reflected beams yields most valuable data on the properties of surface layers and the changes caused by adsorption and catalytic action. In metals, light penetrates to a depth of 0.001-0.003 nm. Optically transparent materials can be studied in transmitted beams (see below). Of special interest here are the Faraday effect (rotation of the polarization plane in a magnetic field) and the Cotton-Mouton effect, also called the Voigt effect (birefringence due to magnetization).

In studying the surface phase, the reflectance must be utilized. Vast information about the energy spectrum of the subsurface phase in ferromagnets and dielectrics can be obtained with the aid of the magneto-optical Kerr effect [2.63,69]. According to the magneto-optical Kerr effect, a plane-polarized light beam, which falls normal to the magnetic surface, becomes elliptically polarized upon reflection. The polar and meridional Kerr effects consist in the rotation of the polarization plane and elliptical polarization of the initially (prior to reflection) plane-polarized light. The equatorial Kerr effect

consists of à change in intensity and a phase shift after reflection. By studing the dispersion of the equatorial Kerr
effect it became possible, for the first time, to obtain information concerning the surface magnetism and surface structure
of magnets.

2.2.6 EPR Spectra

The electron paramagnetic resonance (EPR) technique is currently one of the most informative and sensitive methods for
studying the transition-metal ions [2.70,71] and is widely employed in experiments on adsorption and catalysis [2.72,73]. It
easily detects unpaired electrons (if any) in the system, and is
useful for studying the ions with the configurations $d1$
(Mo^{5+}, Ti^{3+}, V^{4+}, W^{5+}), d^9 (Cu^{2+}), and others in oxide matrices. At
the same time, the ions $Mo^{4+}(d^2)$ and $Mo^{3+}(d^3)$ cannot be
observed at ordinary temperatures because of line broadening
associated with spin-lattice relaxation. The overall intensity
of the EPR spectrum can be related to the concentration of paramagnetic ions.

The theory of EPR spectra provides the basis for exact measurements of the degree of oxidation and coordination of transition-metal ions. The electron structure of the paramagnetic
center is deduced from the value of the g factor, its anisotropy, fine and hyperfine line structure, and the shape and the
width of the line.

The g factor allows one to determine the nature of the particle's paramagnetism: whether the magnetic moment is due
solely to the spin or is aided by the orbital moment. In the
case of purely spin magnetism the value of the g factor equals
that for a free electron: $g = g_e = 2.0023$. The difference
between the actual values of g and g_e reflects the contribution
of the orbital moment to the total magnetic moment in the
ground state. For one electron in a nondegenerate d orbital
(d^1) or one hole in an occupied d shell (d^9), the value of the g
factor in either direction is given by

$$g = g_e + (n\lambda/\Delta E) , \qquad\qquad\qquad (2.9)$$

where λ is the constant of spin-orbital interaction; ΔE denotes the splitting between the orbital having an unpaired electron and the excited orbital (as found from optical spectra); $n = 2$ for $d_{xy}-d_{xz}$, $d_{xz}-d_{yz}$, $d_{xz}-d_{x^2-y^2}$, $d_{yz}-d_{x^2-y^2}$, $d_{xy}-d_{yz}$; $n = 6$ for $d_{xz}-d_{z^2}$ and $d_{yz}-d_{z^2}$; and $n = 8$ for $d_{x^2-y^2}-d_{xy}$.

The constant of spin-orbital interaction λ is positive ($g>2$) if the unpaired electron goes over from an occupied orbital to a half-occupied orbital, and negative ($g<2$) if the unpaired electron goes from a half-occupied orbital to a free orbital. These two types of centers are called the hole-type center and the electron-type center, respectively. Simple relations of this kind may fail when the electron is delocalized in a solid [2.74]. Similar relations can be obtained for other transition-metal ions. For 3d transition metals the value of λ is about 10^2 cm^{-1}, and is much less than ΔE (10^3–10^4 cm^{-1}).

For low-symmetry structures the values of the g factor along different axes are not the same (g-factor anisotropy). For example, in a d^9 ion (Cu^{2+}) in a tetragonal configuration with an elongated z axis, the unpaired electron occupies the $d_{x^2-y^2}$ orbital (Fig.2.4). The value of the g factor along the z axis (g_p) differs from those along the y axis (g_s):

$$g_p = g_z = g_e + \frac{8\lambda}{\Delta E(d_{x^2-y^2} - d_{xy})} ,$$

$$(2.10)$$

$$g_s = g_x = g_y = g_e + \frac{2\lambda}{\Delta E(d_{x^2-y^2} - d_{xz}, d_{yz})} .$$

The values of ΔE can be found from optical spectra.

The expressions (2.9,10) have been are derived in the crystal-field theory approximation. The ligand-field approxima-tion, which takes the effects of covalence into account, would require putting a coefficient $K < 1$ before the second terms on the right-hand sides of (2.9,10) to reflect the reduced participation of the d orbitals of the central ion due to creation of a molecular orbital with ligands.

If the transition-metal ion has a nucleus with nuclear spin J, the spectrum exhibits n = (2J+1) hyperfine splitting (HFS) lines of equal intensity. The separation between the lines depends on the spin density of the electron in the immediate proximity of the nucleus; it is called the constant of hyperfine interaction and is denoted by A. This provides an easy way of determining for which nucleus the electron is localized. For instance, all commercial samples of MgO contain a small quantity of Mn^{2+} ions, which give six lines of equal strength in the EPR spectrum. The nuclear spin of ^{55}Mn is 5/2; hence n = 2·5/2 + 1 = 6.

The orbital of an unpaired electron can embrace several nuclei. If the electron is found with equal probability at either of two nuclei of the same kind, each having nuclear spin J, the EPR spectrum will dispaly 4J+1 lines. For instance, the spectrum of $>Co^{3+}OCo^{2+}<$ shows 15 HFS lines (the nuclear spin of ^{59}Co is 7/2). If, however, the unpaired electron interacts differently with the two nuclei of one and the same element, the EPR spectrum will display HFS of (2J+1)(2J+1) lines of unequal strength. For instance, a complex EPR spectrum of supported partially reduced V_2O_5/MgO catalysts displayed 64 anisotropic lines, which were attributed to VO_2^+ ... V^{5+} centers (for the ^{51}V nucleus J=7/2) [2.75].

When the electron is delocalized over several nuclei, HFS is usually not observed. If in such a case the unpaired electron's orbital does not overlap with the wave functions of other unpaired electrons, the EPR lines usually undergo broadening. Another cause of EPR line broadening is the spin-lattice relaxation. When the delocalization of unpaired electrons is considerable, their orbitals overlap with one another. Then a narrow EPR line is observed ("exchange narrowing") without HFS.

The surface nature of the paramagnetic ion is usually judged by the broadening (up to complete disappearing) of the EPR signal upon physical adsorption of O_2, H_2O molecules etc. on the surface (owing to dipole-dipole interaction), or by drastic changes in the signal (again up to complete disappearance) upon chemisorption. In recent years research in the field of catalysis has been assisted by the spin-echo technique as applied to the EPR of paramagnetic ions. This technique has proved quite helpful in the study of diluted systems (Cr^{3+} ions on silica gel) [2.76]. TSUNEKI et al. investigated the effects of adsorption of water and ammonia on the relaxation times of paramagnetic ions.

2.2.7 Nuclear Magnetic Resonance

Unfortunately, investigations of adsorption and catalysis have not yet enjoyed the full benefits of nuclear magnetic resonance technqiues (NMR) available today [2.77]. Two main reasons may be cited in this connection: Firstly, a lower sensitivity as compared with EPR, which requires the use of extra-high-dispersion adsorbent powders and makes measurements possible only when the concentration of resonant nuclei is high enough. Secondly, the chemisorptive complexes have a very low mobility, which results in the resonance lines being quite broad - nearly as broad as the lines of the solid and much greater than the values of chemical shifts. This reflects distortion of the electron structure of adsorbed particles. Consequently, the structure investigations of adsorptive complexes and catalytic transformations have been lagging behind the liquid-phase high-resolution NMR spectroscopy. In the sixties and seventies the research was mainly concerned with studying (by conventional NMR spectroscopy) the mobility of adsorbed molecules at comparatively high occupancies [2.1].

As in solids, the broad lines in NMR spectra of adsorbed molecules arise from strong dipole-dipole interactions between the resonant nuclei. The problems of high-resolution NMR spectroscopy in solids could be successfully dealt with if it were possible to simplify the Hamiltonian of dipole-dipole interactions H_d. If H_d tends to zero, the NMR linewidth is reduced by a factor of 10^3, whereas the chemical shift is merely be reduced by a factor of ($3^{1/2}$. Some improvements of the spectral resolution was achieved by making use of a rapid exchange between the chemisorbed and the physisorbed molecules – the so-called medium-resolution spectra [2.77]. This approach to diluted systems has already been employed in studying transition-metal complexes in solution [2.78]. Applications of this technique to the study of chemisorption will be considered in Sect.4.5.2. Another way of reducing the local magnetic fields consists of spinning the specimen at a very high speed at a magic angle θ_m with respect to external magnetic field (θ_m = arccos $3^{1/2}$ = 54°44'08") – so called MASNMR (magic angle spinning nuclear magnetic resonance). MASNMR was employed for studying adsorbents and catalysts [2.79-82], too.

Outstanding success in improving the resolution of NMR spectroscopy of solids was reported by a team of physicists working at the Massachusetts Institute of Technology (MIT) [2.83]. The spin-Hamiltonian of interacting nuclei was found capable of being substantially modified when the system of nuclear spins was irradiated by a coherent train of microsecond radio-frequency pulses of appropriate amplitude, phase, pulse rate, and duration. This technique is especially promising for studying heteronuclear systems, e.g., the resonance of C and H nuclei. By exerting independent control over each nuclear system, it is possible to pump energy into the lower-concentration spin system, thereby improving the sensitivity of NMR spectroscopy. Using this technique the MIT physicists obtained high-resolution NMR spectra of C^{13} naturally occurring in organic monocrystals. Recently this technique was successfully employed for studying the surface phases.

2.2.8 Mössbauer Spectroscopy

Mössbauer (gamma resonance) spectroscopy is based on the Mössbauer effect, the emission and absorption of gamma quanta by certain nuclei, bound in crystals, without loss of energy through nuclear recoil, with the result that radiation emitted by one such nucleus can be absorbed by another. Currently, this technique is widely employed for studying the surface phenomena of adsorption and catalysis [2.84–86].

The probability for the Mössbauer effect to occur is

$$f = \exp(-4\pi^2 \langle x^2 \rangle \lambda^2) \tag{2.11}$$

where $\langle x^2 \rangle$ is the mean square amplitude of oscillations in the direction of emission of gamma quantum; λ is the latter's wavelength.

According to (2.11), the probability is greater, the longer the wavelength, that is, the softer the radiation. The investigations of surface phenomena employed the nuclei (absorbers) ^{57}Fe, ^{119}Sn, ^{121}Sb, ^{125}Te, and ^{151}Eu. The use of this technique is restricted by its sensitivity: one has to choose systems with highly developed surfaces (hundreds of square meters per gram). However, the sensitivity of this technique can be raised by up to three orders of magnitude by sprinkling the surface with radioactive atoms (e.g., ^{57}Co), which serve as a source of resonant gamma radiation.

As follows from (2.11), the Mössbauer effect can be used for studying the movement of atoms on the solid surface. The mean-square displacement $\langle x^2 \rangle$ of oscillating atoms is greater on the surface than in the bulk; accordingly, the value of f on the surface is smaller. With increasing dispersivity of the solid f decreases. The diffusion of atoms on the solid surface can be judged by the width of Mössbauer lines: the coefficient of diffusion is proportional to broadening.

The Mössbauer effect has also been employed for studying the electron structure of surface atoms. The electron cloud at the nucleus interacts with the positive charge of the nucleus leading to a change in the relative energies of the ground and the excited nuclear states: the so-called isomeric Mössbauer shift. Since it is only the s electrons that have nonzero density near the nucleus, the isomeric shift will characterize the formation of bonds at the expense of s electrons. The surface atoms may differ from the bulk ones in bond ionicity and extent of mutual screening of the s, p, and d electrons. So far, studies of this aspect are sparse. Many more data have been reported on correlations between various surface characteristics and the quadrupole splitting ΔE_Q [2.87]. The quadrupole Mössbauer splitting arises from an interaction between the nuclear quadrupole moment and the electric field gradient in the vicinity of the nucleus. When the field gradient is zero, the excited levels are degenerate and $\Delta E_Q = 0$. The quantity ΔE_Q reflects the deviation of nuclear symmetry from spherical symmetry. The surface atoms occur in lower-symmetry fields than the bulk atoms, which ought to result in an increase in ΔE_Q. Indeed, the increase in ΔE_Q with the rise in dispersivity has been observed experimentally.

An important feature of Mössbauer spectra is the hyperfine magnetic interaction. In a magnetic field the spin degeneracy of nuclear energy levels is removed. The magnetic field can be both external and internal (the latter only with ferro- and antiferromagnets). For example, for ^{57}Fe in ferromagnetic (α-Fe) and antiferromagnetic (α-Fe$_2$O$_3$) states, the selection rules predict a symmetric spectrum of six lines. Small particles (smaller than a magnetic domain) – in the case of α-Fe$_2$O$_3$ this means smaller than 15 nm across – have a Mössbauer spectrum typical of paramagnetic ions: a quadrupole-splitting doublet. Such substances are called superparamagnetic. The hyperfine magnetic interaction has been successfully used for studying supported ferromagnetic catalysts.

2.2.9 Vibration Spectroscopy

The main research on surface compounds relies heavily on infrared spectroscopy [2.88-91]. The mediocre sensitivity of the
IR technique is counterbalanced by adequate resolution and the
convenient capability of being usable at high pressures, thus
bringing experimental conditions close to the actual ones for
catalysis. Note that the above-described photoelectron techniques work only in high vacuum. The IR technique is usually
used to explore the spectral range from 400 to 4000 cm^{-1},
where electromagnetic radiation is absorbed in transitions
between quantized vibrational levels. The vibrational movement
of atoms in a molecule is sensitive to changes in their surroundings, which makes it possible to use IR spectra for studying the structure of adsorptive complexes and surface compounds.

The intensity of the IR signal is proportional to the squared
derivative of the dipole moment with respect to the normal
coordinate. It it especially easy to detect vibrations of polar
bonds (O-H,N-H,C-O). However, in order to detect even such
bonds on the surface, it is necessary to use very high-dispersion powders with a specific surface of 10^5-10^6 cm^2g^{-1}. Samples
are usually prepared in the form of compact tablets, whose
thickness corresponds to 10-40 mg·cm^{-2}, or in the form of high-
dispersion powders deposited onto a transparent support. In the
past the adsorbents were often compacted into tablets together
with silica or KBr powder, in order to reduce scattering in a
disperse medium. Today this routine has been abandoned, because
silica or KBr powder tends to interact with many adsorbents and
adsorbed molecules. The IR spectra of molecules adsorbed on
small metallic surfaces can be recorded using the technqiue of
multiple internal reflection [2.92]. The sensitivity is greatly
improved in modern IR spectrometers coupled with a Fourier
transformer (FTIR). These instruments allows one to obtain
strong spectra of adsorbed substances using powders with a
specific surface of 10^{-3}-10^{-4} cm^2g^{-1} [2.93].

The processing of IR spectra consists of calculating the frequencies of vibrations in polyatomic molecules influenced by the field of the adsorbent or active center on the surface. The theory can be checked using isotopic replacement: since the frequency of vibrations, ν, is inversely proportional to the square root of the reduced mass μ, the replacement of one of the adsorbate atoms by an isotope of different mass must result in a shift in frequency of vibration. For example, the frequency of stretching vibrations of an O-H bond is reduced by factor of $(\mu_D/\mu_H)^{1/2} = 1.334$ when H is replaced by D (Chap.4).

More often, however, the frequencies are analyzed by empirical comparison of experimental IR spectra of surface compounds with the known spectra of bulk structures of solids or homogeneous complexes. The identification of surface compounds can also be based on the concept of characteristic vibrations, whose frequency is assumed to show little dependence on the surroundings. Such characteristic frequencies are found for vibrations of, e.g., O-H, C-H, N-H, and C=C bonds.

The IR absorption intensity can be derived from the concentration of adsorbed substance, provided that the extinction coefficient is known. Some extinction coefficients are listed in LITTLE's classic monograph [2.88], but they are seldom used today. The width and shape of the IR band can be utilised in analyzing the rotational movements of adsorbed molecule.

To a certain extent, IR transmission-absorption spectroscopy is supplemented by Raman spectroscopy. Raman spectra are used for measuring the frequency shifts in the visible region. These shifts correspond to vibrational quanta. In contrast to IR spectra, where the band intensity is associated with the change in dipole moment, the intensity of Raman spectra reflects the change in polarizability arising from vibrational movement. The use of Raman spectra in adsorption started more than 20 years ago; first attempts, however, yielded little success because of the low intensity of the spectra obtained [2.94-96]. Recently the situation has changed thanks to the development of commercial Raman spectrometers using powerful visible-light laser sources (laser Raman spectrometers) and the discovery of enormous coefficients of extinction for substances adsorbed on surfaces of Cu, Ag, and Au [2.97]. The intensity of Raman spectra can be larger by a factor of 10^5–10^6 than that customarily observed in conventional Raman spectroscopy.

In [2.1,2] we have mentioned another type of vibrational spectroscopy: inelastic electron tunneling spectroscopy (IETS). The advantages are the exceptionally high sensitivity (capable of detecting as few as 10^{10} atoms in a sample) and sufficient resolution (about 10 cm^{-1}). The drawbacks are that IETS can be only used with very thin ($\simeq$2 nm) dielectric objects, whose surfaces must be metallized to become the electrodes; what is more, the spectra must be taken at liquid-helium temperature.

The high resolution low-energy electron-loss spectroscopy (HREELS) uses a monochromatic beam of electrons with an energy of 10–20 eV to excite surface vibrations [2.98,99]. The reflected beam exhibits energy loss corresponding to the value of vibrational quanta (100–300 meV or 800–2400 cm^{-1}), which can be monitored with adequate resolution (within 10 meV). The advantage of HREELS is its capability of measuring the vibration spectra of substances adsorbed on minute surfaces of single crystals. This feature is especially useful in studying hydrogen on metallic surfaces, since the hydrogen structures cannot usually be detected by LEED [2.100]. A disadvantage of this technique is that it requires high vacuum, like photoelectronic techniques.

HALLER [2.91] argued that none of the novel vibrational spectroscopic techniques (Raman, IETS, HREELS) are worthwhile, if it is possible to use conventional IR transmission-absorption spectroscopy.

A new method of vibrational spectroscopy has recently been worked out in the Institute of Spectroscopy and in the Institute of Chemical Physics of the Academy of Sciences of the USSR [2.101].It is the so-called method of surface electromagnetic waves (SEW) or polaritons. The method has both high sensitivity and high resolution. It allows one to measure very small coverages of a single crystal by an adsorbate.

3. The Surfaces of Transition-Metal Oxides

The starting point of our discussion of adsorption and catalysis on transition metals will be the surface of transition-
metal oxides. This is because these surfaces are more convenient for studying the principal features of active sites than
the transition metals themselves. In the case of oxides, the
interaction between the transition-metal atom and oxygen is
essentially similar to the interaction between the central atom
and the ligands in complexes. For metals the interaction
between the adjacent atoms has certain peculiar features, which
will be discussed in Chap.5. Here we shall first consider dilute
oxide systems, and then deal with ordinary nondilute systems.

3.1 Surface Atoms of Transition Metals in Dilute Oxide Systems

Cleavage of a crystal in the ideal crystallographic plane
reduces the coordination number of the surface ions. Figure 3.1
shows the coordination of a metallic ion on the surface of
simple MO (e.g., NiO) oxide (a cubic NaCl-type crystal) in the
cleavage planes (100), (110), and (111). As a result of atomic
relaxation, the metallic atom retreats a bit beneath the
surface, as shown in Fig.3.1. On the (100) face of the NaCl-type
crystal one cation is surrounded by five anions. As a result it
is somewhat displaced below the surface, and its coordination
corresponds to a square-based pyramid. On the (110) face, the

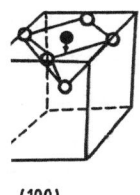 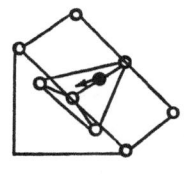 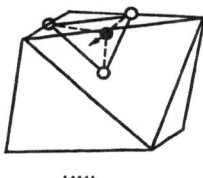

(100)   (110)   (111)

ig.3.1. The coordination of metallic ion on the faces of cubic
ᴐ oxide

ιtion falls short of two anions, and the symmetry becomes
ιtrahedral rather than octahedral, and plane triangular on the
11) face where three anions are lacking. Symmetric coordina-
.on is further distorted on crystal edges and defects.

If the surface ion is a transition-metal ion, all our previ-
ιs considerations regarding stabilization of its energy and
:ructure by the crystal field apply in full. It is clear from
ιble 2.2 that reduction in symmetry (lowering of the coordina-
ιon number) is accompanied by the decrease in the CFSE
:rystal-field stabilization energy) value. Consequently, the
ransition atom on the surface must be expected to take advan-
ιge of various opportunities to raise its coordination number.
ιch an opportunity is provided, in particular, by adsorption,
hich will be discussed in detail in the next chapter.

Dilute oxide systems provide a convenient example for study-
ιg the electronic structure of atoms of transition metals,
ιeir coordination, degree of oxidation, and ability to form
ᴐmplexes with different ligands or adsorbates. Oxide systems
ιth ions of transition metals deposited on the surface are
ιso widely used as catalysts.

Dilute oxide systems, in which a transition-metal oxide is
ᴐntained in a nontransition oxide matrix, have been the subject
ᵬ a large number of research reports [3.1-49]. These investi-
ιtions were, in particular, aimed at obtaining complexes of
ransition metals in a coordination corresponding to the
ιtrix. Such complexes enable one to study the effects of tran-
ιtion-ion symmetry in catalytic and adsorptive activity,

checking the validity of crystal-field and ligand-field theories. Another aim is analysing the relative importance of collective and local interactions in transition-metal compounds. At high dilutions the d-d interactions vanish, and what we get is isolated transition-metal ions in a dielectric matrix. Of crucial importance for adsorption and catalysis is the problem of cation distribution in dilute oxide systems: whether the cations are segregated or distributed evenly throughout the entire volume.

The majority of papers cited deal with the state of transition ions in the bulk rather than on the surface of oxide; the results, however, are used for reaching conclusions regarding certain surface phenomena, which is not always justified.

The dilute oxide systems fall into three main groups:

1) <u>True solid solutions,</u> where the transition-metal ion has the same valence and the same (or nearly the same) size as the original metal atom in the matrix (e.g., Ni^{2+}, Mn^{2+}, Co^{2+} in MgO matrix; Cr^{3+} and Fe^{3+} in an Al_2O_3 matrix).

2) <u>Systems in which the transition-metal ion differs in valence, size, or orbital symmetry from the original metal ion of the matrix</u> (e.g., Cr^{3+} or Cu^{2+} in MgO; V^{5+}, Cr^{5+}, Mo^{6+} in SiO_2, Al_2O_3, and the like). In this case the foreign ions can (a) enter the solid solution with the charge being compensated by other foreign ions (e.g., the extra charge of Cr^{3+} in MgO can be counterbalanced by Li^+), (b) assume a different oxidation state, (c) be restricted to the surface, or (d) form a chemical compound.

3) <u>Systems in which the transition-metal ion is of the same chemical nature as the original ion, but occurs in a different state of oxidation</u> (e.g., Ti^{3+} in TiO_2).

Let us begin by considering the systems of the first type.

3.1.1 The NiO-MgO System

The ionic radii of Mg^{2+} and Ni^{2+} are about 0.074 and 0.072 nm, respectively. The lattice constants are also close: a = 0.421 nm for MgO and a = 0.419 for NiO. Both oxides have face-centered-cubic NaCl-type lattices and produce a continuous series of substitutional solid solutions. Heat treatment above 950°C gives the solid solution $Ni_xMg_{1-x}O$, where the distribution of cations approximates most closely a uniform, statistical pattern [3.6-20]. The proportion of isolated ions under such a distribution is given by

$$x_0 = x(1 - x)^m , \qquad\qquad (3.1)$$

where x is the overall molar fraction of Ni ions, m is the number of cation centers adjacent to the chosen one, $x_0 = n_0/N$, n_0 being the number of isolated Ni ions, and N is the overall number of cation centers in the sample.

The fraction of ions bound in Ni-O-Ni pairs is given by

$$x_1 = mx^2(1 - x)^{m-1} . \qquad\qquad (3.2)$$

Visible-light spectra of $Ni_xMg_{1-x}O$ specimens heat-treated at high temperatures display bands corresponding to d-d transitions of Ni^{2+} ions in an octahedral configuration [3.16]. The EPR spectrum exhibits a singlet line with g=2.226 at x<0.01, which also corresponds to isolated Ni^{2+} ions in the bulk [3.13]. The surface ions generally defy spectroscopic examination, since the surface area of strongly calcinated materials is usually small. There are indications, however, that heat treatment at high temperatures equalizes concentrations of Ni^{2+} in the bulk and on the surface. This was demonstrated in [3.16] for chemisorption of NO, which has the convenient property of being selectively adsorbed by Ni^{2+} ions with a stoichiometric ratio NO:Ni=1:1.

The equilibrium distribution of cations between bulk and surface depends on temperature. Unfortunately, this reality is often overlooked in investigations of solid-oxide solutions. Solid solutions prepared at high temperatures become nonequilibrium after cooling. The equilibrium then does not mean even distribution of cations between surface and bulk. The surface usually has an excess of the larger-radius cation; in the $Ni_xMg_{1-x}O$ system this will be the nickel cation [3.19]. For this reason, of great interest are $Ni_xMg_{1-x}O$ specimens heat-treated at low temperatures (400–600°C). The oxide that has been treated at such temperatures has a well-developed surface (up to 100 m^2/g), and a good share of cations are found on the surface. In this case the distribution of Ni^{2+} ions in the MgO matrix is no longer random, and already at $x \simeq 0.005$ there are pairs or clusters of Ni^{2+} ions interacting in the first and the second coordination spheres [3.10,20]. The conclusion regarding segregation of Ni^{2+} ions in a $Ni_xMg_{1-x}O$ solid solution is supported by data [3.14] on magnetic susceptibility and diffuse X-ray scattering as functions of the concentration of Ni^{2+} ions. The intensity of the EPR spectrum decreases with an increase in x. At $x \simeq 0.1$ the conductivity of oxide exhibits a sharp rise, and polaronic conduction is observed. At $x \simeq 0.4$ antiferromagnetic interactions set in: the Néel temperature, initially equal to about 0°C, grows steadily and approach the T_N value for pure NiO.

On the surface of $Ni_xMg_{1-x}O$ specimens, heat-treated at not too high temperatures, Ni ions are easily detected by optical methods. Their electronic configuration and coordination differ from those of the bulk ions: the latter are usually Ni^{2+} ions in an octahedral configuration. Diffuse reflection spectroscopy demonstrated [3.16] that some of the Ni^{2+} ions on the surface are found in tetrahedral coordination. Other ions of nickel on the surface are Ni^{3+}, apparently in a strong-field configuration with paired spins. Their appearance on the surface is due to charge stabilization by the hydroxyl groups OH^-. The wide singlet in the EPR spectrum pertains to Ni^{2+} clusters, where the exchange interactions are lower because of the proximity to the surface [3.10,11]. These ions have a lower symmetry as compared with the bulk ones, which increases their contribution to the net spectrum.

BICKLEY and STONE [3.8] considered the surface structure of $Ni_xMg_{1-x}O$ doped with Li_2O at low temperature. The Ni^{3+} ions might be assumed to correspond in quantity to the number of Li^+ ions introduced into the surface. It turns out, however, that scarcely any Ni^{3+} ions are formed. This results from intrusion of Li^+ ions into the surface layer with one vacancy formed per every two Li^+ ions. Implantation of Li^+ ions at high temperature ultimately results in a solid solution, with the occupied anion and cation vacancies where Li^+ ions are compensated by Ni^{3+} ions.

3.1.2 The CoO-MgO System

The ionic radius of Co^{2+} is 0.078 nm; the lattice constant of CoO is a = 0.426 nm. These values are also close to those of MgO, so CoO and MgO form substitutional solid solutions with the lattice parameter being a linear function of composition [3.7,12,17-35]. Such linearity is observed even at low heat-treatment temperatures (400-500°C), where hydroxides of Co and Mg are not completely transformed into oxides.

In an attempt to explain the dependence of the magnetic properties of the solid solutions $Co_xMg_{1-x}O$ on the concentration x many researchers have made detailed calculations of the possible number of clusters of Co ions, connected via oxygen and capable of exchange interactions, assuming a random distribution of Co^{2+} ions. Actually, however, especially at low heat-treatment temperatures, the Co^{2+} ions in the lattice are distributed unevenly, and segregation is observed even with low x values. It was demonstrated [3.25] that an increase of the concentration of CoO in MgO leads to an increase in the ratio of the intensity of higher indices to intensity of lower indices in the powder photograph. This can be attributed to the inhomogeneity of the distribution of Co^{2+} ions in the lattice and to clustering of Co^{2+} ions. The same conclusion is suggested by measurements of the magnetic susceptibility κ: the values of κ even at low concentrations of CoO are lower than anticipated, assuming a chaotic distribution of Co^{2+} ions. The calculated values for the magnetic moment μ of the Co^{2+} ion vary from 5.3 Bohr magnetons at low concentration to 4.4 at concentrations of CoO above 8%. The higher value pertains to Co^{2+} in an octahedron; the lower value corresponds to a tetrahedron or square-based pyramid. The electrical properties of the system suffer a change in more or less the same range of concentrations ($x=0.08-0.15$): polaronic conductivity arises [3.17].

The magnetic data agree with the optical, diffuse-reflection spectra. The spectra of all $Co_xMg_{1-x}O$ specimens, heat-treated at both high and low temperatures, exhibited an absorption band at 520 nm corresponding to Co^{2+} in the octahedral coordination [3.24,27,29]. This band has a low intensity and, for specimens heat-treated at no more than 600°C, is disguised by other absorption bands. The latter include bands at 550, 600, and 755 nm, which arise from d-d transitions in fields having distorted octahedral symmetry or that of a square-based pyramid C_{4v} [3.26,33].

Such centers with lowered symmetry are formed on the surface of $Co_xMg_{1-x}O$. It was shown (using optical spectroscopy) that the band at 755 nm virtually disappears after adsorption of ammonia, which indicates that it is associated with a surface site. Thermodynamic calculations of the cation distribution between bulk and surface lead to the conclusion that the surface of the $Co_xMg_{1-x}O$ solid solution is probably highly enriched with Co ions [3.19].

A comparison of magnetic, optical, and adsorptive (Sect.4.7) properties of solid solutions points to the existence of several types of Co^{2+} coordination (Fig.3.2): one which may be isolated (type 1), and two others displaying indirect-exchange interaction between Co^{2+} ions (types 2 and 3). The Co^{2+} ions on the surface are assembled in large clusters, totalling as many

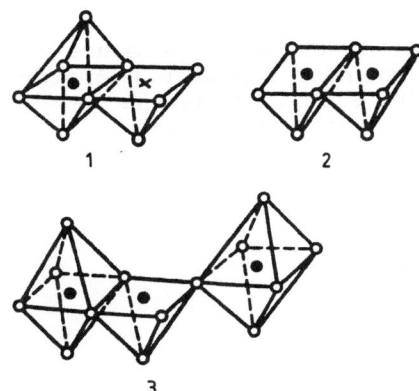

Fig.3.2. Different types of coordination of Co^{2+} on the surface of $Co_xMg_{1-x}O$

as several dozens of atoms [3.25,27,28]. A cluster exhibits a hyperexchange interaction: $Co^{2+}-O^{2-}-Co^{2+}$. When heat treated the cluster size is smaller at high temperatures than with low temperatures. At very high temperatures the ions of the transition metal tend to assume a statistical distribution, isolated rather than in clusters.

An analysis of the cluster formation from the thermodynamic viewpoint indicates that the equilibrium in the crystal may have the form [3.28]

$$X_1 \rightleftarrows \frac{1}{2}X_2 \rightleftarrows \ldots \rightleftarrows \frac{1}{n}X_n ,$$

n being the number of atoms in a cluster. The concentration of clusters with different numbers of atoms X (e.g., Co^{2+} in MgO) is defined not by the statistical distribution (3.1,2) but depends rather on the specific interaction between the ions. The concentration can be expressed as

$$[X_n] = K_n \left[X_1 \right]^n , \qquad (3.3)$$

where K_n is the equilibrium constant. The concentration of large clusters grows with the increase of the overall concentration [X] of Co^{2+} ions and decreases with the rise in heat-treatment temperature. A schematic phase diagram for two cluster types is illustrated by Fig.3.3. At low heat-treatment

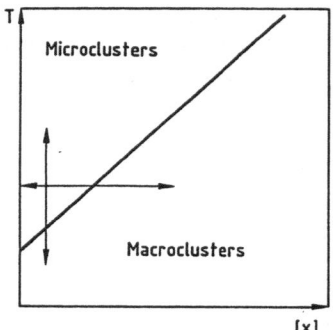

Fig.3.3. Schematic phase diagram of composition versus temperature for a solid solution containing clusters of varying sizes, [x] is the concentration of transition atoms

temperatures aggregation is thermodynamically more advantageous, while at high temperatures the entropy factor facilitates "evaporation" of the macroclusters, which tend to become "micro" instead. The two extremes are segregation of the solid solution into separate phases (low temperatures) and statistically disordered solid solution (high temperatures).

The $Co_xMg_{1-x}O$ systems under study displayed, apart from an anomalous behavior of the magnetic susceptibility with low CoO concentrations, a drastic change in magnetic properties when the concentration of CoO became about 60%. At higher concentrations antiferromagnetic interactions start to take place. As x increases from 0.65 to 1 the Néel temperature grows steadily from $-120°$ to $+120°C$ [3.29].

If a cobalt salt is impregnated into MgO with subsequent heat treatment at a low temperature (400°C) in air, most of the Co atoms will be on the surface in the form of low-spin octahedral Co^{3+}, as indicated by the optical, diffuse reflection [3.24,35]. The Co^{3+} ions here belong to the spinel-type surface compound $MgCo_2O_4$.

From Table 2.1 it is clear that the low-spin octahedral Co^{3+} (electron configuration d^6) exhibits the highest crystal-field stabilization energy (CFSE), as compared with any other possible d^n configurations of ions in oxygen surroundings. Consequently, if a system includes an electron acceptor, the Co^{2+} ion will give away another electron and be stabilized in the form of an octahedral Co^{3+} ion[1]. The excess positive charge of Co^{3+} on the surface is counterbalanced by the surface hydroxyl groups OH^-. Dehydration of the surface with the rise in heat-treatment temperature destroys this balance, and Co^{3+} becomes Co^{2+} in a tetrahedral (at low concentrations) or an octahedral (at high concentrations x) configuration. Anion vacancies $\square_O$ are formed on the surface with a simultaneous giving off of oxygen:

$$2\,Co^{3+} + 2\,OH^- \longrightarrow 2\,Co^{2+} + \square_O + H_2O + \tfrac{1}{2}O_2 .$$

Doping of the $Co_xMg_{1-x}O$ system with Li_2O results in an increase of the Co^{3+} content on the surface [3.125]. The difference from $Ni_xMg_{1-x}O$ is due to Co^{3+} having greater solubility (3.5%) in MgO than Ni^{3+} (0.2% at 1200°C). The segregation of Co^{3+} ions on the surface in the presence of Li^+ also increases, and a $LiCoO_2$ phase is formed when the excess of Li is great [3.22].

3.1.3 The Cr_2O_3-Al_2O_3 System

The Cr_2O_3-Al_2O_3 system [3.17,18,23,36-46] has been the subject of investigation due to the practical importance of aluminum-chromium oxide catalysts for dehydration, dehydrocyclization, and polymerization of olefins.

The ionic radii of Al^{3+} (0.057 nm) and Cr^{3+} (0.064 nm) are close, and the oxides of Al and Cr may form solid solutions. Some difficulties arise from the existence of several modifications of Al_2O_3; the most wide-spread modifications are α-Al_2O_3 (rhombohedral corundum lattice: a=0.513 nm, α=55.16°) and γ-Al_2O_3 (cubic spinel structure). Chromium oxide α-Cr_2O_3 (corundum lattice: a=0.535 nm, α=55.1°) is isomorphous with only one of these, and can form solid solutions in the entire range of concentration. However, α-Al_2O_3 is obtained only after heat treatment at high temperatures (1000°-1200°C). For this reason

---

[1] For the stabilization of Co^{3+} on the MgO surface and spinel formation, see Sect.3.1.4.

heat treatment of a mixture of Cr_2O_3 and Al_2O_3 at moderate temperatures (500°–600°C), including heat treatment of specimens obtained by depositing chromium salts on γ-Al_2O_3, yields the gamma modification with Cr_2O_3 (at low concentrations) mostly confined to the surface.

Let us consider first the bulk properties of the $Cr_xAl_{2-x}O_3$ system. The analysis of EPR spectra [3.36–42] suggests to the existence of several paramagnetic formations in the Cr–Al oxide system. According to O'REILLY [3.38], δ, β, γ, and α phases can be distinguished. The EPR signal with g≈4 and ΔH≈1600 G pertains to isolated Cr^{3+} ions and seems to correspond to some component in the poorly defined fine structure. This signal prevails at concentrations of chromium below 1% and relates to the δ phase. The increase of concentration of Cr^{3+} ions gives rise to a wider line (ΔH≈3400 G) with g≈2. This line somewhat resembles the line of Cr^{3+} in the spectrum of pure chromium oxide, observed above the Néel point. It may therefore be attributed to a magnetoconcentrated system (β phase), whose spectrum exhibits broadening due to a dipole–dipole interaction. Nevertheless, it differs from the spectrum of α–Cr_2O_3 since (first) the linewidth of $Cr_xAl_{2-x}O_3$ is several hundred Gauss larger than that for α–Cr_2O_3, and (second) the solid solution does not show a Néel point. The difference between δ and β systems in optical, diffuse-reflection spectra hardly shows up at all; however, they differ in the spectrum of α–Cr_2O_3.

Figure 3.4 shows the structure of the surface layer of the solid solution $Cr_xAl_{2-x}O_3$ [3.36]. The diagram illustrates the difference between magnetoconcentrated (β-phase) and magneto-dilute (δ-phase) systems. Magnetic measurements [3.38] indicate that the Néel point T_N and a truly antiferromagnetic interaction below T_N (α phase) appear at x≈0.3; T_N grows with the increase of x and reaches 33°C for pure α–Cr_2O_3.

Oxidized $Cr_xAl_{2-x}O_3$ specimens display the so-called γ phase, which is formed exclusively on the basis of γ-Al_2O_3 (EPR

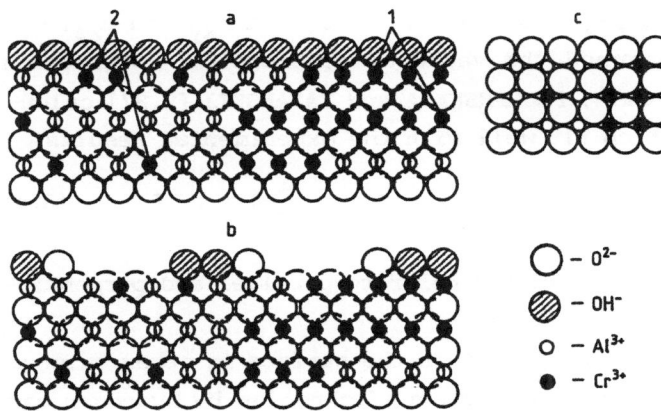

Fig.3.4. Schematic structure of the supported aluminum–chromium catalyst [3.36]: (1) β phase, (2) δ phase; (a) $(Cr_xAl_{1-x})_2O_3$ after the adsorption of water (normal section), (b) the same after desorption, (c) top view (note the second layer of oxygen atoms). In (a) and (b) the rear layer of oxygen atoms is shown by dotted circles

linewidth: $\Delta H \simeq 50$–100 G, $g \simeq 2$) [3.38]. The signal intensity grows with the rise in dispersity of aluminum oxide. The EPR signal from the γ phase is strongest for a concentration of Cr_2O_3 of about 2%. COSSEE and VAN REIJEN [3.39–41] have assumed the EPR signal from the γ phase to correspond to the Cr^{5+} ion, stabilized in an octahedron with the formation of chromyl ions on the surface in the coordination of a truncated square pyramid:

$$-O-\overset{\displaystyle O}{\underset{\displaystyle O\ \ O}{\overset{5+}{Cr}}}-O-$$

PECHERSKAJA and KAZANSKIJ [3.37] have demonstrated that optical and EPR spectra corroborate this assumption. The optical spectra from oxidized supported Cr–Al oxide specimens exhibit the transitions $\Delta E_1 = 18.200$ cm^{-1} (between d_{xy} and

$d_{x^2-y^2}$ levels) and ΔE_2 = 14.000 cm^{-1} (between d_{xy} and d_{xz}, d_{yz} levels). Having assumed the value of spin orbital interaction constant λ = 150 cm^{-1}, PECHERSKAJA and KAZANSKIJ [3.37] calculated g factors which fit well with the experimental results

$$g_z = 2(1 - \frac{4\lambda}{\Delta E_1}) = 1.93 \; , \qquad g_{x,y} = 2(1 - \frac{\lambda}{\Delta E_2}) = 1.97 \; , \qquad (3.4)$$

where the subscripts z and x,y refer to parallel and perpendicular, respectively.

Chromium ions may be present on the surface in other valence states, too. For example, reduction of chromium by hydrogen can lead to the appearance of Cr^{2+} ions on the surface of Al_2O_3. Most researchers reject the possible presence of Cr^{4+} on the Al_2O_3 surface, since the interaction between Cr^{4+} ions in concentrated systems should result in ferromagnetism, which actually is not detected. The Cr^{3+} ions on the $\alpha-Al_2O_3$ surface can form pairs, which, by redistribution of charge, further separate into Cr^{2+} and Cr^{4+} [3.44,45]. When the concentration of Cr^{3+} is below 5% in the Cr-Al-O system, isolated ions prevail on the surface. In the 5 - 10% range there exist paired centers $Cr^{3+}-O-Cr^{3+}$, which convert into $Cr^{2+}-O-Cr^{4+}$. At concentrations above 10% cooperative interaction sets in, and isolated chromium ions disappear as do the paired centers.

Thus EPR and optical spectroscopy enable one to study the coordination and oxidation state of a number of surface chromium atoms. The XPS technique supplies information about all oxidation states of the surface ion at the same time. In [3.43,46] the XPS technique was used to study chromium catalysts supported on $\gamma-Al_2O_3$ with a well-developed surface (235 m^2/g). It was found that, on the surface of catalysts heat-treated at 500°C in air, chromium (at low concentrations of chromium oxide) appears in the form of Cr^{6+} ions. Beginning at 5% concentration the XPS spectra also show Cr^{3+}, which prevails when the concentration reaches 20%. Even slight reduction by out-pumping at 300°C leads to the appearance of Cr^{5+} on the surface. In the range of concentrations of chromium oxide between 1% and 5% the Cr^{5+} ions are the prevalent form of chromium in these slightly reduced specimens.

The acidity of the surface of mixed oxides of the $Cr_xAl_{2-x}O_3$ type is known to be possibly much higher than that of the individual oxides. STONE [3.23] used this system to explain that phenomenon through the crystal-field stabilization energy (CFSE) of the transition-metal ion. It appears that the CFSE is very often responsible for the reconstruction of the surface structure with respect to the bulk. If we are to visualize schematically the surface formed after cleavage of the oxide crystal $M^1{}_xM^2{}_{1-x}O$, the rule of electroneutrality requires equal numbers of M^1 and M^2 cations covered and not covered by oxygen O^{2-} ions. Suppose further that M^1 is a transition-metal ion (e.g., Cr^{3+}) and M^2 a nontransition one (e.g., Al^{3+}). Then the initially random distribution of O^{2-} ions on the surface (Fig.3.5a) yields to rearrangement. As a result, the M^1 ions, due to CFSE, become more shielded by oxygen ions (Fig.3.5b). At the same time the number of nontransition-metal ions M^2 decreases. If the nontransition-metal ion has a large charge-to-radius ratio e/r, it acquires the ability to coordinate donor-type molecules. In other words, the Lewis acidity increases [3.50].

Other 3/2 oxides interact with Al_2O_3 in much the same way as Cr_2O_3. For instance, Ti_2O_3, V_2O_3, Fe_2O_3, and Mn_2O_3 may also form solid solutions with alumina. It is interesting to note that, in all cases, antiferromagnetic interactions set in at one and the same concentration of transition-metal oxide: at $x \simeq 0.3$ in the

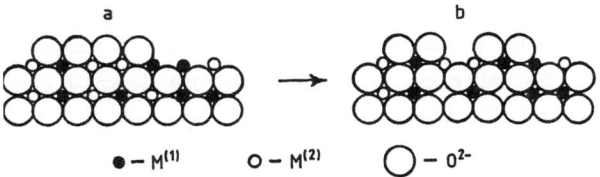

$\bullet - M^{(1)}$ $\circ - M^{(2)}$ $\bigcirc - O^{2-}$

Fig.3.5. The rise in Lewis acidity on the surface of the oxide system M^2O in the presence of transition metal M^1

$M_xAl_{2-x}O_3$ system [3.47]. The interaction between the transition-metal cations, however, starts at much lower concentrations. An analysis of the intensity of EPR spectra from $Fe_xAl_{2-x}O_3$ solid solutions as a function of concentration of Fe^{3+} ions reveals that already at $x < 0.01$ these ions are not distributed at random in the solid solution, but rather they tend to form clusters which display ferromagnetic interactions [3.48]. In the $V_xAl_{2-x}O_3$ system the magnetic moment (per one vanadium atom) decreases linearly with the increase in x in the range $0 < x < 0.04$. This is attributed to the formation of ionic pairs, in which the interaction is already observed at low concentrations [3.49].

The valences of the dissolved ion and the matrix ion being the same, the resistance of the former against oxidation and reduction is the highest. For example, Mn^{3+} readily forms a solid solution with Al_2O_3, whereas Mn^{4+} does the same with rutile TiO_2, because the lattice of $\beta-MnO_2$ (pyrolusite) is isomorphic with rutile. The ions on the surface are less stable: in magnesium oxide the Mn^{3+} ion is easily oxidized to Mn^{4+}. An alternative valence state (namely, Mn^{3+}) can be stabilized by adding Li_2O [3.7].

The 3/2 oxides Fe_2O_3 and Cr_2O_3 can be dissolved in magnesium oxide to get a true solid solution of Cr^{3+} or Fe^{3+} in MgO, if at the same time an equivalent quantity of Li^+ ions is introduced. In the $Li_2O-Cr_2O_3-MgO$ system the increased proportion $[Li^+]/[Cr^{3+}]$ results in a linear reduction of the lattice parameter of MgO and a decrease in magnetic susceptibility κ of the Cr^{3+} ions. In the absence of chromium or iron, Li_2O is insoluble in MgO. The Cr and Fe ions in such cases can exist on the surface in a different oxidation state. Experiment indicates that at $[Li]/[Cr] > 1$ the surface is dominated by deeply oxidized ions: Cr^{5+} and Cr^{6+} [3.51,52]. With the $Li_2O-Fe_2O_3-MgO$ system, heat treatment in a reductive environment results in the appearance of Fe^{2+} ions on the surface.

62

Let us now consider the dilute oxide systems of the second type, in which the transition-metal ion differs significantly from the matrix ion.

3.1.4 Spinel Formation

If the charge of the transition-metal ion introduced into the matrix differs from the charge of the native lattice ion, then the formation of a solid solution (in the absence of defects) is hindered. For example, the Cr^{3+} ion introduced into MgO forms a new phase: magnesium chromite $MgCr_2O_4$ (similarly, $MgFe_2O_4$ with an Fe^{3+} ion). In diluted oxide systems with MgO or $\gamma-Al_2O_3$ serving as the nontransition oxide matrix, spinels are often formed. Crystalline spinel is a dense cubic packing of oxygen ions, with the octahedral and tetrahedral interstices occupied by metallic cations. The general spinel formula is $M^{2+}M_2^{3+}O_4$, where M^{2+} can be Mg, Fe, Mn, Zn, and, occasionally, Co and Ni in the tetrahedral coordination, and M^{3+} is Fe, Al, Cr, Mn, and Co in the octahedral coordination.[2] Gamma-Al_2O_3 also is a defective spinel; it can be described as

$$Al_{\frac{2}{3}+x} \square_{\frac{2}{3}-x} [Al_{2-x} \square_x] O_4$$

with x being the proportion of cation vacancies ($0 \leq x \leq 1/3$). Introduction of M^{2+} cations into such a structure results in the redistribution of vacancies, which has certain effects on the surface properties; in particular, it raises the acidity of the surface [3.6,7,21,22,35,51-61].

The distribution of cations of transition metals in a spinel was explained on the basis of crystal-field theory [3.55,57]. Let us turn to the diagram in Fig.3.6. The term "preference energy" has been used in [3.57] to denote the difference in

---

[2] More precisely, this is direct spinel. In an inverse spinel $M^1[M^1M^2]O_4$ half of the M^1 ions are found in tetrahedral sites, while M^2 ions and the remaining M^1 ions occupy the octahedral sites.

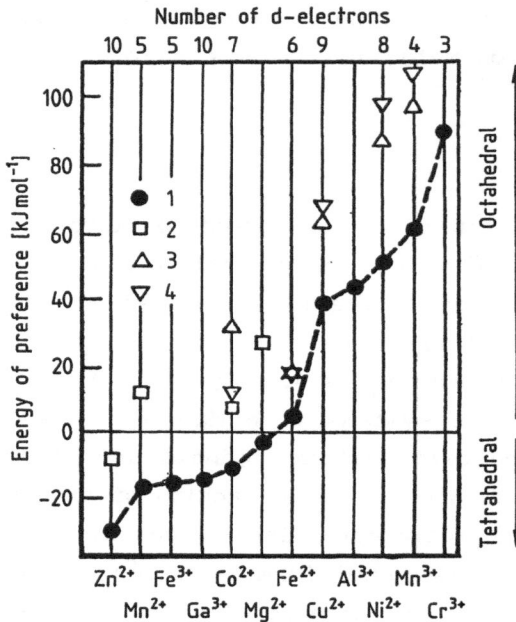

Fig.3.6. The "preference energy" for various cations in the spinel structure. Experimental points for spinel: (1)[3.57]; calculated points: (2)[3.52], (3)[3.49], (4)[3.48]

enthalpies of two ions A and B, equally coordinated with oxygen ions. The zero level corresponds to the enthalpy of the non-transition Al^{3+} ion in tetrahedral coordination. The transition of this ion into octahedral coordination, as seen in the figure, results in an energy gain of about 40 kJ/mol. On the right are Cr^{3+}, Mn^{2+}, and Ni^{2+} ions, which have high octahedral "preference energy"; on the left are Zn^{2+} and, possibly, Cd^{2+} and In^{3+} ions, which tend to occupy tetrahedral sites. Solid circles represent experimental points, while the open markings pertain to data calculated on the basis of crystal-field theory (high-spin complexes) [3.56,57]. These points agree with the general trend.

The diagram in Fig.3.6 was used in [3.55] to explain the optical, diffuse reflection spectra and the structure of supported Co/Al_2O_3 and Co/MgO catalysts. For instance, the spectroscopic data detected Co^{2+} in a tetrahedral configuration in the Co/Al_2O_3 catalyst. From Fig.3.6 it is clear that Co^{2+} tends to occupy tetrahedral sites, whereas Al^{3+} exhibits the tendency to take octahedral sites, so there is a considerable difference in their "preference energies". Hence it follows that $CoAl_2O_4$ is a direct spinel, and the cobalt ion is stable in the Co^{2+} form (heat treatment in air at up to 800°C has no effect on the spectrum). According to XPS [3.59], the surface monolayer of spinel is composed of $CoAl_2O_4$, while further deposition of cobalt oxide results in a Co_3O_4 spinel.

We have already mentioned that diffuse reflection spectra detect Co^{3+} in supported Co/MgO catalysts, in contrast to $Co_xMg_{1-x}O$ solid solutions. This implies formation of spinel-type surface compounds $Mg[Co_2O_4]$ or $Co_3O_4 = Co^{2+}[Co_2^{3+}O_4]$.

In the Co/MgO system the Co^{2+} and Mg^{2+} ions are located near the line which separates the tetrahedral and the octahedral sites (Fig.3.6). Their "preference energies" are close. This accounts for the fact that CoO and MgO form solid solutions in a wide range of concentrations (octahedral coordination). However, for the Co/MgO system there is another possibility: stabilization of the Co^{3+} ion in an octahedral coordination – even the high-spin Co^{3+} state has a much greater octahedral "preference energy" than Mg^{2+}. As regards the low-spin Co^{3+} state, it ought to fit the diagram on the far right (Fig.3.6). Consequently, if the system includes an acceptor of electrons, the Co^{2+} ion must give away an electron and stabilize in the form of an octahedral Co^{3+} ion, forming a spinel $Mg[Co_2O_4]$. Indeed, the optical spectra of Co/MgO reveal Co^{2+} and Co^{3+} ions in octahedral configuration [3.55].

Much work deals with the structure of supported Cr, Mo, V, and W oxide systems, in which the cited metals are found in the extreme oxidation states. The support is made up of MgO, Al_2O_3, SiO_2, and silica-alumina. Because of the considerable difference in oxidation states of the transition ion and the matrix cation, the transition ions do not form solid solutions, but rather occur in surface compounds. Supported systems of this kind are widely used as adsorbents and catalysts.

3.1.5 The CrO_x-SiO_2 System

Surface chromates (3.5a) and dichromates (3.5b) form when Cr_2O_3 is deposited on the surface of silica gel:

$$
\begin{array}{ccc}
\begin{array}{c}
O \diagdown \diagup O \\
 Cr \\
O \diagup \diagdown O \\
| | \\
-Si-O-Si-
\end{array}
&
\begin{array}{c}
O O \\
\| \| \\
O=Cr-O-Cr=O \\
| | \\
O O \\
| | \\
-Si-O-Si-
\end{array}
&
\begin{array}{c}
HO \diagdown \diagup O \\
 Cr \\
O \diagup \diagdown O \\
| | \\
-Si-O-Si- \; .
\end{array} \\
a) & b) & c)
\end{array}
\qquad (3.5)
$$

Even in the course of preparing the specimen, weak reduction leads to the appearance of a small quantity of Cr^{5+} (structure 3.5c).

The optical spectra display one strong band pertaining to the transition ΔE ($t_{2g} \rightarrow d_{x^2-y^2}$) = 13.000 cm^{-1}. Making use of this value and the value of λ for aqueous solutions of Cr^{5+} salts ($\lambda \simeq 160$ cm^{-1}), the authors of [3.37] obtained good agreement between the calculated and the experimental values of g_z and $g_{x,y}$. For Cr^{5+} ions on the silica gel surface the EPR spectrum had the following values for the g factors

$$
g_z = 2 - \frac{8\lambda}{\Delta E} = 1.91 \; , \qquad g_{x,y} = 2 - \frac{2\lambda}{\Delta E} = 1.98 \; . \qquad (3.6)
$$

The Cr^{5+} ion is assumed to replace the protons in surface OH groups, forming the structure of type (3.5c).

Study of the CrO_x-SiO_2 system with the aid of XPS [3.62] and optical, diffuse-reflection spectroscopy [3.63,64] indicates that the rise in heat-treatment temperature of supported specimens (1-9% Cr by mass) results in the decrease in Cr^{6+} content with the appearance of the reduced forms Cr^{2+} and Cr^{3+}. When CrO_x-SiO_2 is subjected to reduction by carbon monoxide at 623 K, Cr^{2+} is formed, which is identified by four d-d bands in the 800-2000 nm range and a charge transfer band at about 240 nm. The Cr^{3+} ions occur on the SiO_2 surface apparently in microcrystals of Cr_2O_3. The intermediate oxidized states Cr^{4+}

and Cr^{5+} could not be detected. Evidently, their concentration is low.

3.1.6 The MoO_x-Al_2O_3 and MoO_x-MgO Systems

The Mo^{6+} state on the surface of MgO, γ-Al_2O_3, and α-Al_2O_3 has been thoroughly studied using diffuse-reflection spectroscopy and other methods [3.24,55,56,65-72]. When Mo^{6+} in tetrahedral coordination T_d goes into octahedral coordination O_h, the starting point of the intensive absorption band, corresponding to charge transfer from the π orbital of O^{2-} ion to the 3d orbital of the Mo ion, exhibits a characteristic shift: absorption in the O_h coordination occurs at longer wavelengths. At low concentrations (<6% by mass) the Mo^{6+} ions are found on the surface of MgO and Al_2O_3 in the tetrahedral coordination. In other words, there is a molybdate $(MoO_4)^{3-}$ ion in the surface monolayer. As the concentration of Mo increases, Mo^{6+} in the O_h coordination appears on the surface. Raman spectroscopy indicates that a polymolybdate phase is then formed, i.e.,

Bulk molybdate $MgMoO_4$ is formed on magnesium oxide. Only after completion of stoichiometry $(MgMoO_4)$ does the phase MoO_3 start to appear on the surface.[3]

Reduction of supported molybdenum-oxide catalysts leads to the appearance of molybdenum ions in lower valence states. Reduced systems of this kind are widely used as catalysts of the industrially important reactions of hydrogenation, polymerization, metathesis of olefins, and hydrodesulfurization. They have been investigated in detail using the EPR technique.

---

[3] Systems of this kind (molybdate + molybdenum oxide) are successfully employed as soft-working selective catalysts in the oxidation of hydrocarbons.

After reduction of the Mo^{6+}/Al_2O_3 system, two EPR lines are observed [3.4,5,65], which have been attributed to Mo^{5+} ions in tetrahedral (T_d) and square-pyramidal (C_{4v}) configurations. Of greater intensity is the signal with $g_z = 1.90$ and $g_{x,y} = 1.94$, pertaining to the C_{4v} coordination. After reduction, one observes signals from Mo^{5+} with $g_1 = 1.920$ and $g_2 = 1.925$. It was found that there exists a certain minute concentration of supported ions, leading to the reduction of the specimen. This gives an EPR signal from the reduced forms of molybdenum. Specifically, this minimum concentration constitutes 0.01% MoO_3 by mass, or 10^{12} Mo^{6+} ions per square centimeter for Al_2O_3 and MgO. The existence of a concentration threshold leads to the conclusion that the reduction of Mo^{6+} ions at 500°C and their subsequent stabilization require formation of binuclear or polynuclear cluster-type centers.

The lower temperature threshold at which heat treatment has reductive effects (400°C for Mo/MgO and 300°C for Mo/Al_2O_3) may be related to the temperature of decomposition of hydroxides of Mg and Al (causing further dehydration of the surface). Apparently the stabilization of reduced forms is considerably promoted by mobile bulk defects. In particular, such defects may be represented by vacancies, which counterbalance the charges of ions intermediate between molybdate (Mo^{6+}) and molybdite (Mo^{4+}). The defects of this kind can be formed due to decomposition of bulk hydroxides. The Mo^{5+} ions in the course of reduction are not stabilized on low-defect α-Al_2O_3.

EPR and diffuse-reflection spectroscopic techniques has helped to specify the early stages of reduction of Mo^{6+} on Al_2O_3[3.69]. At low concentrations (below 4% MoO_3 by mass) the molybdenum oxide interacts with the most basic centers of the carrier, resulting in the formation of oxyanion forms $(MoO_4)^{3-}$. As the concentration of MoO_3 increases to 4-10%, the acidity of the surface grows. Interaction of acid OH groups with Mo oxides results in anion forms $[HMoO_4]^{2-}$ and $[Mo_2O_7]^{4-}$, and molybdenyl cation forms $[MoO(OH)]^{2+}$ and $[Mo_2O_3]^{4+}$. In molybdenum-rich regions, both on the surface and in the bulk, polyoxyanions are formed (containing Mo^{6+} and Mo^{5+}), as well as oxycations MoO^{3+}.

Based on data obtained on reduction of the Mo/Al_2O_3 system [3.4,65], MASSOTH [3.70] proposed a scheme of surface reduction, accounting for the creation of binuclear centers

$$(3.7)$$

In this scheme the binuclear center of two Mo^{5+} ions further separates into Mo^{6+} and Mo^{4+}. A similar scheme has been proposed in [3.72].

Indeed, in a number of works the EPR signal from Mo^{5+} passes through a maximum in the course of reduction of supported Mo/Al_2O_3 catalysts [3.71]. However, Mo^{4+} gives no EPR signal at the temperatures used, while the optical spectra which might have detected molybdenum with a valence less than five are ambiguous. A direct proof of the formation of Mo^{4+} in the course of reduction of MoO_3/Al_2O_3 was obtained with the help of XPS [3.66]. After calibration (using pure MoO_3 and MoO_2 oxides), the bonding energies of molybdenum were measured in the $3d_{3/2}-3d_{5/2}$ levels in the course of reduction of the Mo-Al-O system. Figure 3.7 shows the spectra of partially reduced catalysts. The bonding energy of the $3d_{5/2}$ level is 233.05 eV for Mo^{6+} and 230.02 eV for Mo^{4+}; the intermediate state is ascribed to Mo^{5+}. Investigations indicate the ease of transitions

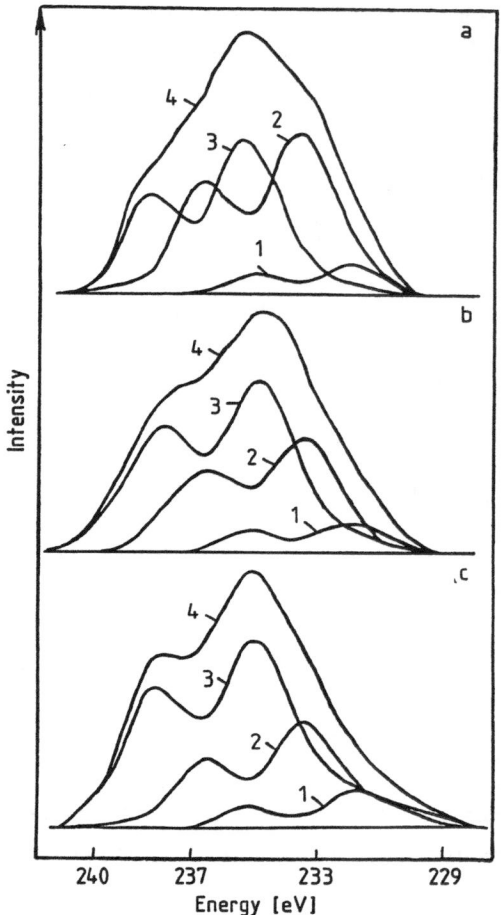

Fig.3.7. The XPS spectra of $Mo(3d_{3/2} - 3d_{5/2})$ lines for 3.5% MoO_3-Al_2O_3 after reduction and partial oxidation [3.66]: (a) 2 hr in H_2 at 460°C, (b) the same plus 5 min in air, (c) the same plus 20 min in air; (1) Mo^{4+}, (2) Mo^{5+}, (3) Mo^{6+}, (4) total spectrum

between Mo^{5+} and Mo^{6+} at reduction and oxidation, which agrees with the proposed scheme. On the other hand, Mo^{4+} is rather resistant to oxidation and apparently forms a separate phase.

An attempt was made [3.78] to assess the size of clusters on reduced catalysts using adsorption of oxygen (Sect.4.4). The cluster of supported MoO_x/MgO catalysts after reduction was found to comprise 50 - 100 Mo^{5+} ions linked by oxygen bridges. The size of the cluster changes little when the concentration of MoO_3 increases from 1 to 6%: the clusters only increase in number. According to [3.78], the cluster size does not exceed the dimensions of the MgO microcrystal face (20–25 nm). On the other hand, the size of the cluster on Al_2O_3 increased from about 10 to 10^4 Mo^{5+} ions when the concentration of MoO_3 varied from 1 to 6%. In this case the cluster size is not restricted by the dimensions of the microcrystal, and can be significantly larger.

3.1.7 The VO_x-Al_2O_3, VO_x-MgO, and VO_x-SiO_2 Systems

The supported vanadium catalysts are very similar in their properties to the supported molybdenum catalysts [3.4,5,75,78–82]. They also have been studied very thoroughly using EPR and diffuse–reflection techniques. In [3.80,81] it was demonstrated that V^{5+} ions on the surface in T_d, C_{4v}, and O_h coordinations can be distinguished by charge transfer spectra: the higher the coordination number, the lower the energy at which strong absorption begins. For instance, in V_2O_5 the charge transfer band at 280 nm corresponds to T_d coordination, whereas the band at 335 nm pertains to O_h. With the increase in concentration of V_2O_3 deposited on MgO, the ratio $[V_{t_d}]/[V_{O_h}]$ of concentrations (where V_{T_d} prevails) shows no significant change.

Paired or cluster centers have also been discovered in supported vanadium catalysts. In V-Al_2O_3 the EPR signal from V^{4+} ions after reduction is detected only starting with the concentration of $5 \cdot 10^{13}$ ion/cm^2. In this case, however, there is an additional proof of interaction between adjacent vanadium ions. Apart from HFS from V^{5+} nuclei ($A_{x,y}$ = 51 G, A_z = 189 G; $g_{x,y}$ = 1.984, g_z = 1.920), the EPR spectrum displays additional splitting A_s = 16 G, due to interaction of the unpaired electron of vanadyl VO^{2+} with the adjacent nucleus of V^{5+} [3.4]. While in V-Al_2O_3 this splitting could possibly be attributed to interaction between VO^{2+} and the Al nucleus, the only plausible explanation for the V-MgO system, which exhibits similar splitting, is the forming of paired VO^{2+} ... V^{5+} centers. The cluster size for V-MgO and V-Al_2O_3 assessed through adsorption of oxygen

[3.78] turned out to be close to the corresponding value for Mo–MgO and Mo–Al$_2$O$_3$.

Judging by EPR of V^{4+} and Mo^{5+}, the nature of support has little effect on the coordination of these ions. The spectral parameters of VO^{2+} and Mo^{5+} in C$_{4v}$ coordination are about the same for V$_2$O$_5$ and MoO$_3$ deposited (with subsequent reduction) on different carriers: MgO, Al$_2$O$_3$, SiO$_2$, and BeO [3.65]. Paired and cluster-type centers (along with isolated ones) were found in all systems.

An interesting group among supported oxide catalysts comprises the so-called molecular layering catalysts. The molecular layering technique has been pursued by ÖHLMAN and co-workers [3.80] ALESKOVSKIJ [3.83]. This technique consists essentially of the controlled buildup of new solid phases on the surface of an initial solid matrix using reactions between gaseous substances and surface radicals of the matrix. A typical example of such reactions is the interaction between gaseous metal chlorides (including those of transition metals) with OH groups, always present on the surfaces of oxide semiconductors and dielectrics [3.84].

It is possible to produce systems with a predetermined number of monolayers (or a fraction of a monolayer) deposited on the carrier. This method was used, in particular, for producing supported vanadium catalysts. The interaction of VCl$_4$ with silica gel (or MgO) results in a monolayer of vanadium complexes:

$$
\begin{array}{ccc}
& & \underset{\text{O}\overset{\diagup}{\diagdown}\text{V}\overset{\diagup}{\lessgtr}\text{O}...\text{H}}{\overset{\text{Cl}\diagdown\ \diagup\text{Cl}}{}} \\
\underset{|}{\overset{|}{\text{H}}} \quad \underset{|}{\overset{|}{\text{H}}} & & \\
\underset{|}{\overset{|}{\text{O}}} \quad \underset{|}{\overset{|}{\text{O}}} & & \underset{|}{\overset{|}{\text{O}}} \\
-\text{Si}-\text{O}-\text{Si}- \ + \ \text{VOCl}_3 \ \longrightarrow \ & -\text{Si}-\text{O}-\text{Si}- \ + \ \text{HCl}.
\end{array}
\qquad (3.8)
$$

Subsequent reduction results in the monolayer of V=O groups, in which oxygen is bound very tightly (binding energy about 400 kJ/mol). The reduction of such systems goes via a series of two-dimensional phase transitions. The size of a two-dimensional phase "patch" or cluster in supported $V-SiO_2$ catalysts was found to be about 2 nm across [3.80]. Such structures have counterparts in some solutions: the heteropoly acids, complex compounds whose anion is composed of two different oxides. In our case both SiO_2 and V_2O_5 are acidic oxides.

The molecular layering technique was used for obtaining surface layers of metallic oxides on dielectrics (SiO_2) [3.80,83], real surfaces of elemental semiconductors Ge and Si [3.85], oxidized metallic surfaces [3.86] and carbon [3.87].

3.1.8 The $CoO_x-MoO_y-Al_2O_3$ System

Of diluted systems containing two transition-metal ions, the most extensively studied one is the $Co-Mo-Al_2O_3$ system, owing to its being an important catalyst of hydrodesulfurization [3.5,58,59,70,73,74,88–91].

According to EPR and diffuse-reflection spectra, at low concentration (<1% Co or Mo) these ions interact with the Al_2O_3 surface in the same way as in systems containing only Co or Mo. The Co^{2+} ions are arranged in a tetrahedron (i.e., surface structure $CoAl_2O_4$ and, occasionally, Co_3O_4 – see Sect.3.1.4), and Mo^{6+} in tetrahedral coordination (Sect.3.1.6) [3.5,58]. With the increase in Co and Mo concentration, two-dimensional structures of fixed composition begin to form. Studies of the $Co-Mo-Al_2O_3$ system with XPS technique [3.59,88] reveal the existence of two-dimensional surface phases of Mo_4Co, Mo_6Co. These investigations enabled a two-dimensional phase diagram to be plotted for MoO_3 and CoO on an Al_2O_3 surface.

According to [3.74,90], at high occupancies by Mo and Co, molybdenum covers the surface of $\gamma-Al_2O_3$ by a monolayer of Mo^{6+} ions, infiltrated by Co clusters. Heat treatment at 700ºC results in the formation of $CoAl_2O_4$. The reducibility of Mo increases in the presence of Co. Many recent reviews show the complex multiphase structure of $Co-Mo-Al_2O_3$ catalysts [3.88].

A convenient example of oxide systems of the third type, in which ions with d electrons are found in a matrix of the same oxide, occurring in a different nonparamagnetic state, is provided by titanium dioxide.

3.1.9 The Ti^{3+}-TiO_2 System

In studying rutile monocrystals reduced in vacuum or in hydrogen, the EPR signal associated with the bulk superstoichiometric Ti^{3+} ions was detected [3.92]. The signal could be detected only at liquid-helium temperature; it vanishes at $-195^\circ C$. This must be attributed to the ionization of defects (the levels of Ti^{3+} lie just 0.05-0.005 eV below the bottom of the conduction band). Analysis of the EPR g tensor from the bulk paramagnetic defects in rutile [3.93] justifies attributing them confidently to Ti^{3+} ions with a $3d^1$ configuration; the spin-lattice relaxation time is about 10^{-9} s. Even allowing for the Jahn-Teller effect (Sect.2.1), the signal can be observed only at very low temperatures. Subjecting the rutile crystal to vigorous reduction in hydrogen (which, incidentally, may initiate a new phase: Ti_2O_3), BOGOMOLOV and SOCHAVA [3.94] obtained an EPR signal at $-195^\circ C$, attributed by them to F centers in the bulk.

It was already in the early EPR investigations of partially reduced samples of rutile and anatase [3.95] that a characteristic signal with g_{av} = 1.95 was detected, which was observed at higher temperatures as well. Adsorption of oxygen and other acceptors irreversibly destroyed the signal, which confirms the surface origin of the adsorption sites concerned. The surface paramagnetic centers were assumed to be the Ti^{3+} ions [3.95-97] and oxygen vacancies having captured an electron [3.98]. Inelastic electron energy-loss spectra from the rutile surface [3.99] show absorption peaks pertaining to d-d transitions; this verifies the presence of Ti^{3+} ions on the reduced surface. The very fact that EPR from high-dispersion TiO_2 specimens can be observed at higher temperatures than from single crystals implies that the parameters of the surface and the bulk paramagnetic centers are not the same. A similar situation is encountered with reduced tin dioxide [3.100].

An analysis of anisotropic properties of the g tensor reveals that the most probable configuration of Ti^{3+} ions on the partially reduced anatase surface is a tetrahedron [3.96], and a square pyramid (C_{4v}) on the surface of rutile [3.101]. Using experimental values of g (g_z = 1.945, $g_{x,y}$ = 1.956) and spectroscopic splitting Δ=17.500 cm^{-1} (obtained from diffuse-scattering spectra), the expressions (2.9) were employed (Sect.2.2.6) to find λ=120 cm^{-1}. This value is somewhat lower than that for an isolated Ti^{3+} ion on silica gel (λ =154 cm^{-1}). This points to partial delocalization of the electron, that is, to a considerable contribution of the covalent component. For this reason further attempts at specifying the nature of the defect (whether it is a superstoichiometric Ti^{3+} ion or a vacancy) make little sense.

The fact that EPR signals come exclusively from the surface paramagnetic defects makes titanium dioxide a very convenient object for studying the spatial distribution of paramagnetic centers on the surface. Information about the distribution of these centers is drawn from the line shape and width of the EPR signal. Figure 3.8a reproduces the EPR signal from Ti^{3+} at –195ºC. Computer analysis of the line using the technique out-

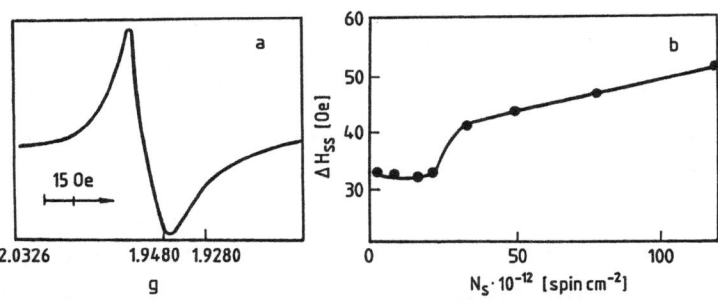

Fig.3.8. (a) The EPR of the Ti^{3+} ion in rutile at –195ºC, (b) the linewidth ΔH_{SS} versus the concentration of Ti^{3+} centers N_S [3.102]

lined in [3.102] reveals that it can be approximated by a superposition of three standard distribution functions: Lorentzian, Gaussian, and anisotropic axisymmetric distributions. With the chosen reduction procedure the linewidth ΔH was found to be independent of temperature in the range from -195^0 to -50^0C [3.103]. This leads to the conclusion that (for the samples studied) the main contribution to the linewidth comes from dipole-dipole broadening. Exchange interactions in systems of this kind may lead only to narrowing of lines.

With a uniform random distribution of paramagnetic centers and Lorentzian line shapes, the linewidth ΔH_{SS}, due to dipole-dipole interaction, is proportional to the concentration of centers N_S: $\Delta H_{SS} = N_S A$, where $A = 1.3 \cdot 10^{-13}$ cm^2. A similar relationship is obtained for the Gaussian line shape.

In [3.103] the concentration dependence $\Delta H_{SS}(N_S)$ was examined. The change in N_S was accomplished by repeated reduction in vacuum of the same rutile specimen for three hours at $T_{cal} = 500-700^0C$ (T_{cal} is the heat-treatment temperature). The results are plotted in Fig.2.8b. For a concentration of $N_S \simeq 10^{13}$ cm^{-2} the theoretical value of ΔH_{SS}^{theor} is about 1 G. As follows from Fig.3.8b, this value is lower, by more than an order of magnitude, than the experimental value of ΔH_{SS}^{exp} for the same concentration, which points to a considerable distinction between the mean concentration N_S, as obtained from experiment, and the local concentration N_S^{loc} which determines ΔH_{SS}^{theor}. In other words, this proves the cluster-like distribution of paramagnetic centers on rutile surface. The local concentration of centers N_S^{loc} within a cluster is $N_S^{loc} \propto N_S(\Delta H_{SS}/\Delta H_{SS}^{exp})$. The average separation between the clusters is dozens of times greater than the average distance between atoms in a cluster. The quantity N_S^{loc} is, naturally, some averaged characteristic of a cluster, since theory [3.104] has so far failed to assess the actual forms of distribution functions for both clusters and atoms within clusters on the basis of the EPR line shape assuming a dipole-dipole mechanism of broadening.

The surface paramagnetic Ti^{3+} ions are formed alongside the oxygen vacancies, whose mobility at elevated temperatures is rather high. KISELEV et al. [3.103] examined the distribution patterns of paramagnetic centers on a rutile surface for different temperatures of heat treatment T_{cal}. Making use of a well-known technique for estimating the coefficients of self-diffusion of defects via the kinetics of electroconductivity variations [3.105], they demonstrated that for disperse rutile specimens after reduction in vacuum ($T_{cal} \simeq 500^{\circ}C$) the value of D is from 10^{-6} to 10^{-8} cm^2/s, which is two orders of magnitude higher than the corresponding values for a single crystal [3.103]. At $T_{cal} = 500^{\circ}C$ the path length for diffusional displacement of defects $\ell_0 = (2Dt)^{1/2}$ (t being the time of reduction) exceeds the size of rutile crystallites by four or five orders of magnitude, so the vacancies may be assumed to occur with equal probability at any point on the surface. The correlations between coordinates of vacancies at such temperatures are weakened, first, because of the high thermal-motion energy of defects, and second, because of the strong screening action of free electrons, whose concentration at $500^{\circ}C$ is about 10^{19} cm^{-3}. In this case the distribution of paramagnetic centers on the surface is accurately approximated by the Poisson function. The probability that an ensemble of m elements will form on surface area element δ, according to Poisson's statistics, is

$$W_m = (N_S \delta)^m / (m!) \exp(-N_S \delta) , \qquad (3.9)$$

where N_S is the mean concentration of centers.

At lower temperatures of reduction the correlative effects can no longer be neglected. Associations of vacancies, similar to clusters of paramagnetic ions, will be more thermodynamically advantageous (Fig.3.3). Their concentration can also be estimated by the law of mass action. A much more laborious task is finding the specific distribution function of clusters according to their size and dimensions. A general outline of the mathematics for the statistical theory of bulk defects, based on the theory of correlation functions, was given in [3.106]. We have indicated above, however, that the distribution of active centers on the surface often differs from the statistical one and results in the emergence of ordered two-dimensional phases.

So far it is possible to form only a qualitative notion about the distribution of "quenched" surface defects in specimens subjected to rapid cooling from T_{cal} to the temperature of the experiment. At relatively low T_{cal} (about 350°-450°C) the surface - because of dehydration due to the removal of closely spaced OH groups and reorientation of incomplete titanium-oxygen tetrahedrons - becomes rather distorted, which may give rise to defect formation. Because of the low mobility of vacancies at such temperatures (about 10^{-8} cm^2/s), the displacement in the course of experiment is small, and thermodynamic equilibrium between the defects is not obtained. As follows from Fig.3.8b at $N_S \leq 2-3 \cdot 10^{11}$ cm^{-2} ΔH = const. and hence N_S^{loc} = const. The rise in N_S is due mainly to the increase in the number of clusters. As T_{cal} and N_S go up, the value of H_{SS} begins to grow steadily (Fig.3.8b), and the difference between ΔH_{SS}^{exp} and ΔH_{SS}^{theor} decreases. In strongly reduced specimens the value of N_S^{loc} exceeds the upper limit of the density of surface defects ($\approx 10^{15}$ cm^{-2}, allowed from the viewpoint of crystal chemistry). The surface defects are then due mainly to the migration of bulk defects into the subsurface region.

The approaches to investigation of spatial distribution of active adsorptive surface centers, discussed here with reference to a model system, can, in principle, be extrapolated to nontransition-metal oxides. This would require developing a technique of controlled seeding of the surface with paramagnetic centers in the necessary concentration: either by irradiation, or by introducing spin markers (ions or ion radicals).

3.2 Transition-Metal Atoms in Zeolites

The introduction of transition ions into zeolites has drawn the attention of many researchers as a possible candidate for obtaining isolated transition-metal atoms devoid of any kind of collective interaction.

The structural properties of zeolites have been discussed by us elsewhere [3.84]. Zeolites possess a skeleton structure, built up of silicon-oxygen and aluminum-oxygen tetrahedrons. Zeolites of type A and faujasite (types X and Y) have a structure of cuboctahedrons. Each cuboctahedron (sodalite cell) contains Al or Si atoms at vertices and oxygen atoms about the center of each face. The sodalite cells are interconnected by oxygen bridges. In type-X and type-Y zeolites the cubooctahedrons are linked by hexagonal prisms and form a diamond-type lattice. The ratio $[SiO_2]/[Al_2O_3]$ equals 2 for A-zeolites, 2.2-3 for type X, and 3.2-5.0 for type Y. The sodalite cells enclose the so-called big cage, about 1.2 nm across in zeolites of types A, X, and Y. The big cages are interconnected by openings (0.9 nm in diameter for zeolites X and Y, and 0.42 nm for A). The space inside the sodalite cell is called the small cage; its diameter is 0.65 nm.

A remarkable recent achievement was the development of a new class of zeolites: the so-called ZSM-5 zeolites. Their structure is characterized by the presence of cylindrical passages 0.56 nm in diameter, intersecting at right angles. The $[SiO_2]/[Al_2O_3]$ ratio in ZSM-5 zeolites is in the range 40-50. Thus the number of acidic centers in the zeolite (which depends on the number of $[AlO_4]$ tetrahedrons) is much lower than that in zeolites of types A, X, or Y; however, the strength of these centers is almost as high as that of strong mineral acids. Zeolites of the ZSM-5 type with high silicon content have permitted catalytic reactions requiring high stereoselectivity, e.g., synthesis of engine fuel from methanol [3.107].

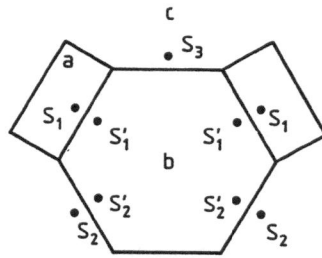

Fig.3.9. Cation sites in the faujasite structure (zeolites X and Y): (a) hexagonal prism, (b) small cage, (c) large cage. Points denote the location of cations

The cations which neutralize the charge of $[AlO_4]$ tetrahedrons may take any of the three main positions (Fig.3.9). In S_1 sites, located centrally in hexagonal prisms, the cations have octahedral coordination with respect to oxygen atoms in the skeleton. In S_2 sites the cations are in the centers of six-sided oxygen passages, interconnecting the big and the small cages. In S_3 sites the cations are in the centers of four-sided oxygen rings, interconnecting the big cages. If the site in the center of a six-sided oxygen window is offset toward the small cage, it is denoted by S_2'. Similarly, the site in the center of the opening between the sodalite cell and hexagonal prism biased toward the small cage is denoted by S_1'.

The transition ions can be used as modifiers of the zeolite properties; they can also serve as a probe for studying the distribution of cations of a given valence in the zeolite's inner space. Especially useful in this respect are the EPR and optical spectroscopic techniques, which do not work well with ions of nontransition metals. The use of EPR and optical spectroscopy for surveying the distribution of cation centers in zeolites has been reviewed in [3.108,109]. Studied most extensively were the X- and Y-zeolites, containing ions of Mn, Co, Ni, Cu, Cr, and Fe.

A large number of EPR investigations were carried out on manganese-containing X- and Y-zeolites, mostly because signals from Mn^{2+} ions are readily detected at any temperature and for arbitrary crystal-field symmetry. A clear-cut theory has evolved for identifying the EPR spectra of Mn^{2+} ions. This theory provides a basis for using the Mn^{2+} ions as a "paramagnetic probe" for studying the arrangement of other (nonparamagnetic) cations in a zeolite cell [3.110,111].

After exchanging Na^+ for Mn^{2+} in X- and Y-zeolites and A, the EPR spectrum displays a line with g = 2.000 and a HFS constant A = 96 G. This line corresponds to fully coordinated hexaquacomplexes $Mn^{2+}(H_2O)_6$, "floating" in the water contained within the big cage. TIKHOMIROVA and coworkers [3.110] denoted them as $Mn^{2+}[(H_2O)_6]_2$, to indicate that the hydrate envelope consists of no less than two layers of water. As the temperature of dehydration is raised to 100-200°C, X- and Y-zeolites give two lines (g = 2.000, A = 87 G; g = 2.007, A = 89 G), pertaining to Mn^{2+} ions localized at sites S'_1 and S'_2. The symmetry of these sites corresponds to a shortened tetrahedron (C_{3v}). The Mn^{2+} ion is linked to three oxygen atoms in the zeolite skeleton and on the hydroxyl group:

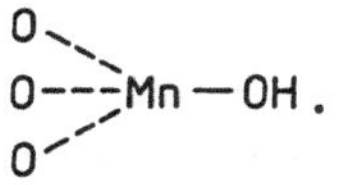

$$O \diagdown$$
$$O \text{---} Mn - OH \, .$$
$$O \diagup$$

Further rise in the heat-treatment temperature of X- and Y-zeolites (to 300-400°C) results in the appearance of an EPR line with g = 2.000 and A = 95 G. It pertains to Mn^{2+} ions in the octahedral configuration, located in S_1 sites inside the hexagonal prism (Fig.3.9).

Similar results were obtained for other transition ions. For example, the optical spectra of Co zeolites of types X and Y after exchange coincide essentially with the spectra of aqueous solutions of Co^{2+} salts. Thus we may conclude that the cobalt ions after exchange and hydration are located in the big cages in the form of a $[Co(H_2O)_6]^{2+}$ complex. Variations in optical and EPR spectra, caused by heating, were attributed to localization of cobalt ions in the form of trigonal complexes at S_2 sites, similar to the manganese complexes depicted above. The EPR signal with g_z = 2.00 and $g_{x,y}$ = 2.75, observed after heat treatment of CoNaY zeolite at 400-500°C, was attributed to Co^{2+} ions [3.109], localized in the plane of a six-sided oxygen passage (halfway between S_2 and S'_2 sites – Fig.3.9). It was, however, demonstrated [3.112,113], on the basis of measurements of magnetic susceptibility of CoNaY zeolites with subsequent calculations of effective magnetic moments of Co^{2+} ions and comparison with the EPR and optical spectra, that at high heat-treatment temperatures the Co^{2+} ions occur in octahedral coordination. Apparently, like manganese ions in Y-zeolites, the Co^{2+} ions are localized at S_1 sites inside hexagonal prisms [3.114].

Notice that fixation of transition cations at S_1 sites is assisted not only by the Madelung potential, but also by the crystal-field stabilization energy. As indicated above, the value of the CFSE is greatest for the octahedral ligand field. Nevertheless, the CFSE values can be rather high also for transition cations in other configurations (except Mn^{2+} and Fe^{3+}: d^5 configuration). This circumstance is apparently responsible for the fact that the transition-metal containing zeolites after exchange of Na^+ for M^{2+} ions, do not achieve complete occupany of S_1 sites (16 sites per cell), as opposed to Ca and Mg zeolites. In fact, X-ray diffraction analysis of NiNaY zeolites [3.115] reveals that the maximum occupancy of S_1 sites by Ni^{2+} ions is restricted to 12 sites per cell. The remaining Ni^{2+} ions localize in other sites.

The general rule of occupation of zeolite sites by transition-metal ions is that dehydration results in consecutive taking of S_2, S'_2, S'_1, and S_1 sites. The actual pattern of occupation, however, depends strongly on the conditions of dehydration and the nature of cation. For example, at slow heating rates the cations tend to localize at S_1 sites in faujasite [3.109].

In Cu-containing X- and Y-zeolites the $[Cu(H_2O)_6]^{2+}$ ions were also detected after accomplishing the exchange for Cu ions and before heating. After heat treatment and partial dehydration at

100–300°C the EPR spectrum displayed two types of signals: $g_z = 2.38$, $g_{x,y} = 2.08$, A = 126 G; and $g_z = 2.32$, $g_{x,y} = 2.05$, A = 162 Ge. The second line corresponds to the Cu^{2+} ion in the square pyramidal or plane-quadrangular coordination (C_{4v}) [3.116]. These ions seem to be located in S_3 sites. The distinction between Cu^{2+} and other ions with 3d configuration arises from peculiar properties of the d^9 configuration and the existence of the Jahn-Teller effect (Sect.2.1). Dehydration at 400–500°C leads to a strong suppression of the EPR spectrum of Cu^{2+} ions, which in [3.116] is attributed to their transition into the plane of a six-sided window, where the field is trigonal. The EPR signals may then vanish altogether. According to other researchers [3.117,118], the drop in intensity of EPR spectra, observed with high heat-treatment temperatures at a small degree of exchange of Na^+ for Cu^{2+} (about 2%), is due to a reduction of Cu^{2+} ions, where the EPR spectrum may disappear because of strong exchange interaction.

Clustering of transition-metal ions is especially favored at higher levels of exchange (>50%). Exceeding this threshold, the ions are no longer restricted to S_1 sites. More cations go to S_2 and S_3 sites, located in the big cages. Paired ions MOM^{n+} may enter the sodalite cell at S'_1 and S'_2. The formation of paired and cluster-type centers was observed by the EPR technique, magnetic-susceptibility measurements, etc. For example, the drop in intensity of EPR spectra at high temperatures for CrNaY zeolites was ascribed to the formation of cation pairs of type Cr^{3+} ... O^{2-} ... Cr^{3+}. When the exchange reaching a high level, these pairs aggregate, giving rise to the Cr_2O_3 phase and destroying the zeolite lattice [3.119].

The status of Fe ions in $Fe^{2+}NaY$ zeolites was investigated with the aid of Mössbauer spectroscopy [3.120]. Two types of iron atoms were identified, one of which was to Fe^{2+} ions in S'_2 sites. Bringing the zeolite in contact with oxygen at 350°C gave rise to reversible changes, attributed to the formation of oxygen bridges in the sodalite cell:

$$2 Fe^{2+}_{S_1} + \frac{1}{2}O_2 \longrightarrow Fe^{3+} \ldots O^{2-} \ldots Fe^{3+}$$

Ferrous ions, initially located at S_1 sites, are drawn by oxygen into adjacent S_1' sites in the same sodalite cells with simultaneous oxidation to ferric ions Fe^{3+}. Other ligands are evidently capable of transferring cations from one site to another.

The enhancement of the static field caused by replacement of Na^+ by the transition ions M^{2+} and M^{3+} results in an increase in zeolite's acidic properties. The acidic properties of X and Y zeolites, exchanged for transition-metal ions, were studied in [3.121]. The proton acidity was assessed via the appearance of a 1545 cm^{-1} band in the IR spectrum of zeolite-adsorbed pyridine ($\rightarrow C_5H_5NH^+$); the Lewis acidity was estimated via the bands between 1435 and 1450 cm^{-1} ($\rightarrow C_5H_5N \ldots M^{n+}$). When Na^+ is exchanged for transition ions, the acidity always grows according to this process:

$$(3.10)$$

At the same time the intensity of the valence vibrations of the OH groups (≈ 3640 cm^{-1}) goes up. The acidity increases, according to [3.121], because of the action of the electrostatic field. The acidity was highest for CrY and CuY zeolites and lowest (of those studied) for CdY, which happens to have the greatest cation radius. The acidity of X-zeolites was lower than that of type Y. No correlation could be found with the ionization potential of the cation. Apparently, the acidity is not directly influenced by the cation's capability to exist in different states of oxidation.

It is noteworthy that some zeolites, subjected to partial ion exchange for M^{n+} ions, were more acid than those which experienced complete exchange. It is possible that at a high degree of exchange the OH groups trade off the proton for the less acidic M^{n+} ion.

3.3 The Surface of Simple Transition-Metal Oxide

As we have demonstrated in Sect.2.1, the coordination of transition atoms on an oxide surface is, to a large extent, determined by the crystal-field stabilization energy. Evidently, the surface properties in the row of transition metal compounds will vary in a manner more or less similar to the bulk properties. Figure 3.10 illustrates the oxygen bonding energies on the surface of oxides of fourth-row transition metals [3.122]. The bonding energy was assessed via the direct calorimetric technique, through measuring the temperature dependence of equilibric oxygen pressure or with the aid of thermal-desorption curves. The extreme pattern in Fig.3.10 resembles the above-mentioned two-peak CFSE curves (Tables 2.1,2; Figs.2.5,6).

At high densities of transition atoms their interaction may give rise to additional structural effects. In common nondiluted oxides this interaction may involve the entire lattice. Many transition-metal oxides are antiferromagnetic. The Néel points for antiferromagnetic oxides of 3d transition metals are $-157°C$ for MnO, $-75°C$ for FeO, $19°C$ for CoO, and $250°C$ for NiO. The magnetic structure of these oxides is built up of parallel

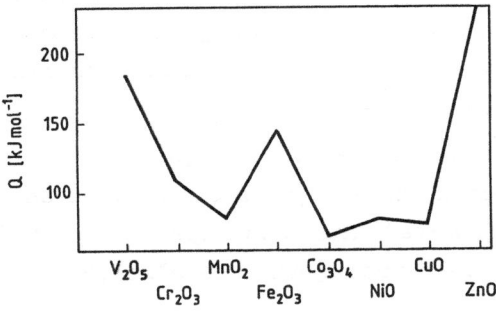

Fig.3.10. The bonding energy of oxygen on the surface of fourth-row transition metals' oxides in the normal valence state [3.122].

(111) planes with alternately parallel and antiparallel spins. The effects of magnetic orientation deform the crystalline lattice. For instance, in MnO and NiO the lattice is compressed in the direction normal to the "magnetic" (111) plane and becomes rhombohedral, while in CoO the compression takes place in the (100) direction and the lattice becomes tetragonal. Only above the Néel point is the structure purely cubic [3.123,124].

It would be natural to assume that the magnetic structure must affect the surface properties of transition-metal oxides. The ionic and covalent crystals exhibit a finite transition layer, in which the distorted surface structure tapers off into the ordered bulk structure. In [3.50] we have quoted the results of a large number of experimental findings, which confirm the existence of such a layer in semiconductors and dielectrics. The same considerations apply to the transiton-metal oxides. In some cases such transition layers were observed in atomic images of oxide surfaces by means of electron microscopy [3.125]. The changes in symmetry of the surroundings of the surface magnetic ions must be expected to give rise to the formation of a macroscopic transition layer, responsible for the surface magnetic properties of the crystal.

The discovery of surface magnetism was reported in [3.126]. Magneto-optical examination of nonbasal faces of hematite (α-Fe_2O_3) detected a surface layer less than 100 nm thick, whose magnetic properties were significantly different from those of the bulk phase. The surface hematite layer was found to be similar to orthoferrite. In essence, the surface layer can be likened to the interface between ferromagnetic domains; here, however, the layer separates the multidomain crystal structure from the external environment.

A theory of surface magnetism was devolped in [3.127], where KRINCHIK and ZUBOV also computed the energy of the surface anisotropy of hematite and estimated the participation of the surface in phase transitions in ferromagnet, see also [3.128].

Chemical etching of the surface was found to destroy the surface magnetism of hematite [3.127].

The LEED method [3.129] was used for assessing the spacing between metallic and oxygen ions in the subsurface layer of transition-metal oxides (single crystals) [3.130,131]. On the nonpolar (100) face of cubic single crystals such as NiO, CoO, MnO, the surface atoms show almost no displacement with respect ot the bulk atoms. This implies high covalence of the M-O bond on the surface, since the O^{2-} ions on the surface are unstable. They can stabilize only in the bulk at the expense of the Madelung field and the CFSE. On the polar (111) face of CoO (terminal O atom) the inward shift of surface atoms, according to LEED, amounts to 15 ± 5 %.

Another peculiar property of transition-metal oxides, which has effects on the surface, is the existence of several oxidation states of transition-metal ions which easily convert into one another. As a result, the transition-metal oxides are rarely stoichiometric: they contain either too much or too little oxygen. As regards the electronic properties, the oxides of the first type (Cu_2O, NiO, FeO) are p-type semiconductors, whereas those of the second type (V_2O_5, MoO_3) are n-type semiconductors.

The O^{2-} anions (ionic radius: 0.136 nm) in oxides of 3d metals exceed the cations in size. In p-type semiconductor oxides the excess oxygen atoms supplement the lattice, giving rise to Schottky-type defects (cation vacancy $\square_{M2+}$ for MO oxide) and excess holes $\oplus$

$$(1/2)O_2 \leftrightarrows O^{2-} + \square_{M2+} + 2\ominus . \qquad (3.11)$$

The holes can be captured by adjacent M^{n+} ions, which assume a higher oxidation state $M^{(n+1)+}$. For example, the excess oxygen in NiO can be counterbalanced by the appropriate number of higher-valence cations, specifically Ni^{3+}.

In n-type semiconductor oxides the deficit of oxygen usually results in Frenkel-type defects: interstitial cations (M_i^{2+}) and electrons:

$$M^{2+} + O^{2-} \leftrightarrows M_i^{2+} + (1/2)O_2 + 2e^- \ . \tag{3.12}$$

The electrons can be captured by adjacent metal ions M^{n+} (perhaps by M_i^{2+}), bringing them into a lower oxidation state $M^{(n-1)+}$. For example, MnO_2 is an n-type semiconductor with an oxygen deficiency, replenished with lower-valence ions Mn^{3+}. Equally probable is the capture of electrons by anion vacancies giving rise to F centers.

The distribution of defects in nonstoichiometric oxides has been discussed in detail in a number of monographs [3.132,133]. The conductivity of p-type oxide semiconductors increases with the rise in oxygen pressure according to (3.11), whereas that of n-type semiconductors decreases according to (3.12). As a rule, the higher-valence oxides have a tendency toward partial reduction, which results in the appearance of stoichiometric excess of metal in the lattice and n-type conductivity. The lower-valence oxides tend to partial oxidation, leading to stoichiometric excess of oxygen and p-type conduction. For instance, CrO_2 and CrO_3 are n-type semiconductors, while CrO is a semiconductor of the p type. Cr_2O_3, however, can be either an n-type or a p-type semiconductor depending on the oxygen pressure. More often it is a p-type semiconductor with excess oxygen ($[O]/[Cr] \leq 1.52$). The dependence of Cr_2O_3 conductivity on oxygen pressure was explained by vacancy formation according to the scheme

$$(3/2)O_2 \leftrightarrows 3O^{2-} + 2\square_{Cr}^{3+} + 6\oplus \ .$$

Above 1000°C chromium oxide may undergo spontaneous redistribution without violating the stoichiometry:

$$2Cr^{3+} \rightarrow Cr^{2+} + Cr^{4+} \ .$$

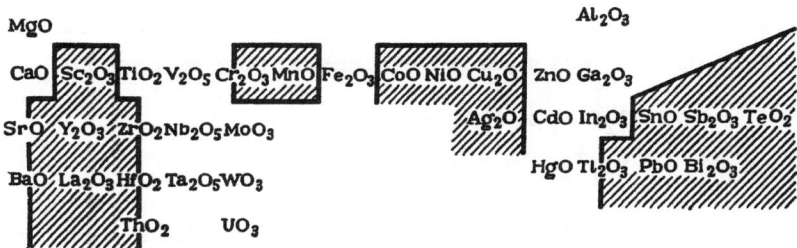

Fig.3.11. The type of conduction of oxides versus the metal's position in the periodic table. The dashed portions denote oxides whose conductivity at low temperatures is predominantly p type; undashed portions denote n-type conductivity

If metals in oxides are considered only with their chief valences, the conductivity type will be found to show a certain regularity in accordance with the metal's place in the periodic table. This point is illustrated in Fig.3.11 for stable metallic oxides, for which data on conductivity are available [3.134]. The oxides of metals opening the long periods usually exhibit n-type conductivity; of those closing the long periods – usually p type. For instance, in the fourth row TiO_2, V_2O_5, Cr_2O_3, MnO_2, and Fe_2O_3 are n type, while CoO, NiO, and Cu_2O are p type. The same regularity is found in the fifth and sixth rows although the data available are not nearly as extensive.

The addition of ions of different valences, replacing the native cation in the lattice, alters the proportion of ions of various valences. Systems of this kind are often employed as catalysts.

To assess the effects of doping with ions of different valences on the electronic properties of transition-metal oxides, the chemists dealing with adsorption and catalysis often use the so-called principle of controlled valence proposed by VERWEY and DE BOER [3.135]. As indicated above, the electroconduction of nonstoichiometric NiO oxide can be caused by production of Ni^{3+} along with the cation vacancy (3.11). According to the principle of controlled valence the number of higher-valence ions (here Ni^{3+}) will rise if the equivalent number of ions of lower valence (e.g., Li^+) is introduced into the NiO lattice. This will be accompanied by consumption of oxygen:

$$Ni^{2+}O^{2-} + (x/2)Li_2O + (x/4)O_2 \rightarrow$$
$$Ni^{2+}Li^+_x \oplus_x O^{2-}_{1+x} = Ni^{2+}_{1-x}Li^+_x Ni^{3+}_x O^{2-}_{1+x} ,$$

where $x \ll 1$. The Li^+ ions were especially often used for doping the transition-metal oxides because of the similarity of the ionic radius of lithium (0.068 nm) to those of a number of transition elements (0.072 nm for Ni^{2+}) and the resulting ability of Li_2O to form substitutional solid solutions with MO oxides.

Conversely, if a trivalent metal is introduced into the lattice of a hole-type MO semiconductor (In^{3+} or Ga^{3+} in NiO), the number of higher-valence ions will be reduced in proportion to the number of implanted M^{3+} ions with a simultaneous release of oxygen:

$$Ni^{2+}O^{2-}Ni^{3+}_x + (x/2)Ga_2O_3 \rightarrow Ni^{2+}_{1+x}Ga^{3+}_x O^{2-}_{1+x} + (x/4)O_2 .$$

Similar relations exist for n-type oxide semiconductors (with excess metal). The principle of controlled valence was subjected to an experimental check, and in some cases it was verified that the number of bulk dope ions (e.g., Li^+) is equal to the number of native lattice metal ions with altered valence (e.g., Ni^{3+} in NiO).

In view of the difficulties encountered with direct methods for studying the surface atomic and electronic structure, many researchers are looking for other properties correlated with dispersity that can be determined more readily. Having established the correlations between the optical spectra and the dispersity of oxide powder, it is possible to recognize the transition of transition-metal ions from higher coordination numbers to lower ones with appropriate variations in CFSE levels. Variations in the hue of transition-oxide powders in the course of comminution can often be observed visually. In Sect.2.2.5 we· have already mentioned Selwood's results concerning the correlations between dispersity and magnetic properties of oxide catalysts.

The surface properties of doped oxides often differ from those in the bulk. Most physical properties of high-dispersion oxide powders (electroconductivity, lattice parameter, electronic work function, etc.) exhibit nonmonotonic changes with increased doping. They also depend on the doping procedure, temperature, etc. All this affects the mechanism of implantation, and the dopant is distributed unevenly between the surface and the bulk. In some cases the solid solutions obtained are substitutional, in others interstitial.

Doping of NiO with LiO_2 with subsequent heat treatment at low temperature (200–250^{0}C) was found not to produce an adequate quantity of Ni^{3+} ions despite the rise in conductivity [3.136,137]. This can be explained by the fact that in the course of low-temperature heat treatment Li^+ is dissolved in the surface layers of NiO, giving rise to anion vacancies $\square^{O^{2-}}$

$$Ni^{2+}O^{2-} + \frac{x}{2}Li_2O \longrightarrow Ni^{2+}Li_x^+O_{1+x/2}^{2-}\square_{x/2}^{O^{2-}}$$

leaving the concentration of Ni^{3+} unchanged. Accumulation of lithium on the surface of $Li_xNi_{1-x}O$ solid solutions was proved directly with the aid of Auger spectroscopy [3.138].

Defects in oxides exist separately – without interacting – at very low concentrations (according to some researchers, below 10^{-5}% by mass). Interaction between defects (attraction and repulsion) results in rearrangement of adjacent atoms and clusters, and in establishment of superstructures (long-range order).

In most oxides the generation of defects and superstructures is promoted by the presence of impurities, and does not occur so easily in pure oxides. This applies especially to oxides with high-symmetry coordination (most often octahedral) and dense packing of oxygen ions, e.g., the oxides of Ni, Co and Fe. Superstructures here are not common.

Another group comprises the oxides of Mo, V, W, Nb, and Ta with irregular structure, asymmetric arrangement of oxygen ions around the ions of the metal, and loosely packed metal-oxygen polyhedrons. These transition metals form a large number of oxides with complicated stoichiometry, each of which is thermodynamically unstabe by itself. The reduction of oxide (at least, at moderate temperatures) goes through a series of phase transitions. For example, the reduction of Nb_2O_5 occurs via the series of transformations

$$Nb_2O_5 \rightarrow Nb_{35}O_{87} \rightarrow Nb_{25}O_{62} \rightarrow Nb_{32}O_{79} \rightarrow Nb_{22}O_{54} \rightarrow Nb_{12}O_{29} \ .$$

Such transformations do not involve the entire crystalline lattice. The reduction follows the so-called mechanism of crystallographic shear [3.139–141]. The oxide MoO_3 consists of layers, which are "packs" of deformed $[MoO_6]$ tetrahedrons with common vertices (structure in (3.13a), see also Fig.3.14). Reduction results in the loss of oxygen and generates vacancies; the Mo atoms come closer to one another and the octahedrons have common edges (structure 3.13b):

$$
\begin{array}{ccc}
\begin{array}{c}
|/ \quad\quad |/ \\
- \text{Mo} - 0 - \text{Mo} - \\
/| \quad\quad /|
\end{array}
&
\begin{array}{c}
\backslash\ |\ /^{0}\backslash\ |\ / \\
\text{Mo} \quad\quad \text{Mo} \\
/\ |\ \backslash_{0}/\ |\ \backslash
\end{array}
& \quad . \\
\text{a)} & \text{b)} &
\end{array}
\tag{3.13}
$$

This results in a structure of blocks, more or less similar, separated by shear planes having a different (although strictly ordered) structure. The shear causes one microcrystal (microblock) to move against another by a small amount, about the distance between adjacent atoms. The energy of crystallographic shear is small. The shift of octahedrons which helps to fill the vacancies is shown schematically in Fig.3.12. This process results in a stoichiometric compound Mo_nO_{3n-x}, n being the number of octahedrons, which remains unchanged, between the

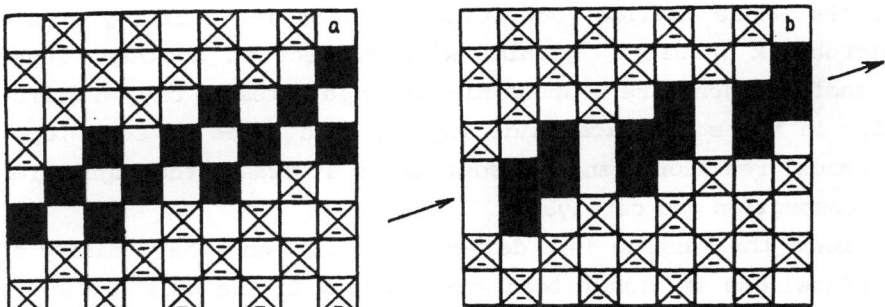

Fig.3.12. Shear structure in the MO_3 oxide crystal: (a) before shear (b) after shear; arrow indicates the direction of shearing force

shear plane. Figure 3.13 shows the block structure of $Nb_{28}O_{70}$ and $Nb_{25}O_{62}$ [3.139]. The double-shear structures are built up of small blocks [the figure shows them as comprising (3x4) and (3x5) octahedrons NbO_6]. The blocks can be either separate or touching at their edges. In the third direction (normal to the plane of the figure) the blocks may either form indefinitely long stacks or terminate somewhere. The block structure can be revealed under the electron microscope using the decoration technique.

Transformations of superstructures, formed from shear structures on nonstoichiometric oxides, often occur very swiftly at oxidation or reduction and exhibit a chain-like or cooperative nature. The energy is spent on the reconstruction

Fig.3.13. Schematic presentation of block structures for (a) $Nb_{28}O_{70}$ and (b) $Nb_{25}O_{62}$ [3.139]. Squares are sections of octahedra

of the whole lattice, but rather just on the shifting of one microblock (cluster, domain) with respect to another. These transformations are responsible for high rates of oxygen diffusion in the subsurface oxide layers, high rates of some topochemical reactions, and (in some cases at least) the high rates of adsorption and catalysis.

Since the surface is a defect by itself, and the mobility of surface ions is higher than the mobility in the bulk, it would be only natural to anticipate the creation of similar surface superstructures, that is, long-range order with regular arrangement of surface defects. Indeeed, a study of the surface of rutile (TiO_2) single crystals using the LEED technique disclosed superstructures of types (1x3), (1x5), and (1x7) on the (100) face after annealing [3.142,143]. Heating does not cause reconstruction of the (110) face, while annealing of the (001) face produces "facets" with (2x1) structure. Heating to 1000°C produces facets on the (114) face, containing titanium with coordination number 4. This structure is unstable and transforms into a stepped (vicinal) surface containing Ti ions with coordination numbers 4, 5, and 6. The superstructures with the longest-range order of (1x7) type are apparently formed due to reduction. These structures are similar to the shift structures in the bulk.

The investigations of surface structures with conventional electron-spectroscopic techniques (high-energy electron diffraction) revealed, in some cases, the distinction between the structure of surface layers and the bulk structure. For instance, the electron-diffraction study of Co_3O_4 films displayed "extra" lines, indicating the existence of a superstructure, these lines do not show up in the spectrum of bulk specimens. The formation of a superstructure has been explained in [3.144] by the presence of excess oxygen in the subsurface layer in the form of O^{2-} ions, neutralized by cation vacancies in the tetrahedral sublattice of Co^{2+} ions. Some of the remaining cations assume a trivalent state

94

$$Co^{2+}_{8(1-x)} \square_{8(x/3)} Co^{3+}_{8(2x/3)} \left[Co^{3+}_{16} \right] O^{2-}_{32}$$

Imperfection of the long-range order causes blurring of some extra lines.

Let us now consider some specific models of transition-metal oxide surfaces commonly found in the work on adsorption and catalysis.

3.3.1 Chromium Oxide

Chromic oxide Cr_2O_3 is commonly used as a catalyst of various reactions [3.37,145-149]. The oxide α-Cr_2O_3 forms a corundum-type lattice, a densely packed hexagonal structure of O^{2-} ions with two-thirds of the octahedral sites occupied by Cr^{3+} ions in a regular arrangement.

BURWELL and co-workers [3.145] put the concept of the dense packing of oxygen ions into the basis of a description of the α-Cr_2O_3 surface. They viewed the hydrated (001) face as consisting of a layer of OH⁻ ions in dense hexagonal packing. Electroneutrality and octahedral coordination of chromium ions are ensured by replacement of the plane of O^{2-} ions in the bulk by an equivalent plane of OH⁻ ions on the surface. The (001) plane contains 9.8 Cr^{3+} ions per square nanometer. Dehydration gives rise to defects: active centers incorporating Cr^{3+} and O^{2-} ions in varying coordination.

However, the conception of a surface consisting of densely packed spherical ions of O^{2-} or OH⁻ hardly gives a true picture of the chromium-oxide surface, covered with oxidized covalence-bound structures. This, in particular, follows from the IR spectra of Cr-O vibrations on the chromium-oxide surface [3.146,147]. Changes in the coordination of surface Cr_2O_3 ions were analyzed with the aid of IR spectra in the region of valence vibrations of the Cr-O bond. Five coordination-unsaturated structures have been assumed [3.147] to exist on the surface of partially dehydrated chromium oxide (3.14a-e). In the presence of oxygen new structures (3.15f-j) are formed. The three downward lines in each scheme represent bonds of Cr atom with underlying oxygen atoms:

$$(3.14)$$

These structures produce five lines in the IR spectrum: a high-frequency doublet (1024 and 1016 cm^{-1}) and a low-frequency triplet (995, 986, and 980 cm^{-1}).

Study of the dependence of the intensity of these bands on the temperature of dehydration or heat treatment in oxygen allows one to assume that the doublet corresponds to Cr=O vibrations near coordination-unsaturated metal ions (structures are a–d,i,j), whereas the triplet stems from Cr=O vibrations in the coordination-saturated octahedron (structures f–h). The lower the coordination number, the higher the frequency of vibrations. These structures do not specify the oxidation state of the surface chromium ions. Most probably, the Cr=O group pertains to Cr^{6+} in structures f–h and Cr^{5+} in the other structures.

Heat treatment in oxygen partially converts the triplet to still lower frequencies (780 and 870 cm^{-1}), attributed in [3.147] to the polymer groups $[CrO_4^-]_n$, where some of the chromium ions are found in a tetrahedral coordination. Regrouping converts the octahedron of oxygen ions into a square pyramid and a tetrahedron. Indeed, the higher oxidation states of chromium (Cr^{6+} or d^0) are more likely to be found in a tetrahedral coordination.

Reduction of Cr^{3+} to Cr^{2+} in the bulk is disadvantageous in terms of CFSE, since the extra electron in Cr^{2+} must take the orbital directed towards two O^{2-} ions. In a square-pyramidal configuration on the surface the reduction $Cr^{3+} \rightarrow Cr^{2+}$ goes more easily. Heat treatment of Cr_2O_3 in oxygen results in partial conversion of Cr^{3+} directly into Cr^{6+} and to enrichment with oxygen. This process is also facilitated on the surface. The intermediate oxidation states Cr^{4+} and Cr^{5+} were detected only in chromium gels heat-treated at low temperatures. At 200°C these ions are found in a very low concentration, while at 250°C there remains only a mixture of Cr^{3+} and Cr^{6+} [3.148]. We see that in contrast to most other transition oxides, the redox processes in chromium oxide are restricted to the surface, while in the bulk the original Cr_2O_3 structure persists.

The EPR spectrum at the initial stage of reduction of CrO_3 to Cr_2O_3 exhibits a narrow line, attributed to the formation of a certain quantity of Cr^{5+} ions [3.37,149]. The persistence of these lines up to quite high temperatures indicates that the degenercy of 3d orbitals is, to a large extent, cancelled, and the Cr^{5+} ion is found in a tetrahedral T_d or square-pyramidal C_{4v} coordination. The broadening of the EPR line of the Cr^{5+} ion at adsorption indicates that this ion occurs on the surface. It is possible that a chromyl-type structure is formed:

$$
\begin{array}{c}
-O \\
-O-Cr=O, \\
-O
\end{array}
$$

where one out of the four Cr-O bonds is shorter than the rest.

Doping of Cr_2O_3 with MgO raises the concentration of Cr^{6+} ions, while doping with TiO_2 increases that of Cr^{3+} ions. Investigations revealed that the impurities remain in the subsurface layers of chromium oxide [3.148].

3.3.2 Molybdenum Oxide

The analysis of the crystalline structure of MoO_3 reveals that the molybdenum atom is surrounded by six oxygen atoms in a highly distorted octahedron. Each MoO_6 octahedron has two common edges and two common vertices with its neighbors, thus forming a lamellar structure (Fig.3.14a) [3.150-158]. The Mo-O bond lengths in an octahedron are highly different: from short $Mo-O_{(1)}$ and $Mo-O_{(3)}$ bonds (0.167 and 0.173 nm) to long bonds $Mo-O''_{(2)}$ and $Mo-O'_{(3)}$ (0.233 and 0.225 nm - Fig.3.14b) [3.151]. It may be assumed that two oxygen atoms in molybdenum oxide are linked to the metal by double bonds, two are linked to metal by a bond which may be counted as double, and the other two are bound more loosely:

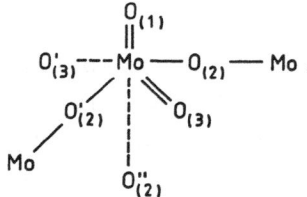

In other words, the structure is based on molybdenum-oxygen tetrahedrons, linked by weaker bonds with other chains -O-$[MoO_2]$-O- . In the IR spectra of MoO_3 the shortest bonds

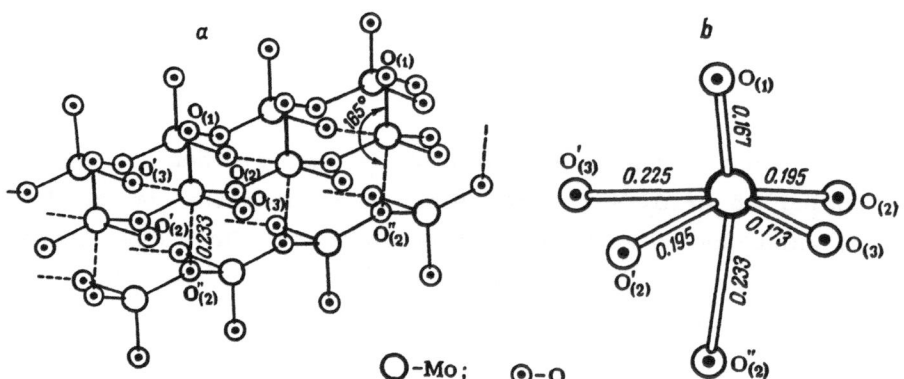

Fig.3.14. Lamellar lattice of MoO_3 (a) and the bond lengths (nm) in an octahedron $[MoO_6]$ of MoO_3 lattice (b) [3.151]

("double" Mo=O bond) show up at 980 cm^{-1}, the Mo–O bonds vibrate at 870 cm^{-1}, while longer Mo ... O bonds produce a wide band at about 600 cm^{-1} [3.152,153]. Such structure facilitates generation of coordination-unsaturated sites on the surface. For example, the removal of oxygen from sites $O'_{(3)}$ or $O''_{(2)}$, opposing the double Mo=O bond in the distorted octahedron, gives rise to a distorted square-pyramidal configuration. Rupture of both Mo ... $O'_{(3)}$ and Mo ... $O''_{(2)}$ long bonds results in a tetrahedron.

It is noteworthy that irregular, asymmetric coordinations of similar kind with varying metal-oxygen bond lengths are typical also for oxides of W, Cr, V, Nb, and Ta. Their structures are also based on a tetrahedron with shorter strong bonds, which is completed to an octahedron by longer bonds [3.151]. Apparently, it is this steric structure and the ease of transition from one coordination to another, together with the ease of oxidation and reduction of these oxides, that are responsible for their outstanding adsorptive and catalytic properties.

Reduction of MoO_3 produces a large number of stoichiometric oxides of type Mo_nO_{3n-x}, where x is small (x = 1, 2, 3 ...), and n may be large. These oxides (e.g., Mo_4O_{11}, Mo_8O_{23}, Mo_9O_{26}, $Mo_{10}O_{29}$, $Mo_{11}O_{32}$, $Mo_{13}O_{38}$, $Mo_{18}O_{52}$, and $Mo_{29}O_{75}$ [3.139, 150,154] can be visualized as a distorted octahedron of oxygen atoms around Mo^{6+} in a block adjoined to another octahedron edgewise rather than vertex-wise as in MoO_3. The Mo-Mo spacing is reduced from 0.32 nm in MoO_3 to 0.2613 nm in $Mo_{18}O_{52}$ (between the atoms in adjacent blocks). This ought to be expected to produce a variety of valence states and coordinations of molybdenum ions on the surface. It was shown in [3.140] that the reduction of MoO_3 by carbon monoxide at 400°C produces a surface phase of Mo_9O_{26}, while the bulk phase MoO_3 is retained.

The Mo^{5+} ion (d^1 configuration) is easily detected by EPR spectroscopy on slightly reduced MoO_3. For instance, partial reduction of MoO_3 gives rise to an axisymmetric spectrum with

$g_z = 1.875$ and $g_{x,y} = 1.933$ [3.156]. This signal does not depend on temperature and persists up to as high as 400°C. This spectrum fits in well with the concept of structure similar to the structure of chromium oxide with Cr^{5+}.

Rapid diffusion of an oxygen vacancy from the surface into the bulk was discovered [3.156] while investigating surface reduction by olefins (RH). The action of propylene on MoO_3 at 300–500°C immediately gives rise to a narrow EPR signal from Mo^{5+} with $g_1 = 1.966$, $g_2 = 1.957$, and $g_3 = 1.894$. This signal exhibits no broadening in the presence of water vapor or oxygen and can be attributed to Mo^{5+} ions in the bulk or in the subsurface region. These ions arise due to a fast migration of an electron produced on the surface at the reduction of Mo_s^{6+} to Mo_s^{5+} with subsequent capture of this electron by a bulk atom Mo_v^{6+}

$$Mo_s^{6+}O_s + RH \longrightarrow \left[Mo_s^{5+}\overset{\overset{\displaystyle H}{\displaystyle |}}{O_s} \right] R^+ \xrightarrow{\;Mo_v^{6+}O_v\;} Mo_s^{6+}\square RO + Mo_v^{5+}O_v + H^+.$$

Then this signal slowly converts into a broad signal from the bulk with $g_{av} = 1.930$ due to attraction of a surface vacancy to this defect:

$$Mo_v^{5+}O_v + Mo_s^{6+}\square \longrightarrow Mo_v^{5+}\square + Mo_s^{6+}O.$$

Migration of a vacancy is accompanied by diffusion of oxygen to the surface. The diffusion of oxygen occurs not at the expense of any interstitial excess oxygen atoms, but rather involves the oxygen atoms in regular lattice sites, and probably follows the mechanism of multiple shifts of microblocks. Thus the surface layer of a catalyst in the course of a reaction is constantly being rebuilt. The oxygen atoms, linked to molybdenum by a double bond and a single bond keep changing places. This perpetual exchange must be taken into account when considering the possible correlations of the surface properties with IR vibrations of Mo=O and Mo-O bonds [3.152,153].

At a high degree of reduction, MoO_3 converts directly into MoO_2 with appropriate reconstruction of the entire lattice, apparently by way of fast disproportionation: $2Mo^{5+} \rightarrow Mo^{6+} + Mo^{4+}$. Only Mo^{6+} and Mo^{4+} were discovered in an XPS analysis of MoO_3 reduction. Vigorous reduction produces metallic molybdenum [3.157,159].

The reduction of V_2O_5 and WO_3 generally follows the same course as that of MoO_3. Slight reduction produces ions in a d^1 configuration: V^{4+} (or vanadyl VO^{2+}) and W^{5+}. Intermediate oxides V_6O_{13} and V_3O_7 were discovered between V_2O_5 and VO_2.

3.4 The Surfaces of Some Complex Oxide Systems

Reliable experimental data on the surface structures of complex oxides are scarer than those available for simple oxides. It must be expected that the variety of valence structure and coordinations of different atoms on the surface will give rise to a diversity of surface structures even greater than in the case of simple oxides. Imagine, for instance, the surface of the complex oxide ABO_x: the metallic ion A with a higher value of the CFSE will be shielded by oxygen to a greater extent than the ion B with a lower CFSE. As a result, the anion vacancies will be concentrated near B, and the coordination number of ion B will be lower.

Let us, by way of an example, consider some data on the bulk and surface structure of molybdates, oxidation catalysts of special importance.

3.4.1 Bismuth Molybdate

Currently bismuth molybdate seems to be receiving more attention than any other oxide catalyst because of its high selectivity in reactions of oxidizing, in oxidation ammonolysis and in oxidation–dehydration of olefins [3.160–172].

Different compositions of bismuth molybdate occur in three crystalline modifications: α phase ($Bi_2O_3 \cdot 3MoO_3 = Bi_2Mo_3O_{12}$), β phase ($Bi_2O_3 \cdot 2MoO_3 = Bi_2Mo_2O_9$), and γ phase ($Bi_2O_3 \cdot MoO_3 = Bi_2MoO_6$). The first has the structure of scheelite $CaWO_4$ and can be defined formally as $Bi_{2/3}\square_{1/3}MoO_4$ or $Bi_4\square_2(Mo_2O_8)_3$, $\square$ being regular cation vacancies. Here Mo is in tetrahedral configuration, and the coordination number of Bi is 8. The coordination numbers of Mo in the β phase are 4 and 6; this modification is sometimes called the Erman phase after the scientist who was the first to investigate it [3.160]. The stoichiometric γ-phase has a koechlinite lattice, which can be loosely described as consisting of alternating layers of Bi_2O_2 and MoO_2 linked through oxygen atoms. In the bismuth layer the Bi^{3+} ion is located a little way above or below the center of the square formed by oxygen atoms. In the molybdenum layer the Mo^{6+} ions are located in the center of the octahedron formed by oxygen ions. The adsorptive and catalytic properties of bismuth molybdate depend on the oxidative and reductive powers of its surface. The processes of reduction and reoxidation of different phases of bismuth molybdate were analyzed in [3.161-170]. In all cases, at temperatures of the catalysis the rate of reduction exceeds the rate of reoxidation by one or two orders of magnitude; in each case, however, the mechanicm of the process are different. In the reduction of the α phase [3.162], the first two steps are very rapid, they are associated with the formation of defects on the surface. Then comes a slow process of bulk reconstruction:

$$Bi_2Mo_3O_{12} = Bi_4\square_2(Mo_2O_8)_3 \longrightarrow Bi_4\square_2(Mo_2O_8)_2(Mo_2O_7\square) + O \,.$$

The two stages ("bulk" and "surface") can also be distinguished in the reduction of the β phase. It has been assumed [3.163] that on the surface the β phase disproportionates into the α and γ phases. Reduction of the β phase may yield a free MoO_3 phase (or MoO_2 with deep reduction):

$$Bi_2Mo_2O_9(surf.) \rightarrow Bi_2MoO_{6-x}(surf.) + MoO_{3-y} + (x + y)[O] \quad,$$
$$Bi_2Mo_2O_9(bulk) + MoO_{3-y} \rightarrow Bi_2Mo_3O_{12-y}(bulk) \; .$$

Reduction of the γ phase goes more slowly and reoxidation faster compared with the α and β phases. The redox processes in the γ phase are usually associated [3.161] with its lamellar structure, in which oxygen occupies three different crystallographic locations: (I) in the MoO_2 plane, (II) in the Bi_2O_2 plane, and (III) in the binding layer:

$$
\begin{array}{ccccccccccccccc}
 & O & & O & & & & O & & O & & & & O & & O & \\
\text{(I)} & O & Mo & O & Mo & O & \xrightarrow[<350^\circ C]{red.} & O & Mo & O & Mo & O_B & \xrightarrow[<400^\circ C]{red.} & O & Mo & O & Mo & \Box_B \\
\text{(III)} & & O & & O & & & & O & & O & & & & O & & O & \\
\text{(II)} & O & Bi & O & Bi & O_A & & O & Bi & O & Bi & \Box_A & & O & Bi & O & Bi & O \\
 & & O & & O & & & & O & & O & & & & O & & O & \\
 & & & (a) & & & & & & (b) & & & & & & (c) & & (3.15)
\end{array}
$$

Reduction of Bi_2MoO_6 just below 350°C first of all removes oxygen from edges of the Bi_2O_2 plane (O_A in the scheme (3.15a) – the surface in each scheme is on the right). The removal of this ion is followed by the fast escape of the bulk oxygen, which diffuses along "self-restoring" rows (II) in the lattice. A vacancy is formed on the surface near the Bi^{3+} ion (3.15b), which is converted into Bi^{2+}. Heat treatment at 350–400°C causes structural rearrangement. The surface becomes dominated by the (111) face rather than the (110) face; the oxygen O_B is removed from molybdenum and fills the vacancies $\Box_A$ (3.15c). The Mo^{6+} ion is apparently reduced to Mo^{4+} in the configuration of a square pyramid. Rapid redox processes are facilitated by the high mobility of oxygen in layers (III), whence it is supplied to the vacancies $\Box_A$ and $\Box_B$. Although the electron is delocalized over Bi-O-Mo bridges, it would be more or less correct to state that at low temperatures the Bi ions are reduced, whereas Mo ions are reduced at high temperatures.

The catalysts $\gamma-Bi_2O_3 \cdot Mo^{18}O_3$ and $\gamma-Bi_2{}^{18}O_3 \cdot MoO_3$, labeled with ^{18}O throughout the bulk, were studied in [3.167]. The analysis of reduction yield indicates that oxygen is removed mainly from layers of $[Bi_2O_2]_n^{2+}$. Reoxidation, on the other hand, occurs at the expense of oxygen migration in the $[MoO_2]_n^{2+}$ layer. These results seem to confirm the role of lamellar structure in the redox processes.

With bismuth-molybdate surfaces XPS investigations are ambiguous [3.168,171]. Fully oxidized structures exhibit the same composition of the surface layer. The surface layer becomes considerably enriched with molybdenum (nearly twofold), the arising Mo^{4+} ions migrate over the surface and form clusters. Reduction of Bi^{3+} occurs at a later stage. Apparently the reduction of bismuth molybdate may follow the mechanism of crystallographic shear, and the compounds of Bi_2MoO_{6-x} form a block structure [3.171]. Such structures evidently promote rapid absorption and rapid release of considerable amounts of oxygen, i.e., they facilitate redox processes. The reduced phase forms clusters, scattered at random in the oxidized phase, and under certain conditions gives rise to superstructure [3.161].

The study of the structure of bismuth molybdate doped with iron oxides utilizing various spectroscopic techniques [3.169,170,172] indicates that the iron oxide replaces Mo in the lamellar structure of gamma bismuth molybdate, thus disrupting the continuity of the MoO_2 layers. According to Mössbauer spectra, however, some of the iron atoms form an ordered magnetic phase (Fe-O-Fe clusters). The proporties of the two forms of iron depend on the particulars of the sample handling. Thorough investigations using complex techniques [3.170] indicate that the oxidized surface of alpha bismuth molybdate doped with iron oxide has the same composition as the bulk. The structure, however, differs from that of the initial phase. A mixed Bi-Mo-Fe phase is formed

$$3 \square Bi_2(MoO_4)_3 + Fe_2(MoO_4)_3 \rightleftharpoons 8MoO_3 + 2Bi_3[FeMo_2]O_{12}.$$

In addition, doping with iron promotes decomposition of the α phase into β molybdate and molybdenum oxide:

$$Bi_2(MoO_4)_3 \rightarrow MoO_3 + Bi_2Mo_2O_9 \ .$$

In the process of reduction, the surface of bismuth molybdate becomes enriched with molybdenum. Mössbauer spectra indicate that the Fe ions occupy different sites in alpha bismuth molybdate: they may either take vacancies or force out Mo ions. This results in expulsion of MoO_3 from the regular lattice into a separate phase. The redox catalytic reaction apparently takes place on the interphase boundary between MoO_3 and molybdate. In a catalyst with addition of Fe ions, the Fe^{2+} ions are more stable in one of the phases while the Fe^{3+} ions are more stable in the other. Oxidation-reduction gives rise to nonequilibrium conditions in which the Fe ions are found in the "wrong" phases. Subsequent relaxation involves exchange of electrons and oxygen in the interphase.

3.4.2 Molybdates of Other Metals

Molybdates of second-group elements crystallize into the scheelite structure: molybdenum has coordination number 4, and the coordination number of the cation is 8. The transition-metal molybdates form different structures, the cation having coordination number 6, while molybdenum, included in the anion, has the coordination number of 4 (in molybdates of V, Cr, Mn, Cu) or 6 (molybdates of Fe, Co, Ni). Phase transitions in some molybdates change the coordination number of molybdenum. There exist, for example, a purple species of $CoMoO_4$ (α phase) and a green one (β phase). The former contains tetrahedrally coordinated molybdenum, while the latter corresponds to the octahedral coordination. The α phase readily converts into the β phase when the powder is crushed in air. Thus the increase in surface area alters the relative stability of phases. Quite probably, the ease of phase transitions in transition-metal molybdates and the readiness of surface atoms to change their coordination are responsible for their outstanding adsorptive and catalytic properties. Doping may help in stabilizing one of the structural modifications.

Detailed investigations of IR absorption spectra indicate the existence of some peculiarities in the region of stretching vibrations of Mo-O bonds [3.153,173]. Scheelite-type molybdates of Ca, Sr, Ba, and Cd display a strong absorption band ν_3 at 750-850 cm^{-1} (Fig.3.15a), corresponding in its location and intensity to Mo-O vibrations in the Mo-O-M structure (M being a cation). Molybdates of V, Cr, Mn, Fe, Co, Ni, Cu, and Zn with a monoclinic wolframite-type structure exhibit absorption bands ν_3 at 750-850 cm^{-1} and ν_1 at 940-990 cm^{-1}, corresponding in location and intensity to Mo-O-M and Mo=O vibrations, respectively (Fig.3.15b). The separation between Mo and O is 0.191 nm in the former and 0.173-0.175 nm in the latter. The spectra of molybdates of Ni, Co, and V, which, like MoO$_3$, have octahedral structure [MoO$_6$], show a wide band at 600-650 cm^{-1}, which seems to be related to the longer Mo-O bonds ($\approx$0.2 nm).

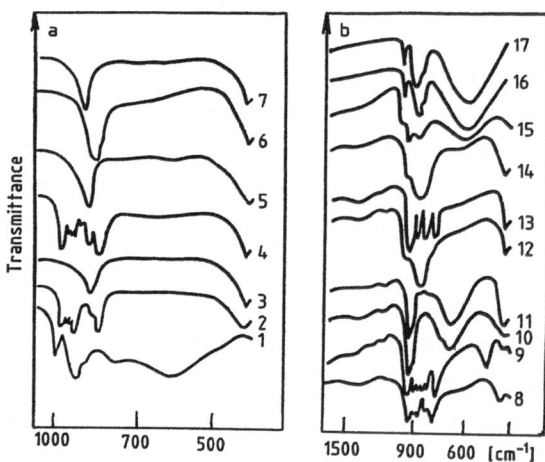

Fig.3.15. Infrared absorption spectra of (a) molybdates of second-group metals, and (b) molybdates of transition metals [3.153]: (1) Be, (2) Mg, (3) Ca, (4) Zn, (5) Sr, (6) Cd, (7) Ba, (8) Zn, (9) Cu, (10) Ni, (11) Co, (12) Fe, (13) Mn, (14) Cr, (15) V, (16) Ti, (17) MoO$_3$

The catalytic and adsorptive action of molybdates and vanadates is now commonly believed to depend on the presence of a double bond Mo=O or V=O [3.152,174]. Some researchers try to attribute the bands in the 900-1000 cm^{-1} range to Mo=O or V=O bonds on the surface. According to [3.175], for instance, reduction of molybdates results first of all in the drop of intensity of the ν_1 band (pertaining to Mo=O) rather than ν_3. These results, however, cannot be assumed final, since the exact values of extinction coefficients for these bands are not yet established.

The reduced forms of molybdenum in molybdates have been detected in many experiments using EPR and optical spectroscopy.

A "shoulder" at 33.000 cm^{-1} in the diffuse-reflection UV spectrum of high-dispersion molybdates of Co and Cd was detected in [3.176] and attributed to the surface. At the same time the IR spectrum showed a band of Mo=O vibrations at 940-945 cm^{-1}, not observed in the spectrum of a bulky sample. TRIFIRO et al. [3.176] attributed these bands to Mo atoms on the surface in the configuration of a highly distorted tetrahedron. The Mo^{5+} ions in such a configuration of a slightly reduced cadmium molybdate give a strong EPR signal with $g_{x,y}$ = 1.940 and g_z = 1.885, corresponding to Mo^{5+} ions forming on the surface. At the same time in most molybdates of transition metals as well as in bismuth molybdate with lamellar structure, the Mo^{5+} ions are generated in the bulk owing to fast vacancy diffusion into the bulk (or, equivalently, fast diffusion of oxygen toward the surface along the layers). The incessant reconstruction of the surface and vacancy exchange between the surface and the bulk take place at low enough temperatures.

The XPS technique was utilized in [3.159] for studying the reduction of the α and β phases of cobalt molybdate. Heat treatment in hydrogen at 400°C was found to result in reduction of Co^{2+} on the surface to metallic cobalt, the α phase being reduced faster than the β phase. The Mo^{6+} ion was not reduced.

We have already mentioned the accumulation of excess MoO_3 on the surface in redox processes involving bismuth molybdate. Similar phenomena were observed in other molybdates, too. For example [3.176], the surface of an Co-Mo catalyst, prepared by mixing solutions of Co and Mo salts, exhibited an excess of molybdenum in the form of MoO_3 (according to XPS data). The excess MoO_3 phase is important for the performance of molybdate catalysts.

The multicomponent Bi-Fe-Co-Mo oxide catalysts of selective oxidation [3.162-165,169,172,177,178] consist of three main phases: β-cobalt molybdate, α-bismuth molybdate, and excess molybdenum oxide. Figure 3.16 shows schematically the structure of such a catalyst, according to [3.169,177,178]. Iron molybdate is stabilized on the surface of other molybdates. The crystalline lattice of $Fe_2(MoO_4)_3$ is coherent with the lattice of bismuth molybdate and is even capable of forming mixed crystals. Reduction produces β-$FeMoO_4$ stabilizing on the surface of β-$CoMoO_4$, which has a similar lattice. Special studies indicate that replacement of β-$CoMoO_4$ by $MgMoO_4$ having the same crystalline structure preserves the redox powers of the catalyst. The cobalt ions seem to not take part in the redox process. The required properties are furnished by Fe^{2+} and Fe^{3+} ions and MoO_3-$Bi_2(MoO_4)_3$ phases, and the redox processes take place in the interphase.

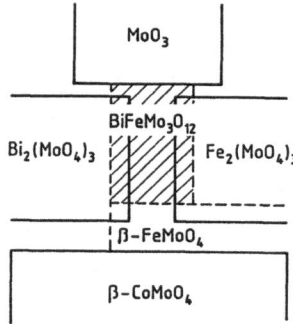

Fig.3.16. The structure of a multicomponent catalyst of selective oxidation of olefins. Solid line denotes phases pertaining to the initial catalyst; dotted line indicates phases produced in the course of catalysis

3.4.3 Magnetic and Ferroelectric Semiconductors and Dielectrics

Many binary oxide systems containing transition-element oxides exhibit peculiar magnetic and electric properties. Application of an external magnetic field to magnetic semiconductors and dielectrics results in resistance changes of by many orders of magnitude. Many magnetic semiconductors show a shift of the intrinsic absorption edge in changes of magnetic order caused by changes in temperature or application of external magnetic field. In recent years a large class of semiconductor and dielectric magnetic compounds has been discovered, having transparency regions in the IR and visible ranges. Such are, for instance, the ferromagnet EuO, $CrCl_3$; the weak ferromagnetic perovskites $RbFeO_3$, $LaMnO_3$; the carbonates $MnCO_3$, $CoCO_3$; the compounds FeF_3 and $FeBO_3$, and the antiferromagnet NiO, etc. The unique properties of magnetic semiconductors and dielectrics have already made them indispensable in many branches of solid-state electronics and laser technology [3.123,179].

Investigations of the surface properties of these magnetic substances have just begun, and seem to be advancing in two major directions. On the one hand, there is the study of effects of adsorption on magnetic order and on electronic properties of the surface layer of a magnet. This information will not only be of academic interest, but will also be very useful in looking out for new materials for solid-state gas analyzers. On the other hand, it seems very important to explore the possibilities of controlling the processes on the surface of magnets by changing the magnetic order in the subsurface region (varying temperature or applying external magnetic fields).

In nonferromagnetic materials with any type of magnetic order the charge carriers (including light-induced nonequilibrium charge carriers) may form ferromagnetic microdomains (several lattice constants across) and restrict themselves to these domains [3.180]. An electron traveling in the crystal will then pull this microdomain along (similar to an electron in a polaronic coat [3.125]). This quasi particle (electron + ferromagnetic domain) has been named a "ferron" [3.180]. Ferrons were detected in certain rare-earth compounds. It was demonstrated in [3.126] that ferronic states are even more likely to be assumed by electrons captured on donor levels of a magnetic semiconductor. On the surface, the role of acceptor and donor states can be played by adsorbed atoms and molecules, which may assist in localizaiton (or delocalization) of surface ferrons.

Systematic investigation of the surface properties of magnetic materials was pioneered by NEEL [3.181], who gave numerical estimates of surface anisotropy for simple faces of ferromagnets with cubic structure. Later, a large body of information (although indirect) regarding the magnetic properties of surfaces was obtained in numerous studies dealing with ferromagnetic metal films.

Similar phenomena are to be expected in ferroelectric semiconductors and dielectrics. A number of ferroelectrics are known to include transition atoms. Ferroelectric properties of crystals are well known to depend crucially on the surface states [3.182]; however, systematic studies of correlations between ferroelectric properties of crystals and their surface properties began only rather recently.

A typical example of a ferroelectric is barium titanate, which experiences a phase transition from the tetragonal to the cubic configuration at about 120°C. In the tetragonal phase the structure of $BaTiO_3$ is built (at the expense of appropriate distortion of the crystalline lattice) of chains of aligned dipoles. This structure facilitates spontaneous polarization of the crystal, and this phase is called ferroelectric. The electric field, produced in a ferroelectric phase and directed along the vector of spontaneous polarization, raises the crystal energy and at each point is proportional to the magnitude of polarization. The cubic phase (above the Curie point) is called paraelectric. A minute displacemant of the Ti ion in the phase transition (by no more than 0.01 nm) produces drastic changes in the electronic structure and conductivity of the crystal.

The existence of spontaneous polarization in an undoped ferroelectric crystal calls for its screening by a space-charge region and results in surface bending of energy bands. Ionized impurities in doped $BaTiO_3$ species take part in screening out the spontaneous polarization, thus reducing the surface charge and straightening the bending of energy bands. This in turn affects the domain structure of the ferroelectric. The changes in the space-charge region of a crystal will affect nucleation. For this reason the bulk doping and the change of surface properties of the crystal will shift the point of phase transition [3.182]. Conversely the change of surface structure due to a reversal of the crystal polarization in an external electric field and the ferro-paraelectric phase transition must contain information about the surface properties of the crystal.

The effects of adsorption on the properties of magnetic and ferroelectric oxides will be discussed in Sect.4.9.

Let us complete our review with a summary of some common features. These include lowered values of the coordination number near the active center (transition atom), the effects of CFSE on structural rearrangements of oxygen atoms around the transition atom; facilitation of valence transitions (i.e., redox processes) on the surface; distinction between crystalline structure of the surface and that of the bulk, resulting sometimes in the formation of clusters and superstructures; and feasibility of structural rearrangements. These peculiar properties of the surface are responsible for the high catalytic and adsorptive activity of transition-metal oxides (Chap.4).

3.5 Energy Levels of Transition-Element Ions in the Bulk and on the Surface of Their Oxides

So far we have been considering the features of the surface structure of transition-metal oxides, the chemical composition of surface, and the distribution of impurities without taking into account the energy spectrum. In Sect.2.1 we discussed the changes in the energy of electrons caused by the stabilizing action of the ligand field (in particular, that of oxygen ions in oxides), but we did not specify the location of the d levels with respect to other energy levels in the solid. In compounds of transition metals d levels are the upper occupied levels. Their energy (counting from the vacuum level or from the middle of the forbidden band, or - in the XPS method - from Fermi level E_F), together with the symmetry of the d orbitals, determines the participation of d electrons in the chemical bonding at adsorption and catalysis.

The energy of surface states (surface levels) is generally different from that of levels and bands in the bulk. One reason is the lattice discontinuity: the termination of the three-dimensional periodicity on the surface. The degeneracy of the d levels is lifted in low-symmetry fields on the surface, and their energy becomes lower than the energy of the degenerate d electrons (t_{2g} or e_g) in the bulk.

Qualitative estimates of the energy of surface states for binary semiconductors were proposed by MARK [3.183]: a one-electron donor-type surface level is located ΔE_s below the conduction band (or above the valence band for acceptor-type levels), ΔE_s defined by

$$\Delta E_s = (1 - \beta) E_g/2 \tag{3.16}$$

where E_g is the width of forbidden band; $\beta = (\Gamma - \mu)/(1 - \mu)$; $\Gamma = M_s/M_v$ is the ratio between the surface and the bulk Madelung constants; $\mu = (I - A)/2zM_ve$; I is the metal's ionization potential; A is the nonmetal's electron affinity; z is the valence, and e is the charge of an electron.

For practical purposes μ can often be neglected so that $\beta = M_s/M_v$. Spectroscopic data obtained for free ions (e.g., for Ni^{2+}) can also be used for transition-metal oxides (e.g., NiO) with appropriate corrections for low-symmetry fields. Such estimates are better made for diluted oxide systems (e.g., $Ni_xMg_{1-x}O$; Sect.3.1.1), where the interaction between the transition-metal ions can be neglected. However, the results derived from spectroscopic data of free ions even in these cases scatter rather far from experiment. In real systems the surface levels are created not simply because of lattice discontinuity. Their formation, as we have emphasized more than once, is accompanied by the creation of defects involving displacement of a number of adjacent atoms (and sometimes even more profound rearrangements) and by production of new surface compounds.

There have been attempts to calculate the energy of local levels using quantum-chemical methods. Calculations of surface band structure in a perfect perovskite-type lattice ($SrTiO_3$) and in ReO_3 and Ti_2O_3 indicate [3.184] that local surface levels arise on the (100), (111), and (110) faces of a single crystal, having t_{2g} symmetry and energy about 1.5 eV above the energy of the t_{2g} band in the bulk. On the (100) face of a cubic perovskite-type lattice these levels overlap, giving rise to surface d band (d_{xz} and d_{yz}). Very promising results have been obtained using the X_α scattered-wave (SW) technique [3.185,186] for calculating the energy-band structure of an oxide surface.

The information about local levels in transition-metal oxides was gained chiefly from experimental research. The investiagation of the energy of local levels in oxides is based on measurements of conductivity, diffuse-reflection spectroscopy, and photoelectron spectroscopy.

3.5.1 Electrophysical Data

In monocrystalline semiconductors the conductivity σ usually grows exponentially with temperature

$$\sigma = \sigma_0 \exp(-E_\sigma/kT) , \qquad (3.17)$$

where E_σ is the activation energy of conduction, k is the Boltzmann constant, and T is temperature.

The slope of the log σ against 1/T curve allows one to find the transition energy from one level to another. At low temperatures the slope is small and corresponds to extrinsic levels with $E_{ex} = 2E_\sigma \propto 0.01$–$0.1$ eV; at high temperatures the slope is steep and pertains to electron transitions across the forbidden band; in oxides $E_g = 2E_\sigma \approx 2$–$10$ eV [3.134].

VERWEY and DE BOER [3.135] were the first to point ou that a d band is not necessarily formed in transition-metal oxides, because cations are separated by anions and their wave functions do not overlap. In oxides such as Fe_2O_3, CoO, and NiO the 3d electrons occupy levels localized by cations, and at room temperature the stoichiometric oxides behave as insulators. In other words, the 3d electrons in transition-metal oxides comply with the London-Heitler model of localized electrons rather than with the band model.

As demonstrated by MORIN [3.187], a 3d band exists in the oxides of early first-row transition metals because of strong cation-cation interaction. Oxides such as Sc_2O_3, TiO, and CrO_2 (as well as NbO_2, RuO_2, and ReO_3 of heavier elements) have metal-type conduction, their resistivity at room temperature being in the range of 10^{-1} to 10^{-6} $\Omega \cdot$cm and depending linearly on temperature. As we proceed to oxides of later elements, the 3d orbitals no longer overlap: NiO, CoO, MnO, Fe_2O_3, and Cr_2O_3 are typical semiconductors, showing exponential increase in conductivity with temperature in accordance with (3.17). At room temperature their resistivity is rather high (10^3 to 10^{17} $\Omega \cdot$cm assuming stoichiometric composition) and the width of the forbidden band is 3–4 eV (i.e., they must be classified as wide-band semiconductors). Finally, there exist such oxides as V_2O_3, VO_2, Ti_2O_3,. NbO_2, and Fe_3O_4, in which a rise in temperature or pressure produces a metal-to-insulator phase transition. At the transition point the conductivity changes by 6–9 orders of magnitude. This is often called the MOTT transition [3.188].

The overlapping of d orbitals grows with the increase in the ratio between the cation and the anion radii r_c/r_a, and decreases with an increase in the cation charge. For oxides containing cations of transition metals in an octahedral surrounding the d-orbitals overlap and the creation of a d band becomes likely real probability when $n \leq 3$, n being the number of d electrons possessed by the cation ($n \leq 2$ for cations in tetrahedral coordination). The oxides of transition elements in the top right corner of the periodic table are insulators and semiconductors; those of the bottom left corner are "metals" – substances whose conductivity depends little on temperature.

The ability of transition ions to exist in a number of oxidation states allows one to attribute the high conductivity of certain oxides of transition metals to electron transitions such as

$$M^{n+1} + {}^tM^{n+} \rightarrow M^{n+} + {}^tM^{n+1} \quad ,$$

different from band-to-band charge transfer (e.g., from the 2p band of O^{2-} to the metal's d levels). Such a transfer, however, according to the principle of controlled valence (Sect.3.3.) is possible only with doping. For example, by introducing Li^+ ions into NiO to replace Ni^{2+}, we can get a number of Ni^{3+} ions by conservation of electroneutrality.

The local levels in NiO were estimated by MORIN [3.187] on the basis of conductivity measurements. Charge carriers in NiO (pure and doped with Li_2O) are generated in the following reactions:

$$O^{2-}(2p^6) + Ni^{2+}(3d^8) \rightarrow O^-(2p^5) + Ni^+(3d^9) \quad , \tag{3.18a}$$

$$Ni^{2+}(3d^8) + Ni^{2+}(3d^8) \rightarrow Ni^{3+}(3d^7) + Ni^+(3d^9) \quad , \tag{3.18b}$$

$$Ni^{2+}(3d^8) \rightarrow Ni^{3+}(3d^7) + 4s \text{ electron },\qquad(3.18c)$$

$$\square Ni^{3+}(3d^7) + Ni^{2+}(3d^8) \rightarrow \square Ni^{2+}(3d^8) + Ni^{3+}(3d^7),\qquad(3.18d)$$

$$Li^+Ni^{3+}(3d^7) + Ni^{2+}(3d^8) \rightarrow Li^+Ni^{2+}(3d^8) + Ni^{3+}(3d^7) .\qquad(3.18e)$$

In a pure oxide the conduction depends on (3.18a–c): (3.18a) describes the creation of a hole in the 2p band of oxygen and an electron localized in the 3d level; (3.18b) represents the creation of an electron-hole pair in the 3d level; (3.18c) describes the excitation of an electron to the 4s band and generation of a hole in the 3d level. In NiO containing a cation vacancy $\square$ the electron is captured by this vacancy producing a localized hole in the 3d level [Ni^{3+} (3.18d)]. In NiO doped with Li^+ the electron is captured at an acceptor site (Li^+), creating a hole in the 3d level (3.18e). In a pure NiO oxide (as well as in MnO, FeO, CoO, etc.) the conduction depends mainly on 2p holes because of their mobility. The Fermi level in NiO(Li) lies halfway between the 3d level of Ni^{2+} and the acceptor level of Li^+Ni^{3+} (about 0.2 eV from each), a distance much smaller than the separation between these levels and the 2p or 4s band. Figure 3.17 shows the energy-band scheme corresponding to (3.18a–e).[3]

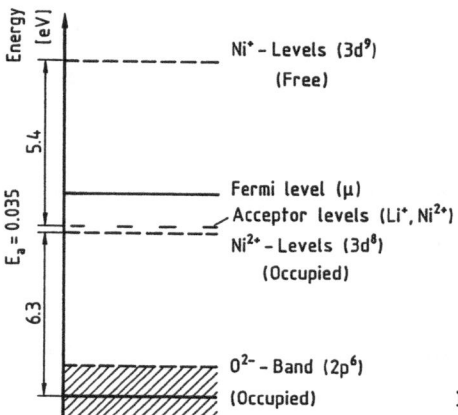

Fig.3.17. The energy diagram of NiO [3.187]

---

[3] New data for the band structure of NiO can be found in [3.189].

According to HEIKES and JOHNSTON [3.190], the activation energy of conductivity E_σ in NiO-type oxides is not related to the energy gap between the donor level and the conductivity band (or between the valence band and the acceptor level). It relates instead to the energy of destruction of polarization around the localized electron (hole). The charge carrier motion results from an activated process of "jumping" of the localized electron from a lower-valence cation to a higher-valence one, e.g., from Ni^{2+} to Ni^{3+} in lithium-doped NiO. The mobility of the electron is defined by

$$\mu = (e/kT) \ r^2 \nu_0 \ \exp(-2\alpha r - W/kT) \quad , \tag{3.19}$$

where r is the minimum distance between cations over which the electron jumps; ν_0 is the phonon frequency which stimulates the jump (10^{12}–10^{13} cm^{-1}); α is a coefficient depending on the wavefunction overlap [$\alpha=0$ and $\exp(-2\alpha r) = 1$ if the overlap is large]; and W is the average activation energy of jumps, which depends on the concentration of impurities.

The value of E_σ (the activation energy of the mobility in the present case) with nearly stoichiometric composition of the oxide is 0.1–0.5 eV; as the concentration of impurities grows, the value of E_σ goes down rapidly to a certain fixed level. At low temperatures there is a probability of an electron jumping to a distant center; at high temperatures the factor $\exp(-2\alpha r)$ provides for a jump to a nearest site.

The value of μ was found to be 10^{-3}–10^{-5} cm^2/(V·s) for electron jumps from cation to cation via an anion, and 0.1–10 cm^2/(V·s) for direct cation-to-cation jumps. The former case may occur with the electron transport between octahedral cations via e_g electrons and anions (Fig.3.20a); the latter is encountered with the transfer of t_{2g} electrons in an octahedron directly between cations (3.20b) [3.191]

$$M^{2+} \quad O^{2-} \quad M^{2+} \qquad\qquad M^{2+} \overset{O^{2-}}{\underset{O^{2-}}{\oslash}} M^{2+}. \tag{3.20}$$

(a)   (b)

The contemporary quantum-mechanical theory of jumps [3.94,188,192-194] takes into account the interaction between electrons and vibrational modes of the lattice. The electron traveling together with its domain of local polarizaiton is called a polaron. The mobility of a polaron is (perhaps hundreds of times) lower than that of an electron, while its energy is lower than the energy of electronic excitations. At low temperatures there exist polaronic bands and conduction depends on nonlocalized polarons. At high temperatures, more common for catalysis and adsorption, the charge carriers are more likely to localize in the form of low-radius polarons [3.193].

According to [3.192-195], in most transition-metal oxides the conduction depends on the low-radius polarons (the case of strong coupling of the electron to the optical-lattice phonons). The energy of the polaron trap W_p, capable of capturing an electron within radius r_0, may amount to 0.3-0.5 eV.

The conduction due to low-radius polarons and the conduction due to ionization of donors can be distinguished with the aid of thermoelectric measurements. Such measurements indicate [3.188,195] that in MnO doped with Li a hole-type band of low-radius polarons is formed at about 0.3 eV from the d^5 levels. In pure NiO the value of W_p appears to be low and the oxide exhibits a band-type conduction (a narrow band about 0.4 eV wide due to indirect overlap of 3d orbitals via anions; see Fig.3.20a). However, a low-radius polaron is created in NiO(Li). This polaron, comprised of Ni^{3+} and raised-up Li^+, is nonsphero-symmetric and behaves as a heavy particle with a mass about 100 times larger than that of an electron. The activation energy of jumps in NiO(Li) is highly dependent on temperature, varying from 0.004 eV at -263°C to 0.2–0.4 eV at high temperatures; this is consistent with the polaronic mechanism.

In complex oxide systems the values of W_p differ for different traps. The electroconductivity depends on random diffusion-type migrations between different valence states and impurities with a drift mobility highly dependent on temperature. The conductivities of amorphous and crystalline solids differ little. For example, vitreous substances (glasses) containing V^{5+} and V^{4+} (or Fe^{2+} and Fe^{3+}) ions exhibit polaronic conduction similar to that in NiO(Li), with an activation energy of 0.4 eV at room temperature and decreasing at low temperatures [3.194].

Various correlations between the locations of extrinsic levels in the band structure of oxides and the nature of these oxides have been considered [Ref.3.134,pp.36–40]. For shallow levels it was shown that the depth of the extrinsic level ΔE_{ex} is proportional to the width of the forbidden band (assuming the impurity to be of the same chemical nature as the metal in the oxide, e.g., M^+ ion in $M^{2+}O^{2-}$ oxides)

$$\Delta E_{ex} = B E_g \; . \tag{3.21}$$

Relations like (3.21) can easily be explained, for example, from the viewpoint of the theory of hydrogen-like local levels. Such relations indicate that the location of extrinsic levels and various properties of oxides modified by impurities can be predicted on the basis of known structures and properties of pure oxides. We are much less familiar with deep levels, which do not fit into the hydrogenic model and require a many-electron treatment.

To provide an example for the use of electrohysical measurement to study the surface levels let us quote from [3.196]. MORRISON used the temperature dependence of conductivity of chromium-oxide powders to identify a donor level 0.3 eV above the valence band which relates to the d band of Cr^{3+} ions. These donor-type levels vanish after bringing Cr_2O_3 into contact with oxygen; thus one attributes them to coordination-unsaturated surface Cr^{3+} ions.

In this connection it ought to be observed that these considerations only apply to monocrystalline semiconductors. In the case of the polydispersed systems commonly used in adsorption and catalysis, the conductivity measurements depend strongly on the magnitude of energy barriers between the particles in a powder. The magnitude of these barriers exhibits an intricate dependence on temperature at which the measurements are taken [Ref.3.50,Sect.4.4]. The conductivity measurements on powdered semiconductors can only supply a qualitative notion about the mechanism of charge transport in polycrystalline systems. For example, when single crystals of NiO instead of powders were used in conductivity measurements, the "electron-jump" model [3.189] turned out to be wrong. Measurements of the Hall effect in pure NiO monocrystals indicate that the mobility of charge carriers remains constant with a rise in temperature, while the number of charge carriers increases [3.123]. Misinterpretation of the earlier results was due to neglect of interparticle boundary effects.

In complex oxide semiconductors (such as transition-metal molybdates, Sect.3.4.2) the number of levels in the band structure is large, whereas their conductivity shows little dependence on the impurity content. Such systems comply with the concepts of the theory of disordered systems, discussed by us in [Ref.3.50,Sect.4.2] in reference to surface phenomena.

In highly doped semiconductors – and such are most complex oxides used in adsorption and catalysis – the extrinsic energy levels will be blurred because of their high concentration and energy fluctuations. The energy levels will depend little on whether the substance is crystalline, amorphous, or even liquid.

Many researchers have based their explanations of adsorptive and catalytic mechanisms on conductivity measurements. However, as follows from the above-developed arguments, the

120

data on the conductivity of transition oxides (especially with high-dispersion samples) supply little information about the energy spectrum of oxides. This has become obvious only recently due to the development of spectral techniques.

3.5.2 Optical Methods

In Sect.2.2.1 we have already mentioned the possibility of assessing the energy of d levels in the oxide band structure with the help of charge transfer spectra. In nontransition-metal oxides like ZnO and MgO, the width of the forbidden band E_g, derived from the location of the edge of continuous absorption in the optical spectrum, corresponds to a charge transfer from the occupied 2p band of O^{2-} into the empty conduction band (e.g., the 4s band of the metallic ion). With the transition-metal oxides, however, it often remains unclear which of the transitions corresponds to the "forbidden bandwidth", determined by the absorption edge.

Two kinds of transitions are possible in oxides with octahedral coordination: from the 2p band either to the t_{2g} level with lower energy or to the higher energy e_g level of the transition metal. Both kinds of transitions are possible (and are observed in some experiments) in oxides of transition metals of the left-hand half of the row (configurations d^0 through d^5). With later elements, the t_{2g} levels are partially occupied, and are completely filled up at the d^8 configuration. In NiO (Fig.3.17) the charge can be transferred from the 2p band of O^{2-} to d^8 levels only if NiO is modified with an acceptor-type dopant (Li_2O). However, the electron may go to a higher level of Ni ($3d^9$) or to the 4s band of oxygen.

From Fig.3.18 it is clear that the width of the forbidden band in the row of transition-metal oxides is widest at the two ends of the first long period and smallest in the middle, the value of E_g for Fe_2O_3 (d^5 configuration) being at the same time a little higher than the values for its neighbors MnO_2 and CoO. Thus we again encounter a vague two-peak dependence, resem-

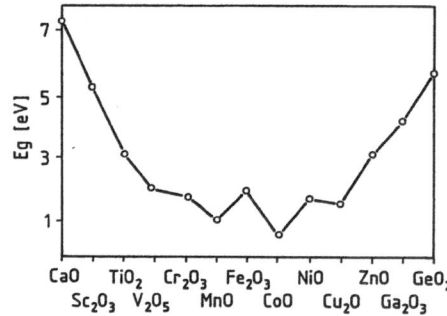

Fig.3.18. The width of forbidden band E_g in the oxides of the elements of the first long period

bling the trend of CFSE mentioned in Sect.2.1 (Figs.2.5,6): E_g is smallest for oxides with d^3 and d^7 configurations and largest for d^0, d^5, and d^{10}. Quite similar trends were exhibited by the energy of ligand-to-metal (L→M) charge transfer in homogeneous ML_n complexes [3.197].

There also are other similarities that can be found between the energy of L→M charge transfer in homogeneous complexes and the width of forbidden band E_g in semiconductors. The energy of L→M charge transfer was shown [3.197] to increase for one and the same metal in complexes with different ligands in the order I < Br < Cl < F. A similar dependence is shown by the value of E_g in solid halides of transition metals.

As shown by JORGENSEN [3.198], the frequency $h\nu$ [cm^{-1}], corresponding to the charge transfer between metal M and ligand L in a homogeneous complex ML_n, is approximately proportional to the difference in electronegativity of M and L. If we express this statement as

$$h\nu = \{x_{opt}[L] - x_{opt}[M]\} \cdot 30.000 \text{ cm}^{-1} , \qquad (3.22)$$

the values of "optical electronegativity" x_{opt} (as Jorgensen calls them) will be close to the Pauling electronegativities. The connection betwen $h\nu$ and Δx_{opt} is quite similar to the connection between E_g and Δx [3.134].

The concept of "optical electronegativity" proved useful for predicting the location of charge-transfer spectral bands for a number of compounds, including solids [3.199]. The energy of these bands in the spectra of solids is lower by ΔE than predicted for complexes in gaseous and liquid phases. The value of ΔE is larger, the greater the overlap of levels of interacting atoms in the solid. Nondiluted solid halides ($MnCl_2$, $FeCl_2$, etc.) are intensively coloured due to the creation of bands. The energy of charge-transfer bands may be 2000–3000 cm^{-1} lower than that in isolated molecules, e.g., linear molecules of $MnCl_2$, $FeCl_2$, $CoCl_2$, and $NiCl_2$ isolated in a cold inert matrix.

The theoretical interpretation of charge-transfer bands was based on the method of molecular orbitals [3.200–203]. In those complexes where the covalent component of the M–L bond is small, the metal and the ligand may be regarded as two essentially different formations with insignificant interaction. Charge transfer proceeds from the higher occupied molecular orbitals (HOMO) to the lower unoccupied ones (LUMO). In order to find the energy of charge transfer one needs a detailed knowledge of the band structure and the energy of the ligand's upper occupied orbitals (e.g., a_g, t_{1u}, e_g) (Fig.2.7) and of d orbitals split by the ligand field for the system with (n+1) d electrons (d^n being the initial configuration). In the case of transition-metal oxides the theory is only valid for oxides with high bond ionicity (NiO).

The charge-transfer energy, found from (3.22), turns out to be rather close to the spectral width of the forbidden band E_g. This corresponds to "donor-type" electron transport from the oxygen ion (2p band) to the metallic ion:

$$M^{n+}O^{2-} \longrightarrow M^{(n-1)+}O^{-}.$$

The "acceptor-type" charge transfer (e.g., from d orbitals of a metal to the higher s band of oxygen) does not occur in oxides, although it may take place in solid fluorides. In lieu of this, the oxides may exhibit donor-acceptor charge transfer between transition-metal ions in different oxidation states.

Some of these transitions for NiO are included in the schemes (3.18).

The interpretation of strong UV absorption by complexes with a more covalent bonding (VO_4^{3-}, CrO_4^{2-}, MoO_4^{2-}, and higher oxides of the same metals) must be based on considering the electron excitation with rearrangement of the electron density of both ligand and metal rather than ligand-to-metal charge transfer [3.200].

The existence of a large number of donor and acceptor levels results in other peculiarities in the spectra of solids. Reduction of the charge-transfer energy or the width of the forbidden band may accompany the generation of polarons, and the creation of extrinsic levels and bands. In many (especially amorphous) semiconductors the absorption coefficient near the edge of the absorption band often exhibits exponential growth (Uhrbach's rule), increasing by several orders of magnitude for a the change in the photon's energy $h\nu$ as small as a few tenths of an electron volt [3.50,204].

Because of these ambiguities, even the spectral data do not provide a sound basis for interpreting such compounds as NiO. According to NEWMAN and CHRENKO [3.205], who studied single crystals of NiO, the rise in absorption starting at $h\nu = 4.4-4.5$ eV is associated with the electron transfer from the 2p band of oxygen to the 3d band of nickel. At the same time, CHERKASHIN and VILESOV [3.206] consider this band to be associated with the electron transfer from the 3d Ni^{2+} band to the unoccupied 3d band of Ni^+ (or to the 3d level of Ni^+] (Fig.3.17) which agrees with the data of the extrinsic photoeffect and the threshold of photoconduction in NiO.

In complex oxide catalysts the overlapping of molecular orbitals of anions and cations causes smearing of discrete levels into bands. The electron transfer between these bands and the capture on local levels of active centers may be responsible for the catalytic action of these substances.

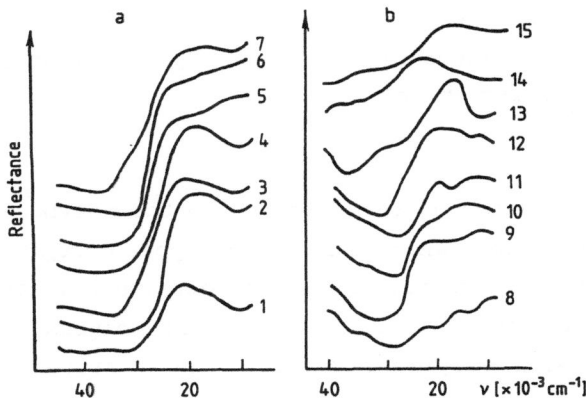

Fig.3.19. Ultraviolet diffuse-reflection spectra of (a) molybdates of second-group metals, and (b) molybdates of transition metals [3.153]: (1) Mg, (2) Be, (3) Ca, (4) Zn, (5) Sr, (6) Cd, (7) Ba, (8) Cr, (9) Mn, (10) Fe, (11) Co, (12) Ni, (13) Cu, (14) V, (15) Ti

Figure 3.19 shows the UV diffuse-reflection spectra of molybdates of the second-group and the first transition period elements [3.153,173,207]. The analysis of UV spectra in the region of the charge-transfer band leads to a conclusion about the formation of common M-O-Mo orbitals (M being the cation). This is especially clear in the case of molybdates of second-group elements. Absorption with charge transfer (20.000-23.000 cm^{-1}) starts at much lower frequencies than in solutions (at 42.500 cm^{-1} for the MoO_4^{2-} ion) or ionic crystals (at 38.000 cm^{-1} for Na_2MoO_4); the higher the atomic mass of the second-group metal cation, the more the charge transfer band is shifted toward longer wavelengths, which cannot simply be attributed to the distortion of symmetry of the molybdate anion. Considering the interaction between molecular orbitals of the MoO_4^{2-} anion and the cation's atomic orbitals, it is clear that highly delocalized π orbitals of molybdate anion (both bonding and antibonding) may interact with either pure or hybridized s and p orbitals of the divalent M cation. Apparently, this interaction gives rise to conduction bands in molybdates, which fit into the energy gap between the oxygen's 2p valence band and molecular

125

orbitals, which essentially are the d orbitals of molybdenum.

The higher the covalent nature of the the $M-(O_4Mo)$ bonding, the lower the charge-transfer energy

$$O^{2-} + Mo^{6+} \rightarrow O^- + Mo^{5+} \ .$$

This facilitates the transition of the electron to the molybdenum d level giving rise to Mo^{5+} and a hole in the upper occupied orbital. The localizaiton of the hole on the oxygen ion results in the creation of a reactive O^- ion. In the case of molybdates of transition metals with varying structure, the picture is complicated by d-d electron transitions. For some molybdates the charge-transfer band exhibits fine structure.

From Fig.3.20 it is clear that there exists a fair correlation between the IR band (ν_3) and the UV charge transfer band in some molybdates (Figs.3.15,19). Evidently, both these bands reflect vibrations and charge transfer in the same structures: bridge covalent bondings, facilitating the long-range transport of electrons and holes:

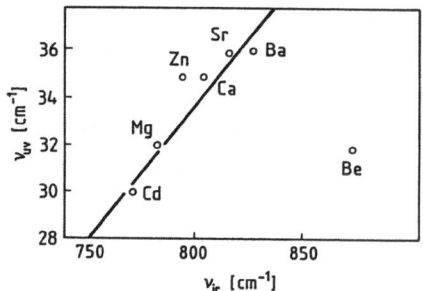

In such systems the energy of charge transfer is lower as compared with the $MoO_4{}^{2-}$ ion, which explains the reduction of Mo^{6+} to Mo^{5+}, the latter being capable of serving as an active site in adsorption and catalysis. At the same time the struc-

Fig.3.20. The correlation between the IR absorption peak ν_3 and UV charge transfer band in the molybdates of second-group metals

126

tures of molybdates of Be and Zn prevent long-range electron transport, which slows down the reduction. For this reason these molybdates do not follow the general correlation.

Charge-transfer spectra have been used to determine the status of surface transition-metal atoms. In [3.208] the visible-light and UV diffuse-reflection spectra of chromia-alumina catalysts were taken while simultaneously measuring the surface coverage by Cr atoms via adsorption of O_2. In addition to the d-d bands in the visible range, corresponding to Cr^{3+} ions, the oxidized Cr/Al_2O_3 catalysts showed strong absorption in the UV range, due apparently to Cr^{6+} ions. The maxima at 27.000, 32.500, and 37.000 cm^{-1} correspond to the charge transfer $O^{2-} \rightarrow Cr^{6+}$. At low concentrations of Cr_2O_3 (<0.1) the intensity of absorption is approximately proportional to the coverage of the surface by chromium oxide, as indicated by adsorption of O_2. Consequently, the spectrum of Cr^{6+} ions is produced on the surface rather than in the bulk.

ASMOLOV [3.24] has studied the charge transfer bands in the diffuse-reflection spectra of molybdenum oxide catalysts supported on γ-Al_2O_3 and MgO. All spectra of MgO-supported molybdenum display an absorption line at 37.000 cm^{-1}. (Maximum absorption corresponds to minimum reflection). The spectra of γ-Al_2O_3 supported molybdenum have maxima at 42.500, 35.000, and 31.000 cm^{-1}. The maximum at about 31.000 cm^{-1} (Fig.3.21) pertains to specimens with high molybdenum content (16.7% of

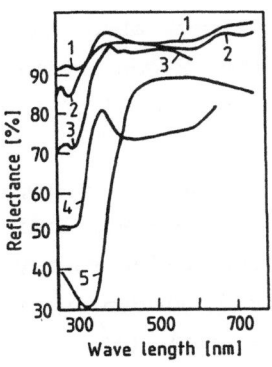

Fig.3.21. The diffuse-reflection spectra of MgO-supported MoO_3 [3.24]. Mass percentage of MoO_3: (1) 0.2, (2) 1.96, (3) 3.85, (4) 9.1, (5) 16.7

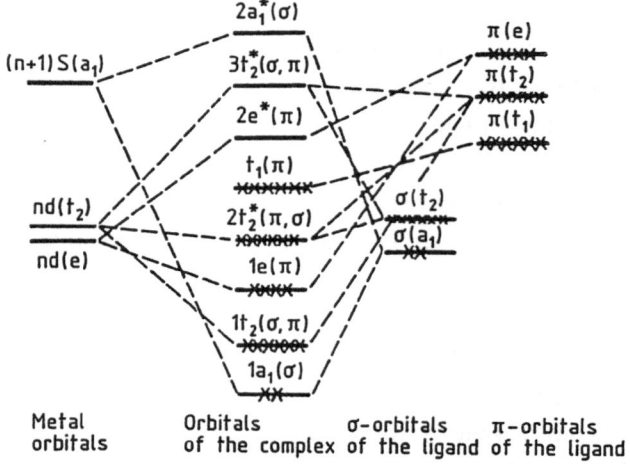

| Metal orbitals | Orbitals of the complex | σ-orbitals of the ligand | π-orbitals of the ligand |

Fig.3.22. The arrangement of molecular orbitals in the complex of $(MO_4)^{2-}$ of T_d symmetry

MoO_3), whereas the specimens with lower molybdenum content display spectral maxima at 35.000 and 31.000 cm^{-1}.

The energy of electron transitions from the atomic or group-molecular orbital of oxygen depends on the symmetry of the field around the molybdenum ion. The scheme of molecular orbitals for the $(MO_4)^{2-}$ complex in a T_d configuration is given in Fig.3.22. The charge-transfer spectra can be explained with the aid of either of two schemes, evolved for transitions in complexes with T_d symmetry: the Wolfsberg-Helmholtz scheme or the Ballhausen-Liehr scheme. The former assumes the transitions $3t_2 \leftarrow t_1$ and $3t_2 \leftarrow 2t_2$; the latter assumes $2e \leftarrow t_1$ and $3t_2 \leftarrow t_1$. Whatever the differences, the first (low-energy) absorption band is due to an electron transition from the oxygen orbital to the antibonding orbital, essentially a pure d orbital of the central atom.

The MoO_4^{2-} anion in a solution is known to have a charge transfer band with a maximum at 42.500 cm^{-1} [3.209]. Ionic solid molybdates have about the same charge-transfer energy, while solid MoO_3 has a maximum at 31.000 cm^{-1}. Evidently, in supported catalysts the maximum at 31.000 cm^{-1} (320 nm), observed with γ-Al_2O_3 species with high molybdenum content, is due to the presence of Mo^{6+} ions in octahedral coordination and pertains to MoO_3, as also indicated by X-ray analysis. The remaining bands find a natural explanation assuming a tetrahedral coordination of the Mo^{6+} ion (37.000 cm^{-1} for MgO and 35.000 cm^{-1} for γ-Al_2O_3).

The diffuse-reflection spectrum of molybdenum oxide, supported on α-Al_2O_3 (where all aluminum atoms are in octahedral coordination), does not differ from the spectrum of γ-Al_2O_3-supported catalyts. This allows one to conclude that at low concentrations molybdenum forms a chemical compound with the carrier: $Al_2(MoO_4)_3$, or $MgMoO_4$ when MgO serves as a carrier. These compounds are confined to the surface and are not detected by either X-ray or electron-diffraction techniques.

In the examples cited, the charge transfer depends chiefly on the immediate surrounding of the transition atom. The spectrum of surface transition atoms in an octahedral coordination does not differ from that of their counterparts in the bulk. However, as indicated in Sect.3.1, the coordination of a large proportion of V, Mo, Co, and other ions on the surface of MgO and Al_2O_3 (in dilute systems) is far from octahedral, as proved by, e.g., EPR spectroscopy. Study of the fine structure of charge-transfer bands in diffuse-reflection spectra of these specimens allows one to draw conclusions about the entire energy spectrum, including the surface local donor and acceptor levels, split off from the band. This method was employed in studying the supported V_2O_5/MgO catalyst [3.65].

The information gained from EPR and diffuse-reflection spectra in the region of d-d transitions (Sect.3.1.7) verifies the assumption that a partially reduced V_2O_5-MgO catalyst contains, along with isolated $V_{T_d}^{4+}$ ions, paired or cluster-type centers of $VO^{2+} \ldots V_{T_d}^{5+}$. Figure 3.23 shows a band scheme of a MgO surface containing a center of this kind. The width of the forbidden band of MgO is assumed to be equal to 10 eV [3.134].

The location of extrinsic levels in the dielectric's forbidden band is deduced from optical spectra on the basis of the following considerations:

The periodicity of the MgO lattice is violated in the neighborhood of the extrinsic centers (VO^{2+} ... V^{5+}). This gives rise to a set of local levels, which are rather smeared because of the presence of fluctuating fields (cf. the theory of disordered systems [3.50,188]). Irregularity of the surface causes a splitting off of acceptor levels from the conduction band and of donor levels from the valence band. The energy gap between the donor and the acceptor levels can be taken to be equal to the gap between the ground level d_{xy} of the VO^{2+} ion and its first excited level $d_{xz,yz}$ overlapping with the lower unoccupied orbitals of the tetrahedral V^{5+} (hyperfine splitting in the adjacent V^{5+} nucleus indicates partial transfer of the electron from V^{4+} to V^{5+}) and (probably) with the levels of the $\square_{Mg}$ vacancies, thus creating a narrow extrinsic donor band.

The Fermi level ought to be located in the middle of the forbidden band, and halfway between the extrinsic bands [3.210]. The extrinsic donor (valence) band is created by d_{xy} levels of V^{4+} in vanadyl. It is located 0.5-1 eV below the Fermi level (about 5 eV from the conduction band of MgO). The remaining levels are arranged in accordance with the absorption bands observed in diffuse-reflection spectra of a V-MgO catalyst. The distance between d_{xy} and the bonding π orbital of V^{4+} in VO^{2+} (transition 4 in Fig.3.23) is taken to be equal to 4.25 eV (34.000 cm^{-1} - charge transfer band to V^{4+} in VO^{2+}). The differences $d_{xy}-d_{xz,yz}$, $d_{xy}-d_{x^2-y^2}$, and $d_{xy}-d_{z^2}$ are 1.4, 1.85, and 2.7 eV, respectively. The appropriate weak bands, typical of the VO^{2+} ion, are observed in the spectrum of V-MgO catalyst. The separation between d levels in $V_{T_d}^{5+}$ and its bonding π orbitals (transition 5) is taken to be 4.6 eV: this is the strongest charge-transfer band.

Transition 6 [the electron transfer from the orbitals of O^{2-} in the lattice (valence band of MgO) to Mg^{2+} orbitals in a

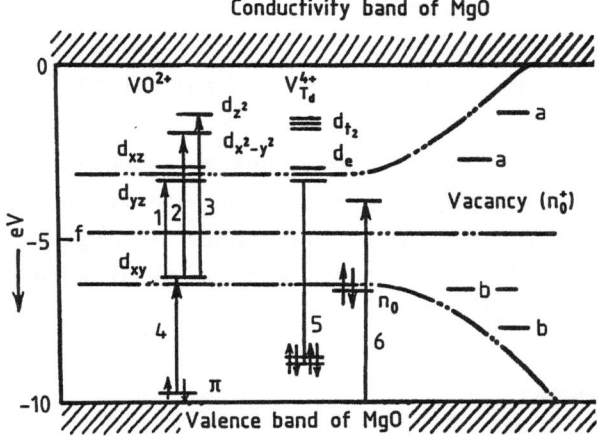

Fig.3.23. The band model of a surface portion of MgO containing a donor V site [3.65]

defect] is assumed to have an energy of 5.5 eV, since both the reduced and the oxidized specimens of V-MgO and Mo-MgO exhibited absorption in this range. The levels of antibonding oxygen orbitals (n_0) are located below the Fermi level more or less symmetric to levels a, arising due to the displacement of Mg^{2+} ions from their regular positions in the MgO lattice caused by interaction with polyanions (vacancy levels in the scheme).

It is clear that this diagram puts the bonding π orbitals of V^{4+} and V^{5+} ions (extrinsic oxygen levels) just a few tenths of an electron volt above the valence band of MgO. This indicates that the energy-band scheme presented in Figs.3.23 gives quite a realistic picture of the levels in V-MgO.

We see that the optical diffuse-reflection spectra, and especially the charge-transfer bands, may be used for assessing the energy of extrinsic levels, which play a crucial role in adsorption, catalysis, and chemical bonding. This applies not only to the bulk bands, but also to the surface bands and levels; and not only to the local electron transitions in isolated complexes (e.g., dilute oxide systems treated in Sect.3.1.), but also to band-to-band transitions. The local and the band-to-band transitions may differ in charge-transfer energy.

In some cases the values of the energy levels obtained from optical spectra agree with the values derived from measurements of conductivity or photoconductivity, but not necessarily so. The local transitions may show up in optical spectra while remaining obscure in electrophysical measurements: the optical spectra carry much more information about the energy structure of the bulk and the surface of solids.

3.5.3 Photoelectron Spectroscopy

Photoelectron techniques (XPS and UPS) were used for studying the composition of transition metal oxides, the degree of oxidation of ions, the binding energy of electrons in intrinsic levels, and the chemical shifts [3.211,212]. Knowledge of low-energy electron levels, which determine the chemical, catalytic, and adsorptive properties of substances, is not so extensive.

An example of the use of the XPS spectrum for studying the band structure of NiO is given in Fig.3.24 [3.213]. The source of radiation was provided by a K_α line of aluminum. Despite the rather low resolution, the spectrum could be interpreted on the basis of correlation with the bands of inner high-energy levels, and with UV and visible-light diffuse-reflection

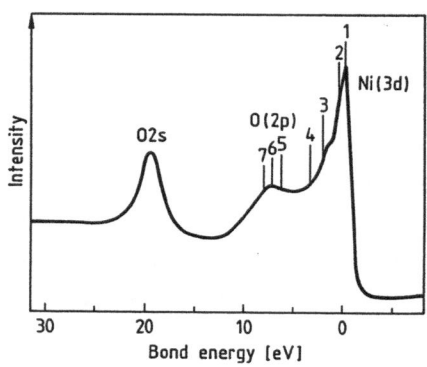

Fig.3.24. The XPS spectrum of NiO in the region of outer levels [3.213]

spectra (by assumption, $Dq = 0.18$ eV). The energy is counted with reference to the peak 1, which is 1.7 eV below the Fermi level of the reference electrode (gold). Peak 2 (0.6 eV) corresponds to ionization of e_g electrons; peak 3 (2.1 eV) to ionization of t_{2g} electrons. The location of the oxygen 2p levels (3.3 eV peak 3) is found from UV spectra. The satellite ("shake-up") peaks 5-7 (6.4, 7.2, and 7.9 eV, respectively) are attributed to charge-transfer transitions from the 2p band of oxygen to the 3d levels of nickel. The satellite structure of this kind is not observed for oxides with occupied d levels (Cu_2O), although it is observed with Sc_2O_3 and TiO_2 (cation configuration d^0). Finally, a strong peak at 19.5 eV corresponds to the oxygen 2s band.

It must be noted that the photoelectron spectra describe the energy of the ionized state (d^{n-1}), whereas the optical spectra account for the energy of the nonionized state (d^n). The photoelectron spectrum of $NiO(d^8)$ must therefore be compared to the optical spectrum of CoO (d^7), while taking into account that the increase in the nuclear charge by one unit reduces the energy of the $O_{2p} \rightarrow M_{3d}$ charge transfer by 1-2 eV. Consequently, the satellite peaks 5, 6, and 7 in the photoelectron spectrum must correspond to the charge-transfer bands 5.5 and 7.5 eV in the optical spectrum of CoO [3.206], which may be attributed to the $O_{2p} \rightarrow Co_{3d}$ transitions. Notice also that the maxima of these bands in Fig.3.24 correspond to transitions from the middle of the oxygen 2p band rather than from the edge of the 2p band, which is responsible for the absorption edge in the UV region.

The interpretation of NiO spectra is open to debate. There exist alternative explanations for the peaks 4-7 in Fig.3.24 [3.214]. The 2p band of oxygen in both XPS and UPS is usually weak and overlaps with 3d bands and satellites.

The XPS spectrum of CoO in the region of valence levels exhibits a wide 2p band of O^{2-} (5.5 eV), overlapping with the d levels of the tetrahedral Co^{2+} (1.9 and 4.0 eV) [3.215]. The spectra of Co_3O_4 show a strong narrow line of octahedral Co^{3+}

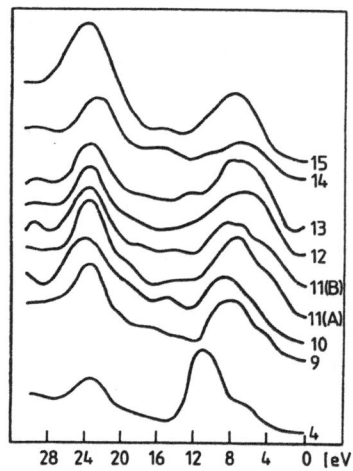

Fig.3.25. The XPS spectra of molyb-dates of second-group and transit metals [3.153]. Enumeration is the same as in Fig.2.19

(1.0 eV) with the levels of tetrahedral Co^{2+} at 2.0 and 5.0 eV, and the 2p band of O^{2-} about 6 eV below E_F.

The low-energy photoelectron spectra of transition-metal oxides are generally characterized by the presence of d levels near E_F (0-3 eV) and the deeper 2p and 2s levels (bands) of O^{2-}. The gap between the 2p and 2s levels is almost always equal to 16 eV, which is close to the corresponding value for a free oxygen atom. This can be explained by the high ionicity of bonding in oxides of transition metals. At the same time the d and 2p levels often overlap, which indicates their hybridiza-tion – that is, the highly covalent bonding.

Figure 3.25 shows the XPS spectra for a number of molybdates in the range of bonding energies corresponding to outer molecu-lar orbitals [3.153]. The maxima on the left stem from the 2s band of oxygen. The maxima on the right have a more complicated structure, from which one can single out the 2p band of oxygen, and the 3d levels as well as other peaks, apparently resulting from the creation of hybridized orbitals. A comparison of Figs.3.19,25 reveals that the band structure according to Fig.3.25 is generally consistent with the charge-transfer bands in Fig.3.19. The spectra in Fig.3.25 also indicate that the ioni-zation cross section (in the spectrum excited by the K_α line of aluminum) is smaller for 2p electrons than for 2s electrons, and smaller still for 3d electrons.

The valence levels of MoO_3 and bismuth molybdate were studied using UPS in [3.216]. The spectra of oxidized MoO_3 specimens display two maxima (5.0 and 7.6 eV), corresponding to the 2p valence band of O^{2-}. The edge of the valence band is about 2.8 eV below E_F. A vacuum heat treatment gives rise to local levels at 0.9 and 2.0 eV, pertaining to the 4d levels of reduced Mo species. Similar levels were observed with reduced Bi_2MoO_6 specimens.

A very important problem is the direct observation of surface states in the photoelectron spectra of oxides. Synchrotron-radiation techniques have helped to discover the surface states in the spectrum, which are extremely sensitive to adsorption and surface contamination [3.217].

The XPS spectrum of NiO displayed a splitting of the bands associated with the oxygen ions [3.214]. One of these split bands was attributed to the surface oxygen ions. The XPS and ELS techniques were used on TiO_2, Ti_2O_3, and V_2O_5; they detected surface levels with an energy in the 0.5–2.5 eV range. These levels were identified as the d levels of surface Ti^{3+} and V^{3+} ions [3.142,218,219].

The number of reports concerning surface levels studied with UPS is increasing rapidly. Especially promising is the use of synchrotron-radiation spectroscopy, due to its high sensitivity and its ability to discriminate different orbitals by changing the energy of the radiation. Progress in theoretical calculations is keeping pace with experimental advances through the development of new quantum-mechanical techniques.

4. Adsorption and Catalysis on Oxides of Transition Metals

In the preceding chapters we discussed the peculiar properties of atoms of transition elements, which determine the structure and the energy spectrum of their compounds. In particular, the symmetry of d orbitals is responsible for certain surface properties of a transition elements' oxides: the crystallography of the surface and the specific oxidation state of the transition ion on the surface. In the following, we shall apply this knowledge to the specific problems of adsorption and catalysis.

4.1 Application of the Crystal-Field and Ligand-Field Theories to Adsorptive and Catalytic Phenomena

Chemisorption of an atom or a cation by a transition metal can be likened to a reaction in a complex resulting in an increase in the number of ligands, one of the ligands being the adsorbed molecule, while the remaining ligands are the anions (electronegative atoms) in the crystalline lattice. The adsorption increases the coordination number, while the desorption reduces it. It would be only natural then to apply the theory of complex formation to adsorption and catalysis on solids containing surface atoms and ions of transition metals. One of the first consistent efforts in accomplishing this was made by DOWDEN and WELLS [4.1], who employed crystal-field theory for explaining the observed regularities at adsorption and catalysis on transition-metal oxides. These concepts have gained wide recognition among the specialists and deserve discussion in greater detail.

The initial experimental observation that suggested the idea of applying the crystal-field theory to heterogeneous catalysis was the two-peak pattern of catalytic activity of oxides of first-row metals in H_2-D_2 replacement, discovered in [4.2]. In Fig.4.1 the first-order reaction-rate constants per unit area (in $min^{-1}m^{-2}$) are plotted against the transition-metal oxides, arranged in increasing order of d electrons (i.e., from left to right of the periodic table). The catalytic activity is seen to be the lowest at the beginning of the period (TiO_2 and V_2O_5, d^0 configuration), in the middle (MnO and Fe_2O_3, d^5 configuration), and at the end (Cu_2O and ZnO, d^{10} configuration – d shell completed). The greatest activity is observed with Cr_2O_3 (d^3 configuration) and Co_3O_4 (d^6 – d^7). As indicated in Sect.2.1, the two-peak pattern is also characteristic of other properties of fourth-row transition metal compounds, the configurations d^0, d^5, and d^{10} being at the one end, and d^3 dn d^7 at the other: namely, the heat of hydration of ions in aqueous solutions (Fig.2.5), the lattice constant of solid oxides (Fig.2.6), the rates of reactions of these ions with other substances, and evidently the catalytic activity of free and complexed ions in solutions. The variations in catalytic activity of various transition ions in aqueous solutions are related in [4.3] to the changes in the coordination of these ions at complex formation and to varying degrees of stabilization by crystalline field.

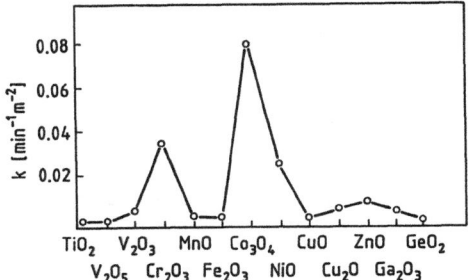

Fig.4.1. Catalytic activity of the fourth-row transition metal oxides in H_2-D_2 exchange [4.2]

Table 4.1. The change in CFSE ΔE_s [kJ/mol] due to the change in coordination in the weak-field approximation according to DOWDEN and WELLS [4.1]

No. of d elec- trons	Ion	$D_{3h} \rightarrow T_d$ (a)	$T_d - C_{4v}$ (b)	$C_{4v} \rightarrow O_h$ (c)	$T_d \rightarrow O_h$ (d)	$D_{3h} \rightarrow C_{4v}$ (e)	$D_{3h} \rightarrow O_h$ (f)
0	Ca^{2+}, Sc^{3+}	0	0	0	0	0	0
1	Ti^{3+}	-17	39	-12	28	15	2.9
2	V^{3+}	-45	74	-23	54	28	5.4
3	V^{2+}	-113	96	29	125	-13	16
3	Cr^{3+}	-145	125	38	162	-18	21
4	Cr^{2+}	-58	121	-50	68	58	9
4	Mn^{3+}	-84	160	-71	92	83	12
5	Mn^{2+} Fe^{3+}	0	0	0	0	0	0
6	Fe^{2+}	-15	24	-7.1	17	8.8	1.8
6	Co^{3+}	-27	42	-12 88*	29	16	3.1
7	Co^{2+}	-28	46	-13	31	17	3.3
8	Ni^{2+}	-70	63	19	80	-8.7	10
9	Cu^{2+}	-58	112	-50	67	60	8.3
10	Cu^+, Zn^{2+}	0	0	0	0	0	0

* Strong-field approximation

DOWDEN and WELLS [4.1] have likewise assumed that chemisorption of a molecule on a surface transition cation is similar to a ligand-gaining reaction of complex formation. For example, the adsorption on a cation on the (100) face of an NaCl-type lattice increases the number of ligands from 5 to 6 and produces changes in coordination (a square pyramid is converted to an octahedron); for the (110) face the coordination changes from a tetrahedron to a square pyramid to an octahedron; for the (111) face from a triangle to a tetrahedron to a square pyramid to an octahedron. Similar changes occur at adsorption on octahedrally coordinated cations in other types of lattices (NiAs, pyrite, CdI_2, corundum, spinel, etc.).

Table 4.1 gives the values of changes in the CFSE ΔE_s caused by the changes in coordination on NaCl-type crystal faces, as calculated by Dowden and Wells in the weak-field approximation using the data of Table 2.2 and experimental values of D_q for oxides and aqueous complexes.

In most cases the increase in coordination number results in further stabilization by crystal field, with the exception of d^0, d^5, and d^{10} configurations. The general trend of ΔE_s is similar to the two-peak pattern in Fig.2.1. For this reason, according to Dowden and Wells, the chemisorption of molecules on cations is usually accompanied by the exothermic effect of crystal-field stabilization. According to [4.1], the correlation between Fig.4.1 and Table 4.1 can be improved by disregarding the data on endothermic effects for Cu^{2+} and Co^{3+} (Column c in Table 4.1): the (100) face is not formed in CuO crystals, while Co_2O_3 (d^6) forms strong-field complexes.

Variations in the CFSE are the greatest in systems with d^3 and d^8 configurations (as well as d^6 in the strong-field approximation). As follows from Fig.4.1, these systems also show the highest catalytic activity.

According to [4.1], the adsorption of a hydrogen molecule may occur in a monomolecular fashion (a, b, c in Table 4.1) in the form of a polarized molecule

$$H^{\delta+} \quad H^{\delta-}$$
$$O^{\delta-} \quad M^{\delta+}$$

Further polarization of the H_2 molecule may result in its breakdown: $MH^+ + OH^-$, increasing by one the coordination number of the cation. The heat of adsorption must change in line with the variations in CFSE in accordance with Table 4.1. Another factor is the increase of polarization and bond strength by higher-valence cations. The relative inactivity of Al_2O_3 (d^0) and Fe_2O_3 (d^5) in adsorption of H_2 (as compared with Cr_2O_3) indicates that the crystal-field stabilization is a factor of greater importance than the polarizing action of the cation.

The electronegative atoms of ligands such as H_2O, NH_3, and H_2S ought to be adsorbed by the cations, which produces a two-peak pattern in the variation of the heat of chemisorption. Molecules chemisorbed on certain oxides are probably ionized (e.g., form complexes H_2^+, O_2^-, NH_2^+); however, the two-peak dependences cannot be explained by considering only the electron transitions without taking into account the change in coordination. Apparently, three factors have to be considered in explaining the catalytic activity and heat of chemisorption: the electron transitions, the polarizing action of the cation, and the CFSE.

Calculations of CFSE indicate that desorption which follows the S_N1 mechanism (that is, simple dissociation reducing the coordination number) must be the slowest from d^3 and d^8 cations in a weak field and from d^6 cations in a strong field.

According to [4.2], the $H_2 - D_2$ replacement goes most easily on semiconductor lattices containing pairs of ions V^{2+}/V^{3+}, Cr^{2+}/Cr^{3+}, Co^{2+}/Co^{3+}, and Ni^+/Ni^{2+}, owing to the possibility of electron transitions

$$
\begin{array}{cc}
H^- & H^+ \\
\vdots & \vdots \\
O^{2-}Cr^{3+}\dots O^{2-}
\end{array}
\;+\; e^- \;\longrightarrow\;
\begin{array}{cc}
H^- & H^+ \\
\vdots & \vdots \\
O^{2-}Cr^{2+}\dots O^{2-}
\end{array}
\tag{4.1}
$$

According to Table 4.1, adsorption is facilitated on the d^3, d^6, and d^8 ions (configuration (4.1a)), while decomposition of the complex (desorption) is facilitated with the d^4, d^7, and d^9 ions (configuration (4.1b)). The activation energy of the net reaction is then the lowest. The direct exchange takes place at desorption of pairs of different atoms in adjacent complexes:

$$
\begin{array}{cccc}
H^- & \dots \;(H^+ \;\dots\; D^-)\; \dots & D^+ \\
O^{2-}Cr^{3+} & O^{2-} \quad Cr^{2+} & O^{2-}Cr^{3+}
\end{array}.
\tag{4.2}
$$

The lowest activity will be observed with the d^0, d^5, and d^{10} systems.

At the decomposition of N_2O the limiting stage is the desorption of oxygen [4.1]

$$2(M^{(n+1)+} \ldots O^-) \rightarrow 2M^{n+} + O_2 .$$

From Table 4.1 it is clear that the lowest endothermicity at these stages (octahedron → square pyramid or square pyramid → tetrahedron) will be observed with the d^0, d^1, d^2, d^5, d^6, d^7, and d^{10} configurations. However, at the limiting stage of catalysis the electron goes from the oxygen ion to the metallic one – this excludes dielectrics with d^0 configuration (CaO and the like). Out of the remaining catalysts, the most active ones ought to be the p-type semiconductors, easily oxidizing to higher oxidation states. In a rather loose accordance with experiment this will give us the following activity series: CuO – Cu_2O > MnO – Mn_2O_3 > CoO – Co_3O_4 > NiO_{1+x} > Cr_2O_3, continued by n-type semiconductors TiO_2, V_2O_5, Fe_2O_3, and ZnO.

DOWDEN, who had proposed the hypothesis about the role of CFSE for catalysis, himself considered this picture to be the first – and a rather rough – approximation [4.4]. It takes no account of the existence of real charges on the surface, of possible σ and π bonding, etc. However, this scheme allows one to derive very simple relationships between the catalytic activity and the configuration of d electrons, which are rather attractive for experimental workers engaged in selecting appropriate catalysts. That is why the Dowden-Wells hypothesis stimulated the appearance of a number of works concerned with checking both the theory in general and its implications.

The expedience of checking the role of CFSE for catalysts having the same crystalline structure was pointed out in [4.5]. For this purpose the H_2-D_2 exchange on Ti_2O_3, V_2O_3, and Cr_2O_3 was investigated. These oxides all have the same corundum-type lattice, are reduction resistant, and have cation configurations d^1, d^2, and d^3, respectively. In these oxides the cation is located in octahedral interstices and surrounded by six oxygen ions; Cr_2O_3 (d^3 configuration) was really found to be a more active catalyst than the other two. According to [4.5], however, the d electrons which form the M-M δ bonding at low temperature (spin pairing of d electrons of adjacent atoms) take no part in catalysis. With the rise in temperature, Ti_2O_3 and V_2O_3 begin to show catalytic activity as the δ interaction vanishes, giving rise to unpaired spins. Further rise in temperature quickly enhances the catalytic activity, as the number of unpaired electrons grows. In Cr_2O_3, on the other hand, the d electrons remain unpaired in the whole temperature range, and the catalytic activity persists even at low temperatures.

It was pointed out in [4.6] that the Dowden two-peak pattern ought to be observed for other reactions as well. In particular, in the reactions of dehydrogenation of cyclohexane and cyclohexene to benzene, the Cr_2O_3 catalyst (d^3) is much more active than $V_2O_3(d^2)$, $TiO_2(d^0)$, and $MnO(d^5)$. A high catalytic activity is also observed in the region of the second maximum (oxides of Co and Ni – d^6–d^7).

Numerous experiments have confirmed (at least in some cases) the two-peak pattern of catalytic activity in the row of oxides of fourth-period metals.

KRYLOV [4.7,8] analyzed the data available on the catalytic activity of oxides of the fourth-row transition metals. When more than one study was done on the same reaction, statistical procedures were applied. Some results of this analysis are presented in Fig.4.2, the ordinate being the "catalytic activity" in appropriate relative units (as a rule, the logarithmic reaction-rate constant related, if possible, to unit area of catalyst surface).

Despite the rather poor reliability of the available experimental data the two-peak pattern is easily traced for a good many reactions. At the same time there are some other peculiar features, not pinpointed in [4.1,2].

1) The **first maximum** is almost always observed with d^3 oxides (Cr_2O_3, MnO_2); the low activity of Cr_2O_3 in certain oxidation reactions is due apparently to its oxidation to CrO_3 in excess oxygen.

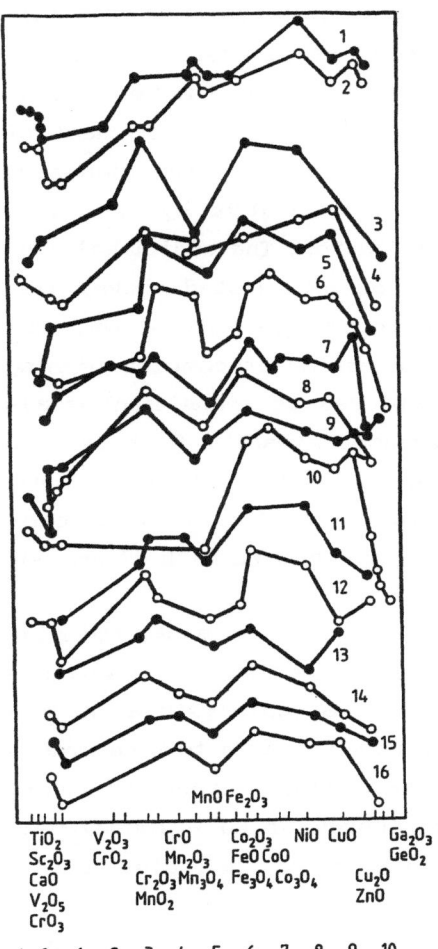

Fig.4.2. Relative logarithmic catalytic activity of the fourth-row transition metal oxides in different reactions [4.7]: (1) dehydrogenation of isopropanol, (2) dehydrogenation of ethanol, (3) dehydrogenation of cyclohexane, (4) O+O recombination, (5) oxidation of H_2, (6) oxidation of CO, (7) oxidation of NH_3, (8) O_2^{18} – O_2^{16} exchange, (9) decomposition of H_2O_2, (10) decomposition of N_2O, (11) decomposition of $KMnO_4$, (12) decomposition of nitroparaffins, (13) oxidation of propylene, (14) oxidation of C_5 hydrocarbons, (15) oxidation of C_6 hydrocarbons, (16) oxidation of benzene

2) The **second maximum** is commonly observed with d^6–d^7 oxides (Co_3O_4), sometimes with d^8(NiO) and even d^9(CuO).

3) The **minimum** with d^5(Fe_2O_3, MnO) is observed in almost all cases, but is not so deep as reported in [4.2]. In any case the activity of the d^5 system is much higher compared with d^0 and d^{10}.

4) The **oxides of metals** closing the period and having the d^{10} configuration (ZnO, Cu_2O) are usually much more active than the oxides with a d^0 configuration (CaO, Sc_2O_3, TiO_2).

143

The regularity turned out to be even more profound than anticipated by Dowden and Wells. According to their hypothesis,
this behavior can be explained by the fact that the net activation energy of a catalytic reaction incorporates the heat of
adsorption (or heat of complex formation) with the sign reversed. This being so, the energy of stabilization by the ligand
field should always be a beneficial factor. The additional regularities in catalytic activity of oxides, listed above, also
fit in with this hypothesis in a more or less natural way. One
must recall, for instance, that the energy of complex forming
minus the CFSE in the row of complexes of transition metals
belonging to the same period increases steadily from left to
right, in line with the increase in ionization potential. Thus,
the higher catalytic activity of the d^{10} configuration as compared with the d^0 and that of d^6–d^8 as compared with d^3 are not
surprising.

Certain quantitative comparisons can also be made. The most
reliable studies report variations in reaction-rate constants
with different transition-metal oxides as catalysts at medium
temperatures (150^0 – 300^0 C) by as many as seven to nine orders
of magnitude. Assuming equal preexponential factors, this corresponds to changes in activation energy of 60 – 100 kJ/mol,
which in general is close to the values of ΔE_s in Table 4.1.

Later DOWDEN [4.9] demonstrated that the model of action of
crystal field at adsorption can be improved by (a) taking into
account the induced charge on adsorbed molecules, e.g.,

$$H^{\delta-} \ldots H^{\delta+}$$
$$O \ldots M \quad \ldots O \ldots M \quad ;$$

(b) employing the strong-field approximation for Mn^{3+}, Fe^{3+},
and Fe^{2+} at certain small distances between the atoms of catalyst and reactant, the latter possibly accounting for the relatively shallow minimum for $Fe_2O_3(d^5)$; (c) using a special model
for CuO considering that the ligands around $Cu^{2+}(d^9)$ are
arranged either in a square or in a highly elongated octahe-

dron, which is not the case with other oxides; (d) taking into account the distortion of coordination at chemisorption of real molecules; and (e) taking π interaction into account.

The change of coordination number and the associated change in CFSE may occur at various stages of catalytic reaction, although only the change in CFSE at the slowest (limiting) stage of reaction can give information on the net reaction rate and thus determine the regularities of catalytic activity. Dowden considers five possible limiting stages in catalysis: (a) adsorption on the transition cation, (b) the reverse process (desorption), (c) replacement of one adsorbate molecule by another via an associative mechanism (implying transient increase in coordination number), (d) reaction between molecules adsorbed on adjacent cations, and (e) reaction between ligands on one and the same cation. In his first study [4.1] Dowden considered only the first possibility (a): adsorption which increases the coordination number. It is this case that conforms with the simple two-peak pattern. Quite the reverse ought to be true for case (b): minimum activity for d^3 and d^8 and maximum activity for d^0, d^3, and d^{10}. The mechanism of associative replacement (the so-called S_N2 mechanism) requires minimum activation energy for configurations d^1, d^2, d^6, and d^7 in the weak-field approximation and for configurations d^1, d^2, d^3, and d^4 in the strong-field approximation. Cases (d) and (e) must be more or less similar to case (a).

The reaction of polymerization of olefins proceeds according to associative mechanism resembling S_N2: the transition-metal ion (Ti^{3+}, Cr^{5+}) must hold simultaneously the growing chain of polymer and the adsorbed olefin molecule (see below). Therefore, the most active catalysts of this reaction are the systems with configurations d^1 and d^2. Conversely, the hydrogenation of olefins follows the dissociative scheme, and the d^3 system is of greater advantage. Experiment indicates that the rate of hydrogenation is higher with $Al_2O_3 \cdot Cr^{3+}(d^3)$ than with $Al_2O_3 \cdot Ti^{3+}(d^1)$.

Table 4.2. The changes in the ligand-field stabilization energy (in relative units) due to the changes in coordination in the weak-field appproximation.

Electron configuration	$D_{3h} \rightarrow T_d$ (a)	$T_d \rightarrow C_{4v}$ (b)	$C_{4v} \rightarrow O_h$ (c)	$T_d \rightarrow O_h$ (d)	$D_{3h} \rightarrow C_{4v}$ (e)	$D_{3h} \rightarrow O_h$ (f)
d^0, d^5, d^{10}	0	0	0	0	0	0
d^1, d^6	−1	−1	−1	−1	−2	−3
d^2, d^7	−2	−2	−2	−4	−4	−6
d^3, d^8	−0.08	−9.67	−3	−12.67	−9.75	−12.73
d^4, d^9	−0.04	−7.33	+1	−3.33	−7.33	−6.37

The above-discussed scheme of coordination variations at adsorption and catalysis was based on the electrostatic crystal-field approximation. The more sophisticated ligand-field theory leads to the same conclusions about consequential occupation by d electrons of two groups of orbitals: the lower t_{2g} orbital (octahedral configuration) and the upper e_g orbital. The former of these is in this case nonbonding and its energy coincides with the nondegenerate d level (Fig.2.7), and the latter is antibonding. Table 4.2 lists variations in the ligand-field stabilization energy (LFSE) for the same variations in coordination as listed in Table 4.1. The true values of the LFSE are taken from work by JATSIMIRSKIJ [4.10], where the LFSE is defined as the change in energy upon transition from the structure with ultimately uniform distribution of electrons over all nonbonding and antibonding orbitals to the structure with the actual electron distribution.

From Table 4.2 it is clear that in almost all cases the increase in coordination number results in more stable complexes, and there are far fewer exceptions than in Table 3.1.

SHOPOV and ANDREEV [4.11] analyzed the energies of a stable molecular orbital, obtained in quantum-chemical calculations of adsorption of a hydrogen atom on an oxide surface cation. The cation's symmetry was C_{4v} prior to adsorption and O_h afterwards. The stability series of MO ran

$Co^{3+}(-15.7 \ eV) > Cr^{3+} > Fe^{3+} > Ni^{2+} > Co^{2+}$ (in CoO) $> Fe^{2+} >$ $Mn^{2+} > V^{3+} > Co^{2+}$ (in $Co_3O_4 - 14.1 \ eV$) ,

in better agreement with the experimental activity of oxides in reactions involving hydrogen [4.2,5,6] than the series con-structed on the basis of crystal-field theory [4.1].

When the ligands are capable of developing π bonds (O_2, NO, CO, olefins, benzene, etc.), the "electrostatic" approach of crystal-field theory fails to predict the actual properties of complexes. Such complexes require account - in addition to the cation's acceptor properties - of the cation's donor proper-ties, the formation of so-called dative bonds (back donation of the metal's d electrons to the antibonding π orbital of the ligand).

The methods of group theory allow one to describe the struc-ture of a complex with reference to the symmetry of atomic d orbitals of the central ion and the molecular orbitals of the ligands. The overlapping of orbitals and creation of a complex is accomplished by combining the orbitals of the central ion and those of the ligands, which have similar symmetric proper-ties with respect to a common reference frame. The overlapping in this case will be the greatest.

The principal idea of creation of a π complex is borrowed from analysis of the structure of complexes of platinum metals, in particular Zeise salt [4.12,13]:

$$K\left[Cl_3Pt \longrightarrow \begin{matrix} CH_2 \\ \| \\ CH_2 \end{matrix}\right] ,$$

in which both carbon atoms are equidistant from platinum; Fig.4.3 illustrates the overlapping of molecular orbitals in the π complex.

The antibonding $π^*$ orbital of olefin is symmetry-allowed to interact with p orbitals of the metallic ion; however, the values of the overlap integrals at the formation of the back-

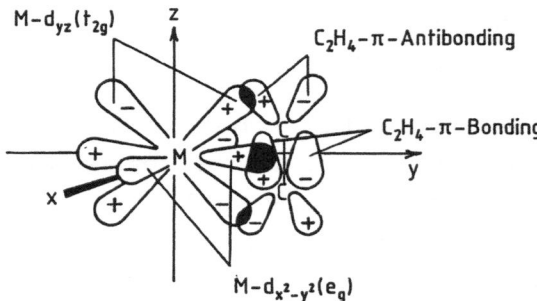

Fig.4.3. Overlapping of orbitals of a transition-metal ion and an ethylene molecule

donation bond involving d orbitals are much higher than those with p orbitals, though the interaction with p electrons strengthens the back-donation bond. Thus the stability of π complexes is ensured, on the one hand, by the existence of vacant d orbitals and the ligand's occupied bonding π orbitals, and on the other, by the availability of occupied d orbitals and the ligand's vacant antibonding π^* orbitals [dative (back donation) or π bond].

The concept of back donation proved to be very useful for developing the theory of complex structure, including complexes of great importance for catalysis. Further improvements allowed removal of almost all restrictions of ionic crystal-field theory, which was replaced by the theory of molecular orbitals. For example, COSSEE [4.14] has employed the concept of π complexing of olefins with a transition-metal atom for explaining the specificity of catalysis of olefinic reactions by transition metals. According to [4.14], the catalytic polymerization of olefins on the surface of Ziegler-Natta catalyst ($TiCl_4$ alkylated by radical R – e.g., C_2H_5) proceeds via insertion of an olefin molecule at the beginning of the growing polymeric chain:

$$
\begin{array}{ccc}
\underset{\substack{| \\ Cl}}{\overset{\substack{R \\ |}}{Cl-Ti}}{\overset{\diagup Cl}{\diagdown}} \cdots \;+\; C_2H_4 \longrightarrow & \underset{\substack{| \\ Cl}}{\overset{\substack{R \\ |}}{Cl-Ti}}{\overset{\diagup Cl}{\diagdown}} \longleftarrow \overset{CH_2}{\underset{CH_2}{\|}} & \longrightarrow
\end{array}
$$

148

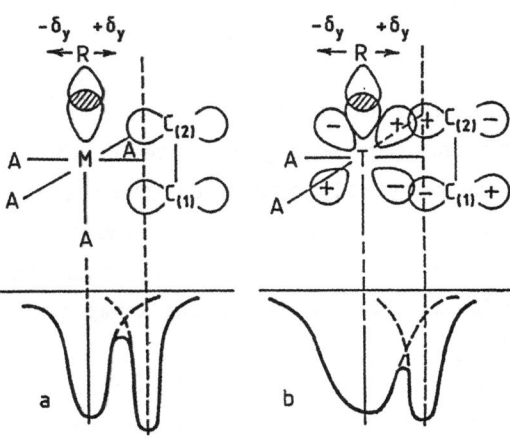

As the reaction proceeds, the vacancy in the coordination sphere of Ti, engaged in π complexing with the olefin, and the growing polymeric chain constantly change places (σ–π rearrangement [4.15]).

Figure 4.4 gives the curves of potential energy for vibrations in octahedral complexes $MA_4R(C_2H_4)$ and $TA_4R(C_2H_4)$ according to [4.14]. Here M is a nontransition metal, T is a transition metal, A is an anion, R is an alkylic group. For the sake of simplicity the potential energies of both complexes are depicted in harmonic approximation by simple parabolas of the same depth. At equilibrium the complex TA_4 is linked to group R via overlapping of alkyl electrons with an axial orbital (e.g., d_{z^2}. The amplitude of vibrations of the alkyl group in the plane of the diagram will be much greater in a complex with a transition metal than in a complex with a nontransition metal, owing to additional overlap with the d_{yz} orbital, such overlapping does not occur in the equilibrium state. Thus, of the two equally strong bonds (assuming equal depth of potential pits), the bond in a transition-metal complex will be more reactive. Large dev-

Fig.4.4. Overlapping of orbitals and potential energy curves for M–R and C_2–C_3 vibrations in octahedral complexes, leading to polymerization of ethylene [4.14]: (a) $MA_4R(C_2H_4)$, (b) $TA_4R(C_2H_4)$

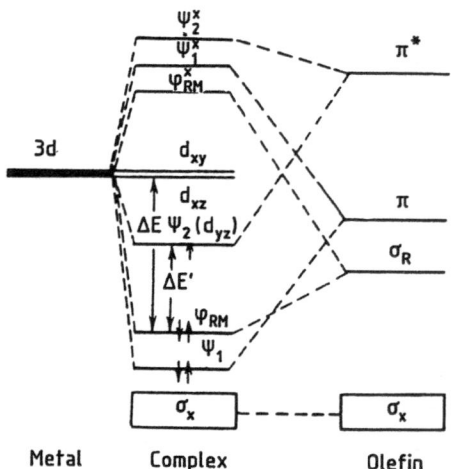

Fig.4.5. Molecular orbitals in the complexes of ethylene with a transition-metal ion in the d^1 configuration [4.12]

iations from equilibrium may lead to the rupture of the M—R or T—R bond and to the creation of a $C_2 - C_3$ bond, C_3 being the carbon atom of the alkyl group, located next to the metal ion. We see that the energy barrier, separating the two potential wells, is lower in the transition-metal complex. Thus the role of the transition metal in the joining of the R group of the olefin consists in supplying its vacant d_{yz} orbital for holding the unsettled R group.

A qualitative diagram showing the creation of molecular orbitals is given in Fig.4.5. Here ψ_1 and ψ_2 are the bonding orbitals. The stabilizing action is due to the possibility of back donation of d electrons from one of the t_{2g} orbitals (d_{yz} on the diagram) to a free antibonding π orbital of ethylene, resulting in creation of molecular orbital ψ_2. For those elements which have no d electrons (or have just a few), the ψ_2 orbital will be empty (or only partially occupied). Then a high stability of π complex is hardly to be expected. However, the energy of the ψ_2 molecular orbital is still lower than the energy of the initial d_{yz} orbital, the latter being comparable with the energies of d_{xy} and d_{xz}.

On the surface of $TiCl_3$ the Ti^{3+} ion occurs in octahedral coordination, four coordination sites being occupied by Cl^-

anions, a fifth site coordinating with the alkyl group, and a sixth forming a π complex with the olefin. Alkylation of the $TiCl_3$ surface by an organometallic compound – e.g., $Al(C_2H_5)_3$ – was verified by special experiments. In Fig.4.5 the molecular orbital, corresponding to σ bonding of Ti with the alkyl group, is denoted ϕ_{RM}. The lowering of the d_{yz} orbital down to the ψ_2 level results in decrease in the energy gap $\Delta E'$ between this orbital and the occupied ϕ_{RM} orbital. This opens the possibility of thermal excitation of an electron from ϕ_{RM} to ψ_2, the latter having the capacity for one more electron. The M-R bond breaks down, and the olefin molecule jams in at the beginning of the growing chain, as shown in the scheme (4.3). Thus the catalytic activity depends on the availability of t_{2g} orbitals, which, on the one hand, must have some d electrons for back donation. On the other hand, these electrons must be not too many (1 to 3) to facilitate the electron transition from the ϕ_{RM} orbital.

On the basis of spectroscopic data, Cossee maintains that the best catalysts are those in which the initial d level of the cation falls between the π and $π^*$ orbitals of ethylene. In the first long row, this condition is best satisfied by Ti, by Zr in the second, and by Ta in the third period. The oxides of transition metals are as active as the chlorides are. Whatever difference there is consists in the mechanism of initiating the polymerization reaction when the active center is alkylated by the R group.

The migration of the alkyl group and its attachment to the coordinated olefin molecule was examined in [4.16] for several intermediate stages in the course of reaction. The calculations by LCAO-MO method were considered alongside the optical spectra. Complexing with C_2H_4 brought the energy gap between ϕ_{RM} and the only occupied d orbital (d_{yz}) down to 11.100 cm^{-1}. This gap may serve as a measure of stability of the metal-carbon bond and thus as a measure of catalytic activity in polymerization reactions. The calculations of molecular orbitals for intermediate stages in the movement of the alkyl group supplied the values of potential barriers for each orbital. These, however, cannot be taken for the activation energy of the reaction, since the orbital energies do not correspond to the total energy of the system.

Some of the later works [4.17-19] challenge the possibility of considerable weakening of the metal-alkyl bond as a consequence of coordination with the olefin.

The calculation by the SCF-X_α technique indicates [4.20] that the geometry of Ti-olefin complexes, as considered by COSSEE [4.12,14,16] is similar to the geometry of isolated complexes of transition elements capable of homogeneously catalyzing the reactions of hydrogenation, albeit the Ziegler-Natta catalysts $TiCl_3 \cdot AlR_3$ are used as heterogeneous catalysts of polymerization. It is pointed out that the formation of a metal-olefin bond requires ligands capable of the "pullout" of electrons. This feature seems to be of special importance for the catalysis by transition-element oxides.

According to some authors [4.4,14], the "cis-insertion" of ligands, discussed above,

$$-\overset{Q}{\underset{|}{M}}-L \quad \longrightarrow \quad -\overset{|}{\underset{|}{M}}-L-Q$$

(L being the ligand attached by the π bond, and Q being the ligand linked to the transition metal M by a σ bond) takes place not only in the polymerization of olefins, but also in many other catalytic reactions of olefins and carbonyl compounds. For instance, the Fischer-Tropsch synthesis is assumed to proceed on oxide catalysts as follows:

152

$$\longrightarrow \quad \xrightarrow{\text{H}_2} \quad \xrightarrow{1/2\,\text{H}_2} \quad \xrightarrow{-\text{H}_2\text{O}}$$

$$\longrightarrow \quad \xrightarrow{\text{CO}} \quad \longrightarrow \qquad (4.4)$$

$$\longrightarrow \quad \xrightarrow{\text{H}_2} \quad \xrightarrow[\text{(as in the beginning)}]{\text{H}_2,\text{CO}} \qquad \text{etc.}.$$

Here the role of the ligand L is played by carbon monoxide, and that of the σ-bound ligand Q by the H atom in the beginning of the reaction and by radical R in the end. The carbonyl compounds MC(O)R experience intermediate hydrogenation.

Similar schemes were proposed for hydroformylation, paraffin oxidation, dehydration and dehydrogenation of alcohols, dehydrogenation of hydrocarbons, and oxidation and methathesis of olefins [4.4,9,14]. These transformations can also be described from the viewpoint of σ–π rearrangement of ligands, an intramolecular process in which the organic group attached to the transition metal goes from being σ bound to being π bound and back again [4.15]. The creation of σ or π complexes was explained through considering the relative gaps between the higher occupied (HOMO) and the lower unoccupied (LUMO) molecular orbitals (e.g., ΔE and $\Delta E'$ in Fig.4.5). The increase of the gap results in stabilization of the complex in a particular form, the decrease promoting its conversion to an alternative form.

The theory of ligand field was further developed to describe interactions of transition-metal atoms with adsorbed ligand compounds, using the so-called cluster methods [4.11,21]. In order to carry out quantum-chemical calculations, a cluster is singled out on the surface – a quasi molecule, which includes the transition-metal ion, surrounded by oxygen anions (sometimes also further cations) in the geometry which approximates the actual arrangement in the oxide, and the adsorbed particle. The size of the cluster depends on the capability of the computer available. The main snag is the question of unsaturated bonds on the boundary of the cluster. This problem is usually overcome by saturating the dangling bonds by imaginary univalent atoms with specially chosen parameters.

4.2 Symmetry and Reactivity of Coordinated and Adsorbed Molecules

We have seen that the structure and the stability of complexes of ligands with transition metals, including those due to adsorption, depend on the symmetry properties of d orbitals. According to some current opinions, the symmetry of d orbitals may also determine the routes of conversion of complexes, that is, the reactivity of the adsorbed and the coordinated molecules.

In Sect.2.1 we considered the first-order Jahn-Teller effect, which determines the nuclear configuration in a complex with degenerate electron states. Of much greater importance for adsorption and catalysis is the second-order Jahn-Teller effect (SOJTE), which determines the electron-vibrational interactions in molecules and complexes with nondegenerate electron states. The second-order Jahn-Teller effect was first formulated by BADER [4.22] and was further developed in reference to the structure and reactivity of complexes by PEARSON [4.23].

Let us assume that the wave function ψ describes the ground (nondegenerate) state of a molecule with the energy E_0, and ψ_k describes the excited state. The energy of the molecule after a small displacement of its atoms (variation in the normal coordinate Q representing the "net" displacement with respect to "reaction coordinate") can be written, according to BADER [4.22], in the second-order perturbation-theory approximation, as

$$E = E_0 + \frac{1}{2} V_{00}Q^2 + \sum_k \left[V_{0k}^2 Q^2 / (E_0 - E_k) \right] , \qquad (4.5)$$

E_k being the energies of different excited states, and V_{00} and V_{0k} being the operators of nuclear-nuclear and nuclear-electronic potential energies:

$$V_{00} = \int \psi_0 (\partial^2 V / \partial Q^2) \psi_0 d\tau , \qquad (4.6)$$

$$V_{0k} = \int \psi_0 (\partial V / \partial Q) \psi_k d\tau . \qquad (4.7)$$

The first correction $(V_{00}Q^2)/2$ is always positive. The second correction, on the other hand, is always negative, since $E_0 - E_k < 0$, and tends to reduce the energy. Most important in the summation in (4.5) is the lowest excited state (or the lowest two or three states, if they are close to one another). Hence, if $V_{00}/2 < V_{0k}^2/(E_0-E_k)$, then the initial configuration is unstable and may go spontaneously into an alternative form. If, however, $V_{00}/2 \simeq V_{0k}^2/(E_0-E_k)$, the molecule may convert into another configuration with low activation energy along the coordinate Q. Thus the term $V_{00}/2$ dominates at the peak of the potential-energy curve while the term $V_{0k}^2/(E_0-E_k)$ dominates at the bottom of the potential-energy curve. The symmetric properties indicate that the integral V_{0k} is nonzero only if the direct product of representations of ψ_0 and ψ_k contains the representation of displacement Q. This is a necessary, though not sufficient, condition for the conversion of the molecule. From (4.5) it is clear that E_0-E_k [that is, the energy of electron transition from the occupied level of the molecule (complex) to the excited level] must be not too high - as a rule, it will be below 3 to 4 eV and will correspond to the visible spectral range. Hence it follows, in particular, that colored substances are more reactive than noncolored ones.

In order to assess the reactivity of a complex with the aid of SOJTE one has (a) to find the symmetry of the one-electron transition between HOMO and LUMO; (b) to find the symmetry of normal vibration which converts the given complex into its alternative (e.g., tetrahedron T_d into square D_{4h}): if the symmetry properties of the electronic and the vibrational transitions comply with one another the initial complex is unstable; (c) to check the difference E_0-E_k: for the reaction to occur it must be not too high; (d) to apply a similar procedure to the resulting complex in order to check for the possibility of further conversion. This technique, naturally, will yield unambiguous results only for complexes simple enough. Incorrect drawing of molecular orbitals may lead to wrong conclusions.

We shall here list the structures of tetracoordinated complexes as calculated by PEARSON [4.23] on the basis of SOJTE (using also the first-order aprproximation for degenerate structures). It appears that a tetrahedron is far from being the most stable coordination for most complexes; the square coordination D_{4h} may be more stable:

	d^0	d^1	d^2	d^3	d^4	d^5
High spin	T_d	D_{2d}	D_{4h}	D_{2d}	D_{4h}	D_{4h}
Low spin	T_d	D_{2d}	T_d	D_{2d}	D_{2d}	T_d

	d^6	d^7	d^8	d^9	d^{10}
High spin	D_{4h}	D_{4h}	D_{4h}	D_{2d}	T_d
Low spin	D_{2d}	T_d	D_{2d}	D_{2d}	T_d

The stability of octahedral complexes MX_6 is due to the fact that the gap Δ between the levels is then the greatest. At the same time the SOJTE technique predicts various departures from a regular octahedron in MX_6 complexes. For example, the conversion of the octahedron into a trigonal prism is facilitated by (1) π-donor ligands with occupied π orbitals, (2) vacancies on the t_{2g} orbital of the metallic atom, (3) high effective charge of atom M, and (4) strong interaction between ligands. The latter is encountered with "bulky" ligands. The conditions (2) and (3) are satisfied with atoms in d^0 and d^1 configurations (Mo, V, W, Re in the highest oxidized states). Indeed departures form octahedral coordination are most frequently observed with solid oxides of Mo, V, W, and Re. The theory also explains the displacement of the metal atom from the center of the octahedron in such complexes.

The SOJTE technique is capable of predicting the geometry of coordinated ligands. For instance, a biatomic molecule may coordinate with a metal ion in any of the three ways: in a line (4.8a), at an angle (4.8b), and normally (4.8c):

$$M\cdots O - O \qquad M\overset{\cdot\cdot O}{\cdots}\diagdown_O \qquad M\cdots\overset{O}{\underset{O}{|}} \ .$$

$$\text{(a)} \qquad\qquad \text{(b)} \qquad\qquad \text{(c)}$$

In complexes of metal M with π ligands CN, CO, N_2, NO, and O_2 the boundary HOMO and LUMO may have either σ or π symmetry. If HOMO and LUMO have the same symmetry (e.g., π and π^*), the transition is entirely symmetric and the initial complex is stable; if the transition between levels of different symmetry (e.g., $\sigma \rightarrow \pi^*$) is energetically advantageous, the complex is unstable. In an O_2 molecule the antibonding π^* orbitals are partially occupied and lie rather close to the π levels (Sect.4.4). Both $\pi \rightarrow \pi^*$ and $\sigma \rightarrow \pi^*$ transitions are possible in the MO_2 complex; the complexes may therefore have either symmetric (linear and rectangular) or asymmetric (angular) structures. The same is the case with M · NO complexes. As regards complexes of metal with CN, CO, and N_2, all electron transitions (including $\sigma \rightarrow \pi^*$) are energy consuming. In according with (4.5) the second-order Jahn-Teller approximation is negligible, and only linear complexes are formed, incapable of transforming into complexes of types (4.8b and c).

In considering bimolecular reactions Pearson applied the symmetry principles to concerted one-stage reactions, which have only one potential barrier (Fig.4.6). The reacting molec-

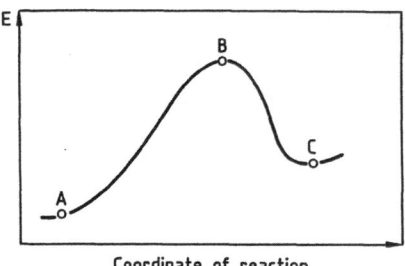

Fig.4.6. The potential curve of a reaction: A – initial substances, B – activated complex, C – reaction products

Coordinate of reaction

ules produce at each point of the potential curve a certain symmetry group, and the nuclear displacements may be considered as normal vibrations of a pseudomolecule of the relevant symmetry. On the basis of the above-developed arguments regarding the configurational instability of the nondegenerate state (4.5), it can be assumed that the reaction coordinate must belong to a fully symmetric representation of all points of the potential curve with the possible exclusion of extremes A, B, and C (Fig.4.6).

The integral V_{0k} (4.7) will be nonzero if ψ_0 and ψ_k (the wave functions of the ground and the excited states) both have the same symmetry. The wave functions ψ_0 and ψ_k are usually replaced by linear combinations of atomic orbitals which form HOMO and LUMO, respectively. For the reaction to take place it is necessary that (a) both orbitals should have the same symmetry, (b) their energies should be close enough , and (c) the orbital ψ_0 should represent the bond that break in the course of the reaction, while ψ_k pertains to those formed anew (this applies to bonding orbitals; the converse is true for antibonding orbitals). The physical picture in general corresponds to the electron flow between the orbitals, provided the overlap between these is positive in the region where the new bonds are formed.

When the difference between energies of orbitals of the same symmetry $E_0 - E_k$ is large, the concerted reaction is symmetry-forbidden; the reaction is symmetry-allowed only when the difference is small.

The reaction between the two molecules requires considerable overlap of their orbitals. As the reaction proceeds and advances along the reaction coordinate Q (Fig.4.6), an intermediate state is attained (Point B), which has different molecular orbitals. The reaction will have a low potential barrier if HOMO (ψ_O) and LUMO (ψ_k) do overlap.

In the simplest bimolecular reaction

$$H_2 + D_2 \rightleftharpoons 2HD$$

the only occupied orbital is the bonding σ_g orbital, and the nearest free orbital is the antibonding σ^*_u orbital. Let us assume that the reaction goes via a four-site intermediate state:

$$
\begin{array}{ccccc}
H-H & & H-H & & H \quad H \\
+ & \longrightarrow & | \quad | & \longrightarrow & | \; + \; | \\
D-D & & D-D & & D \quad D .
\end{array}
$$

The point symmetry group of this intermediate state C_{2v} is the same as in the beginning of the reaction. The molecular orbitals of the intermediate state, comprising the bonding orbitals σ_g, are denoted A_1, while those comprising the antibonding orbitals σ^*_u are labeled B_1 (Fig.4.7). The HOMO of one H_2 or D_2 molecule has zero overlap with the LUMO of another molecule. In other words, a low-molecular excited state of the same symmetry as the ground state does not exist. This implies that the reaction of H_2-D_2 exchange, should it take the four-site route, will have an extremely high potential barrier.

Indeed, the uncatalyzed exchange never follows the one-stage four-site scheme, but rather occurs as a multistage reaction involving single atoms:

$$D + H_2 \rightarrow HD + H \text{ , and so on.}$$

The four-site reactions are almost always symmetry-forbidden, as, e.g.,

$$2NO \rightleftharpoons N_2 + O_2 .$$

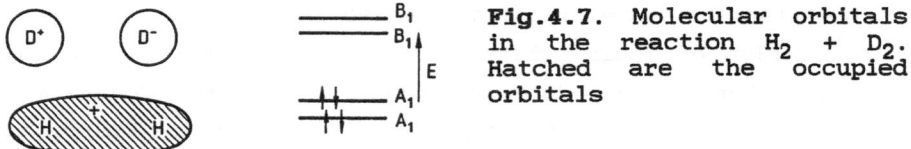

Fig.4.7. Molecular orbitals in the reaction $H_2 + D_2$. Hatched are the occupied orbitals

At the same time the Diels-Alder reaction is symmetry-allowed. The same is true for the reaction of 1.4-hydrogenation of butadiene

$$HC \overset{\text{\Large/}}{\underset{\text{\Large\char92}}{}} \begin{matrix} CH_2 \\ \\ CH_2 \end{matrix} \quad + \quad \begin{matrix} H \\ | \\ H \end{matrix} \quad \longrightarrow \quad HC \overset{\displaystyle CH_3}{\underset{\displaystyle CH_3}{\parallel}}$$

although the 1.2-hydrogenation of olefins

$$C_2H_4 + H_2 \rightarrow C_2H_6$$

is symmetry-forbidden. The action of the catalyst in the case of forbidden reactions is especially effective because the catalyst removes restrictions on symmetry.

WOODWARD and HOFFMANN [4.24] have formulated the rule of conservation of orbital symmetry on the basis of considerations less complicated than the second-order Jahn-Teller effect. This rule states that orbital symmetry is conserved in concerted reactions. The activation energy is low, if the ground states of reactants and products are symmetry-correlated; the reaction is then said to be symmetry-allowed. If the ground state of the reactant and the excited state of the reaction product are symmetry-correlated, then this reaction is forbidden – its activation energy is high.

A brief publication by WOODWARD and HOFFMANN [4.24], which appeared in 1965, has sparked off hundreds of works whose authors applied the rule of conservation of orbital symmetry for predicting wether many various reactions are allowed. Symmetry-allowed, for instance, are the reversible reactions cyclobutene $\leftrightarrow$ butadiene, hexatriene $\leftrightarrow$ cyclohexadiene, and the Diels-Alder reaction between ethylene and butadiene.

Above all, however, we are concerned with those symmetry-forbidden reactions that are capable of being catalyzed. Let us consider the concerted reaction of ethylene dimerization into cyclobutane:

$$\begin{matrix} CH_2 \\ || \\ CH_2 \end{matrix} + \begin{matrix} CH_2 \\ || \\ CH_2 \end{matrix} \rightleftarrows \begin{matrix} H_2C - CH_2 \\ | \quad\quad | \\ H_2C - CH_2 \end{matrix}.$$

The approach of C_2H_4 molecules to each other may only go via an intermediate state, in which both molecules lie parallel to one another while conserving maximum symmetry (Fig.4.8). The combination of π orbitals of the two C_2H_4 molecules gives two bonding π and two antibonding π^* orbitals (Fig.4.9). The combination $\pi_1 + \pi_2$ has mirror symmetry with respect to both planes 1 and 2 and is denoted SS; the combination SA $(\pi_1 - \pi_2)$ is symmetric with respect to plane 1 and antisymmetric with respect to plane 2, etc. The reaction results in replacement of π bonds of ethylene by σ bonds of cyclobutane, whose symmetry is illustrated in Fig.4.10 (the two bonding σ orbitals and the two antibonding σ^* orbitals).

The rule of conservation of orbital symmetry demands conservation of symmetry of each molecular orbital over the reaction coordinate. The so-called correlation diagram (Fig.4.11) shows the HOMO and LUMO of the initial reactants and the final products with their energies. Equisymmetric levels are con-

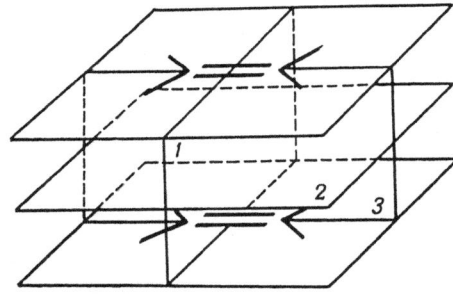

Fig.4.8. The transition state in the dimerization of ethylene: 1, 2, 3 are the planes of symmetry

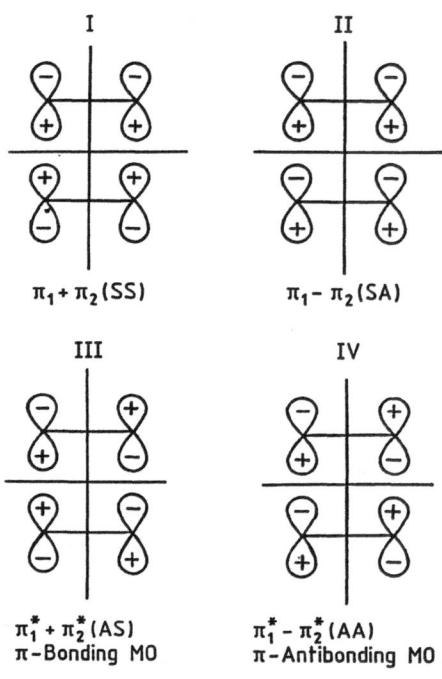

I   II

$\pi_1 + \pi_2$ (SS) $\pi_1 - \pi_2$ (SA)

III   IV

$\pi_1^* + \pi_2^*$ (AS)
π-Bonding MO

$\pi_1^* - \pi_2^*$ (AA)
π-Antibonding MO

Fig.4.9. The combinations of π orbitals of ethylene molecules participating in the dimerization

nected with straight lines. The molecular SS orbitals transform into the orbitals of products with little change in energy and no change in electron occupancy. The AA orbitals remain unoccupied and their energy is about the same. The remaining levels intersect: the combined π orbital SA of olefins becomes the σ^* orbital SA of cyclobutane, and the combined π^* orbital AS of

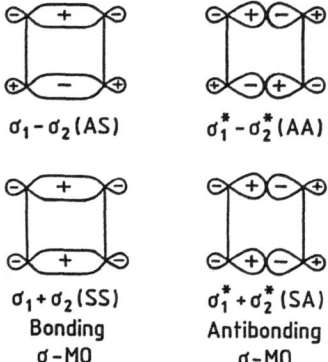

$\sigma_1 - \sigma_2$ (AS)

$\sigma_1^* - \sigma_2^*$ (AA)

$\sigma_1 + \sigma_2$ (SS)
Bonding
σ-MO

$\sigma_1^* + \sigma_2^*$ (SA)
Antibonding
σ-MO

Fig.4.10. The combinations of σ orbitals of the cyclobutane molecule

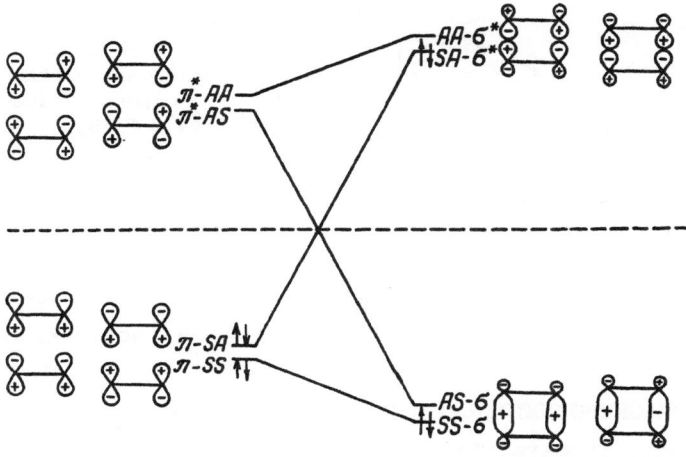

Fig.4.11. Correlation diagram of orbitals in the production of a cyclobutane molecule from two ethylene molecules [4.25]

olefins becomes the σ orbital AS of cyclobutane. The dashed horizontal line corresponds approximately to the energy of the 2p orbital of a free carbon atom. The gap between π and π* levels is about 5 eV.

According to WOODWARD and HOFFMANN [4.25], the existence of intersecting levels and the change in their occupancy by electrons implies that the reaction's activation energy is high. In this case the reaction is symmetry-forbidden in either direction. In order that the cyclobutane bond be formed, the electrons must occupy the bonding σ orbitals SS and AS. From Fig.4.11 it is clear that this is only possible with a high activation energy. The interaction between electrons prevents the reaction from crossing the lines which connect equisymmetric levels on the correlation diagram. Consequently, the transition from one ground state to the other goes gradually, over a high potential barrier. This barrier is of the same order of magnitude as the energy required for the transition of the two ethylene electrons from the occupied bonding orbital to the antibonding one. The lowest excited state of two molecules of ethylene correlates with the first excited state of cyclobutane:

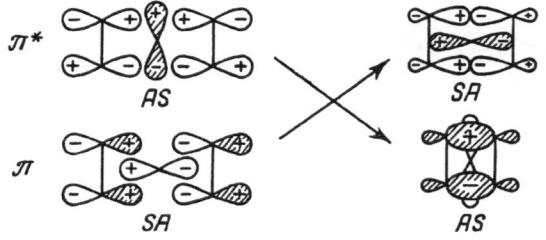

Fig.4.12. Orbitals SA and AS and orbitals of a hypothetical catalyst of dimerization of ethylene [4.26]

this conversion is not forbidden by symmetry. The photochemical reaction $2C_2H_4 \rightarrow C_4H_8$ is possible, although thermal activation is forbidden.

MANGO and SCHACHTSCHNEIDER [4.26] extrapolated the Woodward–Hoffmann rules to catalytic reactions.

As we have seen, the conversion of two ethylene molecules into cyclobutane requires introducing two electrons to the AS orbitals of olefin and removing two electrons from the SA orbital (Fig.4.11). Figure 4.12 shows the orbitals of the hypothetical catalyst capable of doing this. For the sake of simplicity the pairs of electrons are localized (hatched area) in the bonds to be broken and formed. Thus, the catalyst is a kind of "electron switch", which makes a symmetry-forbidden reaction possible. The overall energy of the catalytic reaction partly goes to the transfer of an electron from the SA orbital to the AS orbital.

It is easily proved that it is $d(t_{2g})$ orbitals that have the required symmetry (Fig.2.1). These orbitals must have one or two electrons. The geometry of a possible complex of a transition atom with two olefin molecules is shown in Fig.4.13a. The symmetry of combination of π orbitals (C_{2v}) is the same as for a noncatalyzed reaction. The symmetry planes are the yz and xz planes. The crystal field, produced by olefin ligands (Fig.4.13b), splits the levels of d_{yz} and d_{xz} orbitals. The pair of electrons, if available, will take the lower d_{yz} orbital, thus providing for back donation to the antibonding π^* orbital of olefins (AS). The vacant d_{xz} orbital will accept an electron from the olefin π orbital SA.

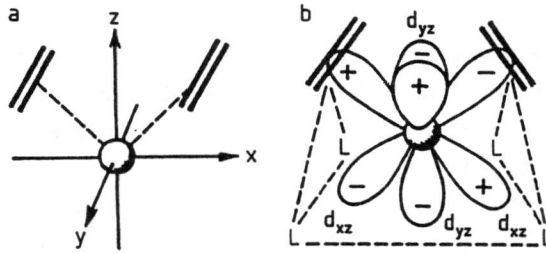

Fig.4.13. Geometry (a) and layout (b) of d_{yz} and d_{xz} orbitals in the complex of metal with two ethylene molecules [4.26]

The correlation diagram of orbitals for the catalyzed dimerization of ethylene into cyclobutane is given in Fig.4.14. The considerable lowering of the first unoccupied AS level, due to complex forming between olefin and the transition atom, ensures its occupation by electrons and results in considerable reduction in activation energy by comparison with the uncatalyzed reaction.

The calculations of the changes in the energy of molecular orbitals at every point of transition from the initial reac-

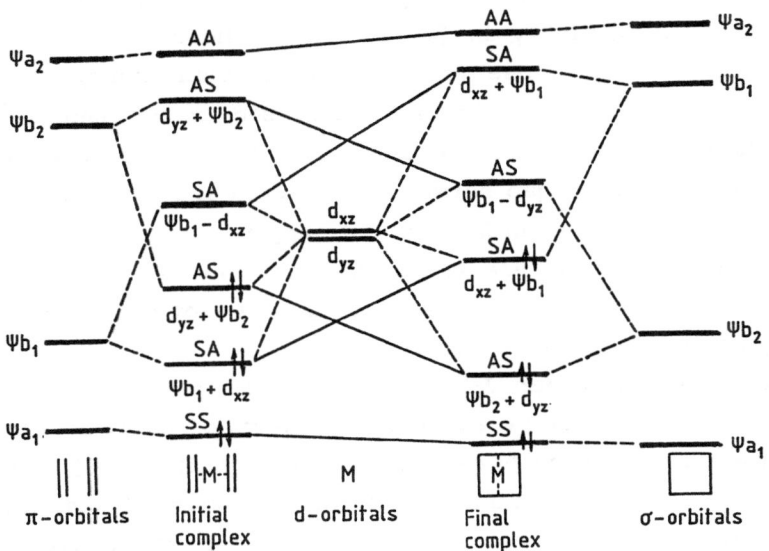

Fig.4.14. Correlation diagram of orbitals for the catalyzed dimerization of ethylene to cyclobutane. Dashed lines indicate the correlation between the orbitals of ligands and metal at complexing

tants to the final products must account for the fact that the electron levels do not necessarily coincide with the straight lines drawn between the initial and the final levels, as shown schematically in correlation diagrams (Figs.4.11,14). The calculation of one-electron levels for the reaction H_2+D_2 in the presence of a transition-metal catalyst indicates that the path from the initial to the final levels is far form being straight. This results in additional lowering of potential energy in comparison with the straight crisscrossing of terms. The overlap of orbitals may be even greater if the reaction involves more than one atom of transition metal. However, reasonable conclusions regarding the lowering of activation energy in reactions catalyzed by transition elements can only be made on the basis of consistent quantum-chemical calculations.

This catalytic mechanism requires that the transition-metal levels be located between the HOMO and the LUMO of the system catalyzed (Fig.4.14). According to [4.27], this applies to all d electron configurations (d^0 through d^{10}), rather than only up to d^8 as assumed initially in [4.26]. This conclusion is confirmed experimentally: close values of activation energy were obtained for ethylene dimerization catalyzed by d^8 and d^{10} systems. The d^{10} system seems to be active in those cases when there is a small energy gap between d orbitals and the nearest unoccupied p orbitals, as in Ag^+ and Cu^+. Apparently, not only d electrons, but also sp electrons can act as an "electron switch", although the d electrons provide more options in both the energy of levels and the symmetry.

The use of second-order Jahn-Teller effect gives the same results as do the simple rules of conservation of orbital symmetry, although in a much more complicated way. In [4.26], and later in [4.27-33], the principle of conservation of orbital symmetry elucidated the mechanism of a number of catalytic reactions: the conversion of acetylene into cyclooctatetraene, isomerization of allylbenzene into β methylstyrene, isomerization of cycloolefins, and oxidation of CO. Most works dealt

with the mechanism of metathesis of olefins, including hetero-
geneously catalyzed reactions.

All these reactions were assumed to follow the concerted
mechanism. Naturally, there exist alternative mechanisms of
catalytic action in symmetry-forbidden reactions. For example,
a catalyst may route the reaction via creation of active intermediate products: radicals, ions excited molecules, etc.
According to recent research [4.34], such processes are quite
common in catalysis. The existence of such active particles may
result in catalytic mechanisms resembling the chain or even
branching chain reactions. In this case the importance of
orbital symmetry is rather subdued if not entirely nullified.
For instance, the reaction of metathesis of olefins, catalyzed
by compounds of Mo, Re, and W, was long assumed to follow a
"quasi-cyclobutane" mechanism

$$
\begin{array}{c}
\underset{\underset{CH_2}{\overset{\|}{CH}}}{\overset{R_1}{\overset{|}{CH}}} + \underset{\underset{CH_2}{\overset{\|}{CH}}}{\overset{R_2}{\overset{|}{CH}}} + M \longrightarrow \underset{\underset{CH_2 \_\_ CH_2}{\overset{|}{M}}}{\overset{R_1 \quad R_2}{\overset{|\quad|}{CH \_\_\_ CH}}} \longrightarrow \begin{array}{c} R_1 - CH = CH - R_2 \\ + \\ CH_2 = CH_2 \end{array} + M ,
\end{array}
$$

which complies with the rule of conservation of orbital symmetry [4.26,35]. More recently, however, it was demonstrated
[4.36] that metathesis follows the mechanism of chain reaction
via the rupture of the double bond and creation of carbene complexes:

$$ 2M + R_1CH{=}CH_2 \longrightarrow M{=}CH_2 + M{=}CHR_1 . $$

$$ M{=}CHR_1 + R_2CH{=}CHR_2 \longrightarrow M{=}CHR_1 + R_1CH{=}CHR_2 , \text{ and so on.} $$

The exemplified use of orbital symmetry conservation rules
relates mainly to reactions of homogeneous complexes in solutions, representing possibly also individual stages in catalysis. The stereochemistry of surface complexes is naturally
more complicated; in most cases, however, the surface atoms
have enough freedom to allow rearrangements similar to those
described above.

MESSMER and BENNETT [4.32] point to the difficulties encountered when applying the rules of orbital symmetry to chemisorption and heterogeneous catalysis due to the lack of detailed knowledge about the actual symmetry of orbitals of
surface atoms. In their view, one of the most simple cases is
the graphite latttice with exactly known symmetry. The HOMO of
reacting molecules must have the same symmetry as the set of
levels in a small range above the Fermi level, whereas the LUMO
must have the same symmetry as the set of levels in a small
range below the Fermi level. The juxtaposition of HOMO allowed
the authors of [4.32] to draw conclusions regarding the localization of chemisorbed H_2, O_2, N_2, Fe_2 molecules and H, N, O, F
atoms on the basal graphite plane.

A detailed study of the electron exchange between orbitals
and the role of electron-vibrational interactions in catalysis
was made by BERSUKER [4.33]. According to him, the most favorable conditions for the initial stage of the catalytic process
(the change in the electron structure of the reactant) are
created when the reactant develops multiorbital bonds with the
catalyst. A large amount of charge may be transferred on each
orbital, while the total charge transfer, in accordance with
the principle of electroneutrality, is small. The multiorbital
bonds are formed with the help of transition-metal atoms. An
example of such a charge transfer in opposing directions was
given above (Fig.4.12) in reference to the mechanism of ethylene dimerization involving d_{xz} and d_{xy} orbitals. The creation
of a π complex (Fig.4.3) is accompanied by electron transfer in
opposing directions (back donation and donor-acceptor bond).
With catalysts that do not contain any atoms of transition elements the creation of multiorbital bonds may involve several
catalyst atoms; however, participation of d orbitals is beneficial in this case too.

According to BERSUKER [4.33], the electron-vibrational (or
vibron) interaction results in the transfer of the charge in
both directions when the bond is formed between the catalyst

168

and the reactant. The treatment here is similar to the treatment of SOJTE. The electron transition on each orbital results in a certain change in the dynamic bond constant. Since the symmetry of the orbitals is not the same, these changes in the dynamic constants produce a certain net force that changes the conformation of the whole molecule and is responsible for the catalytic action.

The contribution of "vibronic" action of the catalyst to the reduction in activation energy of a model reaction $N_2 + 4H = N_2H_4$ was calculated in [4.33]. The potential barrier is reduced by 115 kJ/mol by virtue of one electron being transferred from the σ orbital of N_2 to the catalyst and one electron being transferred from the catalyst to the π^* orbital of N_2. In [4.37] the theory of vibron interactions was used for considering the oxidation of CO by solid solutions $Co_xMg_{1-x}O$, and rationalizing the decrease in activation energy with the increase in the concentration of Co ions in the solid solution.

4.3 Energy Levels of Adsorbed Molecules in the Band Structure of Oxides and Complexes of Transition Elements

Analysis of the relationship between molecular and electronic processes on the surface of semiconductors and dielectrics [4.38] reveals the importance of local levels of adsorbed particles for adsorptive and catalytic action. Therefore, it is necessary to know the location of these levels with respect to the bands and levels of the solid adsorbent.

The local adsorptive levels in the band structure of oxides were considered by BONCH-BRUEVICH [4.39], MARK [4.40], MORRISON [4.41], and DOWDEN [4.9]. For purely ionic bonding correlations were established between the depth of the adsorptive level and the permittivity of the subsurface layer, the value of Madelung's constant in the adsorptive site, the ionization potential of the metallic atom serving as the adsorptive site, and the adsorbate's electron affinity. It must be admitted, however, that the problem of calculating the exact location of the energy levels of adsorbed particles in the forbidden band is far from being solved even for the simplest model systems.

The counterparts of electron transfers between bands in a solid and levels of adsorbed species are - in homogeneous complexes of transition metals - the electron transitions between the metal and ligand, detected by charge transfer spectra. As demonstrated in Sect.3.5.2, the charge transfer spectra of transition-metal oxides are quite similar to the spectra of homogeneous charge transfer complexes of respective transition metals. This justifies the attempts to base analysis of the catalytic and chemisorptive activity of oxides on the known regularities of the charge transfer energy in homogeneous complexes. The chemisorptive activity is commonly assumed to depend on the depth of the local level.

For the description of charge transfer spectra (3.22), JORGENSEN [4.42] employed his concept of optical electronegativity x_{opt} not only to atoms, but also to the molecular ligands. He quotes the following values of optical electronegativity:

	O_2^-	O_2^{2-}	O^{2-}	H^-	Cl^-	F^-	H_2O	NH_3
x_{opt}	3.0	2.8	3.5	2.2	3.0	3.9	3.5	≈ 3.3 .

The Madelung's potential in the oxide lattice affects the value of x_{opt} only slightly.

These figures can be used in first-approximation calculations of the energy of charge transfer from the adsorbed molecule to the transition ion in the oxide. The ligand-to-metal charge transfer energy in this approximation corresponds to the distance from the local donor level of adsorbed molecule (e.g., O_2^-) to the d band of transition ion $M^{(n+1)+}$ (the transition $O_2^- \rightarrow O_2 + e^-$ and $M^{(n+1)+} + e^- \rightarrow M^{n+}$). Since $x_{opt} \approx 2$

for most transition metals, and the energy of charge transfer is proportional to the difference $x_{opt}(L)-x_{opt}(M)$, the higher the value of $x_{opt}(L)$ the deeper the respective level of the adsorbed molecule must be, and the less readily it will give away the electron (i.e., be oxidized).

The absolute values of the L→M electron transfer energy, calculated in this way, are high, and much higher than the activation energy of catalytic reactions; only the relative quantities are reasonable. It must be noted, however, that as the polarizability of the ligand increases (and the adsorbed molecules often are ligands with considerable polarizability), the charge transfer spectrum is shifted toward the visible range - in other words, the energy of charge transfer becomes lower.

The role of charge transfer in the catalytic activity of complexes of transition metals was analyzed by BERSUKER [4.43,44]. In complicated complexes the energy borderline between HOMO and LUMO (Sect.4.2) is quite similar to the Fermi level in a heterogeneous catalyst. By analogy, the borderline between HOMO and LUMO of the reactant may be called the reactant's Fermi level. Simply by comparing the Fermi levels of the catalyst (which may also be heterogeneous) and the reactant, it is possible to determine the magnitude and the direction of charge transfer. The complete calculation must account for multiorbital charge transfers [4.33] (Sect.4.2).

The redox capacity of the system, its capability for electron transfer, and hence its catalytic activity in redox reactions increase with the rise in the concentration of the catalyst's levels near the Fermi level E_F. The MO techniques were used in [4.44] for calculating complexes of ferroporphyrin (catalyst) with hydrogen peroxide (reactant). Equalization of Fermi levels of the catalyst and reactant results in partial transfer of charge from ferroporphyrine to the coordinated H_2O_2 molecule (Fig.4.15). Having accepted the charge, the H_2O_2 molecule becomes unstable and decomposes, e.g., into two OH^- ions: $H_2O_2+2e \rightarrow 2OH^-$. The charge is transferred from the system

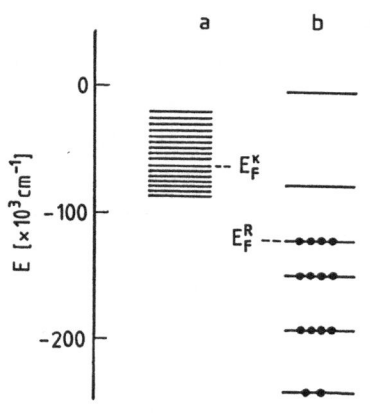

Fig.4.15. The set of one-electron energy levels (to scale) and their occupancy by electrons [4.44]: (a) the complex of ferroporphyrin, (b) hydrogen peroxide molecule

with the higher E_F to the system with the lower E_F. The catalysis here is conditioned by charge transfer, causing instability of the H_2O_2 molecule. Ferroporphyrine serves for the reacting molecules (presently for H_2O_2) as a buffer, which readily drains off the "excess" charge or makes up for its deficit. The capacity of such a buffer is even greater when a charge transfer complex is formed with a heterogeneous catalyst (e.g., a transition-metal oxide), where there are many levels - both free and occupied - near the Fermi level.

When applying the rules of conservation of orbital symmetry (Sect.4.2) to heterogeneous catalysis, one must take into account the actual structure of surface d bands (d levels) and their energy with respect to the energy levels (orbitals) of the reactant. Such a calculation was attempted for the reaction of ethylene dimerization into cyclobutane on the surface of perovskite-type crystals [4.45]. The correlation diagram is given in Fig.4.16. The t_{2g} band of Ti^{3+}, split by the ligand field, is loosely represented by two discrete states d_{xz} and d_{yz} on the (100) face of perovskite. The diagram shows the two extreme cases, which according to XPS, are typical of the surfaces of different perovskites: (a) the lower t_{2g} level lies 8 eV below the Fermi level, close to the π level of ethylene, and (b) the lower t'_{2g} level lies 5 eV below the Fermi level and is

172

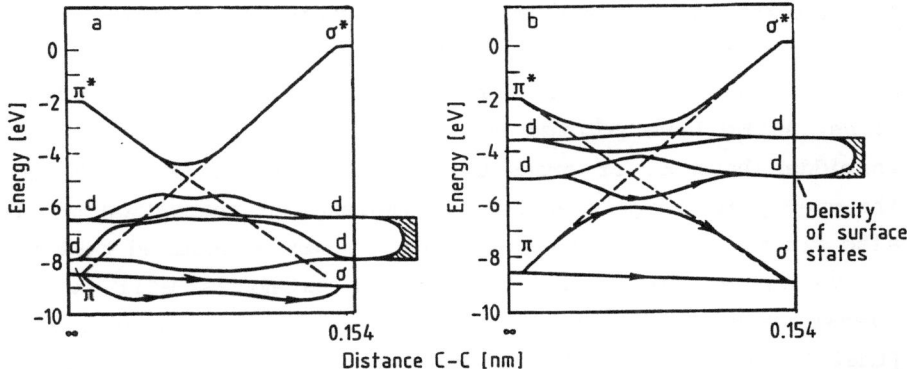

Fig.4.16. Correlation diagram of molecular orbitals for the dimerization of ethylene on the (100) face of perovskite [4.45]: (a) t_{2g} orbital is 9 eV below E_F, (b) 5 eV below E_F

close to the antibonding π^* level of ethylene. In the first case (Fig.4.16a) the potential barrier is virtually removed, while in the latter (Fig.4.16b) the potential barrier still remains [4.45].

However, the potential barrier can be lowered considerably in the latter case too (Fig.4.16b), if the d level is occupied by no more than five electrons. The potential barrier is the lowest when the energy of d levels is close to (slightly above) the energy of ethylene's π level (Fig.4.16a). In the second case (Fig.4.16b) the transfer of charge from the d level to the antibonding π^* orbitals may be obstructed by the Coulomb repulsion due to excessive charging of the molecule. In a solid, however, the Coulomb repulsion can be weakened because of the screening of transferred charge by a large number of surface levels near the Fermi level ("redox capacity", according to BERSUKER [4.44]).

Let us now turn back to the two-peak pattern of catalytic activity of oxides of transition elements (Figs.4.1,2). The reactions which conform with this pattern have one feature in common that must not be overlooked. They can all be generally classified as the redox reactions – that is, reactions involving transfer of electrons. It would be reasonable therefore to look for alternative explanations of the two-peak pattern, which would not rely heavily (DOWDEN's assumption [4.1]) on CFSE concept. Dowden's explanation is not specific for redox reactions. As we have pointed out in Sect.3.5.1, the width of the forbidden band E_g of transition-element oxides (Fig.3.17) and the energy of charge transfer in complexes of transition metals also fit into the two-peak pattern. It can be assumed that the activation energy of adsorption and catalysis will be the lowest when the energy of the electron transition is lowest (that is, with d^3 and $d^6 - d^7$ systems). The energy level of an adsorbate in such systems is close to the d level of the transition element. When considering the actual regularities displayed by the catalytic activity, one must combine the approaches of the coordination theory and the electron theory of adsorption on real surfaces. For nontransition element oxides this was done in [4.38,46].

Notwithstanding that the question of energy levels of adsorbed substances is of crucial importance for adsorption and catalysis, the relevant experiments so far are few. In [4.41,47,48] the energy of adsorptive levels on oxide surfaces was assessed by measuring the temperature dependence of point-contact conduction in compacted powder tablets in the presence of adsorbates. As we have pointed out in Sect.3.5.1, this technique gives unreliable results. The same can be said about other electrophysical techniques (measurements of photoconductivity, field effect, etc.), which are capable of giving correct estimates of the energy of local levels only for adsorption on a monocrystalline surface.

FRANKL [4.49] points out that electrophysical techniques are only indirect tools for studying the surface levels. What is measured is not a direct characteristic of the surface states, but rather some characteristic of the space-charge layer, which is then related to the parameters of the surface states. By contrast, the optical and UPS techniques yield direct information: the useful signal here depends on electrons or holes localized on the surface states. It is clear that the direct techniques in principle allow one to detect a larger number of levels than the indirect techniques do, since the former are sensitive not only to space-cahrge variations, but also to strictly local transitions (e.g., in the coordination sphere of the transition ion). Unfortunately, the absolute sensitivity of the direct techniques is so far inferior to the sensitivity offered by indirect (electrophysical) techniques.

The first experiments using the optical methods for assessing the energy of local levels of adsorbed molecules were conducted by TERENIN and PUTSEIKO [4.50] more than 35 years ago. Many researchers have subsequently used the diffuse reflection spectra for studying various catalysts. However, even today there are not much reliable data on the energy levels of adsorbed molecules. In many works the shift of charge transfer band due to adsorption was used not for determining the level of adsorbed molecule, but rather for measuring the new location of the level of the surface ion of the catalyst [4.51-53]. Some works, for instance, dealt with adsorption of organic substances on molybdenum-containing oxide catalysts. The edge of the charge transfer band was found to shift toward longer wavelengths, pointing to the transition of molybdenum from tetrahedral to octahedral coordination when the adsorbate molecules fill its coordination sphere.

A promising technique for studying the surface levels of adsorbed molecules is photoelectron spectroscopy, especially its version with UV excitation (UPS). The number of investigations in which this technique is employed for studying adsorption is increasing rapidly.

VILESOV and SUKHOV [4.54] have used UPS for studying molecules of aromatic compounds on the surface of oxides. Physisorption was found to lower the potential of naphthalene, antracene, and similar molecules by 0.9 - 1.0 eV.

In [4.55,56] the UPS technique was used for studying the adsorption of oxygen, hydrogen, and water on the (100) and (110) faces of reduced TiO_2 (rutile) single crystals: Ti^{3+} ions produce peaks in UPS spectra at −0.6 eV. Adsorption of O_2 gives rise to peaks at −5.2, −6.8, and −10.3 eV, at the same time reducing the peak at −0.6 eV. Adsorption of H_2 gives rise to peaks at −4.3, −7.6, and −10.0 eV, at the same time increasing the peak at −0.6 eV. Apparently, adsorption of O_2 is accompanied by a reduction of Ti^{3+} ions, and adsorption of H_2 by oxidation of Ti^{4+} ions. Adsorption of O_2 removes the surface d levels of Ti_{3+} ions. Adsorption of H_2O gives rise to three peaks, pertaining to the orbitals of water itself. Two of these have the same energies as the orbitals in a free water molecule, and the third is shifted by 1 eV and pertains to the orbital involved in the bonding with the surface. Adsorption of certain molecules on a ZnO surface was studied in [4.57,58]. Considerable changes in the energy of HOMO were observed on adsorption of CO, benzene, ethylenoxide, methanol, and acetone.

The reduction of a WO_3 surface was detected in [4.59] by the appearance of a peak of d levels in the UPS spectrum. In [4.60] the lowering of the peak of d levels in the UPS spectrum of an NiO single crystal was observed after adsorption of O_2 at 700 and 1450°C, which was attributed to the oxidation of Ni^{2+} to Ni^{3+}. It must be observed that in all the above-mentioned works the change in intensity of d levels at adsorption of O_2 or H_2 cannot be taken for an irrefutable indication of participation of d orbitals of surface atoms in adsorption. Bulk atoms possibly participate in these processes as well, the electron being relayed to them from the surface layer.

Now we turn to considering the specific mechanisms of chemisorption of individual molecules on the oxides of transition metals. We shall be dealing mainly with the adsorption on isolated ions of a transition metal, implanted in a matrix of a nontransition-metal oxide (Sect.3.1), since it is for these systems that the knowledge of the coordination and the electronic structure of surface complexes is the most consistent.

4.4 Adsorption of Oxygen on Oxides

When adsorbed on oxides, oxygen may assume a number of ionic and molecular forms. In the ground state the O_2 molecule occurs in a biradical triplet state $^3\Sigma_g$ with the two unpaired radicals on the antibonding π^* orbitals. The excited singlet states of O_2 have a relatively low energy: $^1\Delta_g \simeq 1eV$ and $^1\Sigma_g^+ \simeq$ 1.65 eV. The nonparamagnetic states of oxygen are also O^{2-}, O_2^{2-}, and O_3, as well as certain aggregated states like O_4 ($=2O_2$). Interaction with the surface may produce paramagnetic ionic states O^- and O_2^-. The simplest reaction

$$O_2 + e^- \rightarrow O_2^-$$

is exothermal (the electron-affinity energy of O_2^- is 0.87 eV [4.61]). The reaction

$$O_2 + 2e^- \rightarrow 2O^-$$

is weakly endothermal ($\simeq 1eV$). Highly endothermal (6.5 eV) is the reaction producing the nonparamagnetic O^{2-} ion; this ion therefore can be produced only in the bulk at the expense of the Madelung potential. The state O^- can be produced either directly from O_2 or from O_2^- via intermediate O_2^{2-} (the reaction $O_2^- + e^- \rightarrow O_2^{2-}$ consumes about 2.1 eV [4.61]). It must be noted, however, that successive conversion ($O_2 \rightarrow O_2^- \rightarrow O_2^{2-} \rightarrow 2O^- \rightarrow 2O^{2-}$) hase almost never been observed so far; the creation of a particular ionic state of oxygen at adsorption depends on the availability of the required number of electrons or surface donor states.

The measurements of conductivity of oxides upon adsorption of O_2 have long been almost the only source of knowledge about the mechanism of adsorption. Generally, the conductivity of n-type semiconductor oxides increases, and that of p-type semiconductor oxides decreases, with an increase in the pressure of oxygen P_{O_2}. Most of these measurements, however, were carried out with powdered specimens and thus are not too useful for drawing conclusions about the details of the process. The majority of researchers who have studied the conductivity as function of oxygen pressure agree that the recharging of chemisorbed forms of oxygen involves not only the free electrons of oxide, but also the local donor-type surface states: metallic cations, Lewis acids, and anion vacancies.

The less numerous works, which used oxide single crystals, are in a way exceptional [4.57-60,62]. For example, a study of the adsorption of O_2 on the (1010) face of a ZnO single crystal [4.62] reveals that on stoichiometric ZnO the O_2^- ion can be produced from physically adsorbed O_2 at the expense of the thermal activation of an electron from the conduction band. Desorption restores the conductivity to its initial values. The process on the surface of nonstoichiometric ZnO involves oxygen vacancies and interstitial zinc atoms. The oxygen (O_2^- or O_2) interacts with vacancies, becoming O^- and then O^{2-}.

The independent production of O^- and O_2^- on ZnO is apparently confirmed by the data on thermal desorption. According to [4.63], for instance, the curves of thermal desorption of O_2 (adsorbed on ZnO at 25°C) display peaks at 180 - 190°C and 280 - 290°C, the former corresponding to a larger amount of oxygen released. The release of oxygen at each of the two peaks is proportional to the duration of contact with oxygen at 25°C. This points to the existence of slow adsorption, resulting in the production of both O_2^- (low-temperature peak) and O^- (high-temperature peak). In our view [4.64], the O_2^- and O^- forms are produced independently at low temperatures, and O_2^- is not covnerted into O^-, because otherwise the prolonged adsorption would have resulted in an accumulation of predominantly the "high-temperature" form. On ZnO, for instance, these two forms compete for adsorption sites:

$$O_2(gas) + Zn^+ \rightarrow O_2^- Zn^{2+} \ ,$$
$$O_2(gas) + 2Zn^+ \rightarrow 2O^- Zn^{2+} \ .$$

The two independent and concurrent processes were observed [4.65] in studying the adsorption of O_2 on Cr_2O_3 at 131–163°C and on supported MoO_3/Al_2O_3 catalysts [4.66]. Apparently, O^- and O_2^- are produced on different sites.

The work function increases upon adsorption of oxygen. Ideally, were the electrons at adsorption transferred only from the solid surface to the O_2 molecule, this would have resulted in a negatively charged surface layer and a positively charged subsurface region, thus preventing further adsorption of oxygen after reaching a small fraction of a monolayer [4.67]. Experimentally, this is not the case, since the surface process is followed by the process in the bulk: the negative oxygen ions diffuse into the bulk, and the lattice cations move to the surface. In some cases these two stages can be studied by measuring the work function. For example, in NiO the adsorption of oxygen below 250°C is restricted to the surface, while above 250°C the process involves both the surface and the bulk [4.68].

Much greater achievements in the study of the mechanism of oxygen adsorption are due to the employment of EPR spectroscopy.

Most thoroughly investigated are the EPR spectra of the molecular ion-radical O_2^-. The structure and the magnetic properties of O_2^- have been studied especially well for the bulk of peroxides NaO_2 and KO_2 [4.69] and alkali halides [4.70]. Analysis of the hyperfine nuclear structure of ^{17}O (nuclear spin J = 7/2) verified that these really are the molecular ion radicals O_2^- with equivalent nuclei.

The calculation of the parameters of EPR spectra of O_2^- (in alkali halides) was first carried out by KANZIG and COHEN [4.71]. The calculation was based on the diagram of levels shown in Fig.4.17a. The unpaired electron is found on the antibonding 2p orbital (π_g^y) of oxygen. The main axes are directed as

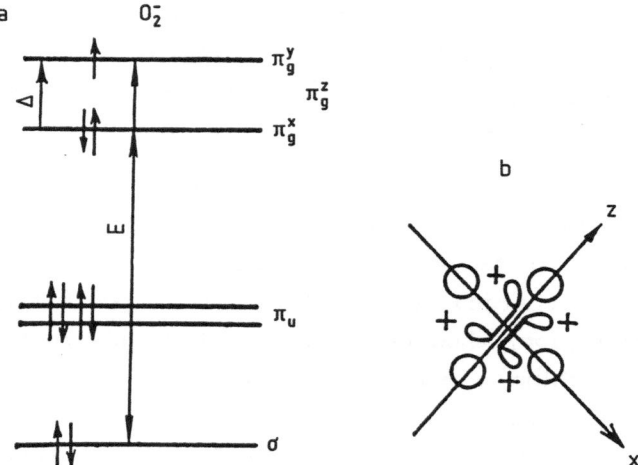

Fig.4.17. Energy diagram of adsorbed ion radical O_2^- (a) and the arrangement of axes and the symmetry of adsorbed ion radical O_2^- (b).

in Fig.4.17b. The following expressions were obtained for the components of the g tensor:

$$g_z = g_e + 2\ell \sqrt{\frac{\lambda^2}{\lambda^2 + \Delta^2}} \quad,$$

$$g_y = g_e \sqrt{\frac{\Delta^2}{\lambda^2 + \Delta^2}} - \frac{\lambda}{E} \left[\sqrt{\frac{\lambda^2}{\lambda^2 + \Delta^2}} - \frac{\Delta}{(\lambda^2 + \Delta^2)^{1/2}} + 1 \right] ,(4.9)$$

$$g_x = g_e \sqrt{\frac{\Delta^2}{\lambda^2 + \Delta^2}} - \frac{\lambda}{E} \left[\sqrt{\frac{\lambda^2}{\lambda^2 + \Delta^2}} - \frac{\Delta}{(\lambda^2 + \Delta^2)^{1/2}} - 1 \right] .$$

The y axis is directed along the p orbital (π) of the unpaired electron. The following designations are used: λ is the constant of spin-orbital interaction (equal to the constant of spin-orbital interaction in a neutral oxygen atom); E and Δ are the magnitudes of splitting (Fig.4.17); ℓ accounts for the action of the effective field on the surface; and g_e is the value of the g tensor for a free electron (g_e = 2.0023). Substi-

180

tution of experimental values for g tensor components for O_2^- in KCl (g_z=2.435, g_y=1.9512, g_x=1.9551) yields λ/Δ =0.23, λ/E = 0.025 and ℓ =1.04. From these calculations it follows that the EPR spectra of O_2^- must obey the following relations: $g_z > g_x$, g_y; $g_x \simeq g_y$ (axial symmetry, $g_z > g_e$). The anisotropy of the g tensor $\Delta g_z = g_z - g_e$ in the first approximation is a linear function of $1/\Delta$, assuming $E \gg \Delta > \lambda$.

Later KASAI [4.72] studied the spectrum of O_2^- on the zeolites NaY and BaY, treated by gamma rays. The calculations conformed with the Kanzig–Cohen model, and simplified expressions for the g tensor components were obtained, assuming $E > \Delta \gg \lambda$:

$$g_z \simeq g_e + \frac{2\lambda}{\Delta} \; ; \qquad g_y \simeq g_e + \frac{2\lambda}{E} - \frac{\lambda^2}{\Delta^2} - \frac{\lambda^2}{E\Delta} \; ;$$

$$g_x = g_e + \frac{\lambda^2}{E\Delta} - \frac{\lambda^2}{\Delta^2} \; . \qquad\qquad (4.10)$$

The calculation of λ/Δ from these expressions, using the experimental values of the g tensor components, indicates that the splitting between the two p π orbitals for Ba^+ is twice as big as that for Na^+.

LUNSFORD and JAYNE [4.73], who studied the chemisorption of oxygen on MgO and ZnO exposed to UV radiation, conclude – on the basis of (4.9) – that the status of the oxide surface has a strong influence on the splitting between the π^*_g levels of the O_2^- ion, although its effect on the gap between the π^*_g and σ_g levels is small. For the two extreme cases of splitting they have obtained the following values of g tensor components: $g_x = 4$, $g_y = g_z = 0$ for $\Delta =0$; $g_x = g_y = g_z = 2$ for $\Delta = \infty$.

When adsorbed by oxides the O_2^- ion is most easily detected with systems containing cations in d^1 configuration (one unpaired electron). On the basis of the above-developed arguments and their own research, concerned with O_2^- ion radicals on the surface of reduced TiO_2 and supported V/SiO_2 catalysts, MIKHEIKIN et al. [4.74] arrived at the conclusion that Δ in (4.9) is an almost linear function of the charge of the cation near which the O_2^- ion radical is stabilized. This is true when the axis of the oxygen molecule is parallel to the surface. For adsorption of O_2 on V/SiO_2 [4.75] the signal from O_2^- displayed hyperfine splitting, caused by [51]V nuclei adjacent to the ion radical O_2^-. The low value of isotropic HFS constant A_{iso} confirms the assumption regarding the high ionicity of the bond between the adsorbed oxygen and the surface.

The ionic nature of bonding between O_2^- and the surface of reduced specimens of SnO_2 and TiO_2 was proved in the study of adsorption of oxygen enriched with [17]O [4.76,77]. An additional signal, whose parameters were close to the parameters of O_2^- (g_1=2.034, g_2=2.004, g_3=1.994), did not exhibit HFS due to [17]O and was therefore ascribed to the electron, localized on the lattice defect at the adsorption of oxygen.

A great diversity of forms of O_2^-, adsorbed on the oxides of rare-earth elements and other M_2O_3 oxides, were discovered in [4.78]. The parameters of signals from O_2^- with low Δ_g coincided with those of peroxide ions produced in peroxide hydrates, while the signals with high Δg were close in parameters to the signals from the bulk of rare-earth element peroxides $M^{3+}(O_2^-)_3$.

Table 4.3 lists figures which support the correlation between the charge of the oxide cation and the parameters of the g tensor of adsorbed ion radicals O_2^-. By the "charge of the cation" we mean here the final formal charge of the cation at which the O_2^- ion radical is stabilized. In the case of TiO_2 the electron was initially donated by the Ti^{3+} ion; in the case of Al_2O_3 it was the electron captured previously by a vacancy on exposure to gamma radiation.

The anisotropy of the g factor is plotted in Fig.4.18 as a function of cation charge [4.64], on the basis of data listed in Table 4.3. Large variations in Δ_g (from 0.0033 for HO_2 to 0.165 for NaO_2) are observed with O_2^- ion radicals stabilized at univalent cations. This may be related to the unequal polarizing power of different cations. Indeed, the calculation for the LiO_2 molecule indicates that the molecular ion of oxygen is

Table 4.3. The parameters of the g tensor of ion radicals O_2^-, adsorbed on the surface of oxides containing differently charged cations.

Adsorbent	Cation charge	$g_1(g_x)$	$g_2(g_y)$	$g_3(g_z)$	$\Delta g = g_3 - g_e$	Ref.
NaY zeolite	1	2.002	2.066	2.113	0.111	4.72
BaY zeolite	2	2.005	2.009	2.057	0.055	4.72
ZnO	2	2.002	2.007	2.051	0.049	4.73
MgO	2	2.0018	2.0089	2.0777	0.0754	4.79
MgO	2	2.0011	2.0073	2.0770	0.075	4.73
MgO	2	2.0019	2.0070	2.0740	0.072	4.80
MgO	2	2.0006	2.0075	2.0894	0.087	4.80
Al_2O_3	3	2.003	2.008	2.037	0.035	4.81
Sc_2O_3	3	2.003	2.007	2.047	0.045	4.78
Y_2O_3	3	2.002	2.007	2.055	0.053	4.78
SnO_2	4	2.004	2.010	2.029	0.027	4.82
TiO_2	4	2.002	2.009	2.024	0.022	4.82
TiO_2	4	2.003	2.010	2.025	0.021	4.76
TiO_2/SiO_2	4	2.003	2.009	2.026	0.024	4.83
V_2O_5/SiO_2	5	2.004	2.011	2.023	0.021	4.83
MoO_3/Al_2O_3	6	2.002	2.009	2.018	0.016	4.84
WO_3/SiO_2	6	2.002	2.010	2.023	0.019	4.85

highly polarized in the field of the ion of lithium, and the electron density of the oxygen's orbitals is strongly biased toward lithium [4.86].

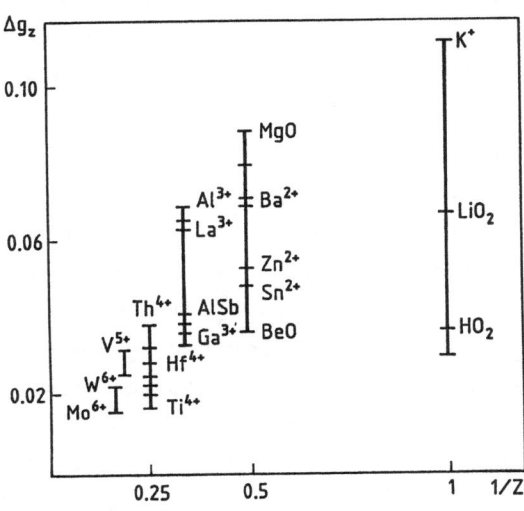

Fig.4.18. The g-factor anisotropy of the ion radical O_2^- versus cation charge

However, the parameters of the g tensor exhibit considerable variation not only with different (equally charged) ions, but also with the same ions in one and the same lattice. For example, the value of g_z for O_2^- on TiO_2 (anatase) varies, depending on the pretreatment of the catalyst, from 2.019 [4.82,87] to 2.030 [4.76], and g_z for O_2^- on ZnO varies from 2.051 [4.76] to 2.039 [4.88]. Evidently, this change in g_z is due to varying coordination or effective charge of the cation. It must be observed also that adsorption of oxygen on the surface in most cases gives rise simultaneously to several forms of O_2^- with slightly different EPR parameters. The possible reorientation of the ion radical, which might tell us about the parameters of the EPR signal, must not be overlooked either.

The deviation of Δg_z from a tentative linear correlation between Δg and the cation's charge may arise from the fact that the theoretical EPR parameters of O_2^- are based on the model of O_2^- stabilization which is rarely encountered with real catalysts. Indeed, by assuming axial symmetry this model takes into account – apart form the ground state of O_2^- – only the parity-allowed states (σ_g for g_z and π_g for $g_{x,y}$) and disregards the state π_u with the closest "free" states $^2\Delta_u$, $^2\Sigma_u$, etc. As a matter of fact, however, the existence of centers of symmetry as high as C_{4v} is hardly to be expected on a real surface. More likely, the coordination is rather distorted, and the real distribution of charges around the central atom must result in polarization of the molecular oxygen ion, which in turn tilts the molecule's axis so that there is a nonzero component of crystal field along the axis of the molecular ion. Combined with polarization, this field component will remove parity restrictions, adding π_u levels (about 4 eV from π_g^*, while σ_g is 5 eV off) to π_g^* levels and $^2\Delta_u$ and $^2\Sigma_u$ levels (removed by 4 – 5 eV from π_g^*). This accords with the assumption of SYMONS [4.89], that the O_2^- in real systems occupies an intermediate position between the peroxide radical and the classic model used by

Kanzig for calculating the parameters of the EPR signal from O_2^-.

SETAKA and KWAN [4.90], who studied the adsorption of oxygen on Al_2O_3-supported oxides UO_3, CeO_2, and CdO, came to the conclusion that with CeO_2/Al_2O_3 and UO_3/Al_2O_3 the principal values of the g tensor of the molecular oxygen radical pertain to peroxide, that is a radical of type M-O-O˙ with the more covalent M-O bond. This structure implies an unequal standing of oxygen atoms in the molecular ion. This conclusion is based on the comparison drawn between the averaged g factor $g_{av} = 1/3(g_1 + g_2 + g_3)$ for the observed signals from adsorbed oxygen, and g_{av} from radical forms of oxygen in mixed aqueous solutions of salts, containing H_2O_2 and an electron-donating transition ion. The "angular" structure of this kind is based, as shown in Sect.4.2, on second-order Jahn-Teller effect as applied to oxygen complexes.

However, the conclusions drawn in [4.90] would seem to us rather hasty. Observe, for instance, that the values of g_{av} for an O_2^- ion on the surface of TiO_2 (2.013) and in the $Ti^{3+}-H_2O_2$ system (2.011) are very close to one another; furthermore, adsorption of oxygen on TiO_2 is known to result in O_2^- ions composed of identical atoms. Nevertheless, the authors of [4.90] are quite right in pointing to the lack of clear-cut correlation between Δg_z and the cation's charge for the systems in question.

Inequality of oxygen atoms in ion radical O_2^- on ThO_2 was detected in [4.91] by using oxygen enriched with the ^{17}O isotope. The angular structure $Th^{4+}-O\diagdown O$ is formed. It is interesting to note that two forms were observed with a highly anisotropic g tensor: $\Delta g = 0.017$ after adsorption at -195°C, and $\Delta g = 0.046$ after adsorption at 25°C.

Angular complexes of the ion radical O_2^- with nonidentical atoms of oxygen seem to be commonly formed when oxygen interacts with ions having more than one d electron. The adsorption of O_2 on RhNaY zeolite (Rh^{2+} in d^7 configuration) gave rise to a signal with g_1 = 2.014; g_2 = g_3 = 1.943. The use of oxygen enriched with ^{17}O isotope enables one to obtain estimates of hyperfine splitting. These particles were assumed to have angular structures. Their bonding is much weaker as compared with ionic forms of O_2^-. Pumping out at room temperature eliminates them [4.92].

Another much studied paramagnetic form of adsorbed oxygen in the atomic ion is the radical O^-. The ground state of a free ion radical O^- is the ^2P state. The crystal field (e.g., the axial field on the solid surface) splits this state – in the first-order approximation – into a singlet level and a doublet level. The expressions for the components of the g tensor of the EPR signal from O^- in the axial field were obtained in [4.93]:

$$g_z = 2C^2, \ g_{x,y} = 2C^2 + 4AC , \qquad (4.11)$$

where $C = 1 + \left[2\lambda^2/(2E+\lambda)^2\right]^{-1/2}$, $A = \lambda C/(2E+\lambda)$, $2A^2+C^2 = 1$, $\lambda > 0$; E is the splitting between the doublet and the singlet.

For $E \gg \lambda$, $g_z \simeq g_e$, $g_{x,y} = g_e + (2\lambda/E)$; for $E \ll \lambda$, $g_z = 2/3g_{x,y} = g_e$. These equations describe the signals from O^- observed with crystals of alkali halides. The signal with $g_z = 2.012$, $g_{x,y} = 2.008$ from $\alpha-Al_2O_3$ exposed to gamma radiation was attributed to the ion radical O^-, produced at the expense of the lattice O^{2-} ion near a cation vacancy, that is, to a V center [4.94]. On this occasion the hole on the p orbital of the oxygen ion is attracted by the cation vacancy, which is charged negatively. The components of g tensor are then

$$g_{x,y} = g_e, \ g_z = g_e(1 - (2\lambda/E)) , \qquad (4.12)$$

where $\lambda < 0$ and thus $g_z > g_{x,y}$.

The EPR data available on the O^- ion radical allow one to conclude that there are two types of orientation of O^- in the axial field: (a) when the unpaired electron is localized on the p_z orbital of oxygen, which is directed along the $M^{n+}-O^-$ bond,

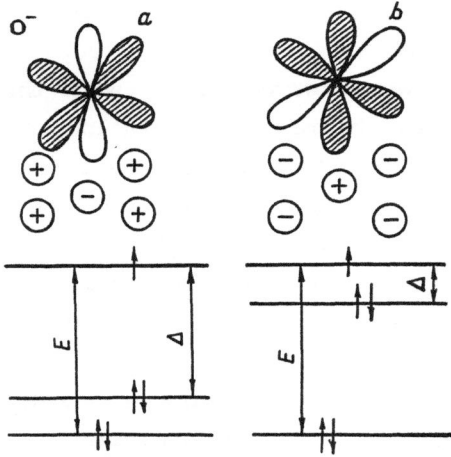

Fig.4.19. Energy levels of ion radical O^-: (a) σ type, (b) π type. The hatched orbitals are occupied by pairs of electrons

O^- being thus a σ-bound radical; and (b) when the unpaired electron is localized on the p orbital (p_x or p_y), which is normal to the M^{n+}-O^- bond, the bonding being thus of the π type. Both configurations and the respective band schemes are shown in Fig.4.19. The first-approximation expressions for the components of the g tensor were obtained in [4.74]:

$$g_y = g_1 = g_e, \quad g_z = g_2 \simeq g_e + (2\lambda/E), \quad g_x = g_3 = g_e + (2\lambda/\Delta) ; \tag{4.13a}$$

$$g_z = g_1 = g_e, \quad g_y = g_2 = g_e + (2\lambda/E), \quad g_x = g_3 = g_e + (2\lambda/\Delta) . \tag{4.13b}$$

The meaning of E and W is shown in Fig.4.19; z is the direction of the axial field.

The effect of crystalline structure on the EPR spectra of O^- can be monitored by taking signals from O^- (V centers) in different lattices. The V centers in the lattices of bivalent metal oxides (as well as in some other lattices) usually have spectra as in Fig.4.19a (σ bonding) [4.95,96]. The value of g_z for σ-bound O^- is about the same for all lattices and close to g_e. The most considerable are the changes in $g_{(x,y)}$; the value of $g_{x,y}$ decreases, as we go from SrO to BeO, from 2.093 to 2.021 - that is, the splitting E and W increase (4.13). This may be related to the increase in the polarizing power of the cation.

The π type of bonding of O^- was observed mainly with the lattices containing anions in tetrahedral configuration, O^- being produced when a hole is localized on one of the oxygen ions incorporated in the anion. It must be noted that with the π type of bonding the values of the g tensor depend little either on the formal charge of metal in the anion or on its radius. For example, $g_z = 2.057$ for WO_4^- in $CaWO_4$, whereas $g_z = 2.059$ for AlO_4^- in SiO_2. The values of g_y vary from 2.014 for VO_4^{2-} in $CaWO_4$ to 2.004 for SiO_4^{3-} in $SrSiO_4$; the values of g_x vary from 2.001 for $[WO_4-WO_3]^{2-}$ in $CaWO_4$ to 2.009 for VO_4^{2-} in the same matrix.

The adsorption of oxygen at relatively low temperatures (<300°C) usually gives rise to spectra of both O_2^- and O^-. In order to obtain a pure signal of O^-, the adsorption of O_2 on the oxide surface can be replaced by adsorption of N_2O with subsequent decomposition. The signal from O^- was observed in [4.97] after adsorption and decomposition of N_2O on partially reduced supported V^{5+}/SiO_2 catalysts. The O^- ion on V^{5+}/SiO_2 catalysts gives a single line with g = 2.025, which displays HFS due to ^{51}V nuclei (I=7/2). Hyperfine splitting indicates that O^- is stabilized near one V^{5+} ion, the low value of HFS pointing to the high ionicity of bonding.

Subsequently, the O^- ion was also detected directly after adsorption of O_2. In [4.83,98,99] the π-bound anion radicals O^- were detected on V/SiO_2, Mo/SiO_2, and W/SiO_2 ($g_z = 2.012-2.024$, $g_{x,y} = 2.020-2.030$). In the case of V/SiO_2 the adsorption of O_2 gives rise to O^- ions apparently at the expense of oxidation of V^{2+} ions. Study of the adsorption of O_2 on reduced Cr/SiO_2 catalyst led to the discovery of π-bound O^- with $g_{av} = 2.028$, produced at the expense of oxidation of Cr^{3+} ions:

$$\frac{\diagdown}{\diagup}Cr^{3+}(\square)_2 + O_2 \longrightarrow O^{2-}Cr^{6+}O^-.$$
$$\diagup|\diagdown$$

These anion radicals O^- are rather stable and endure heating up to 100°C [4.100].

The ion O_3^- was detected at adsorption of oxygen on gamma-irradiated MgO after decomposition of N_2O and appearance of O^- [4.101] or after direct adsorption of O_2 [4.102]. The parameters of the signal from O_3^- (g_1 = 2.0147, g_2 = 2.0120, g_3 = 2.0008) are close to the parameters of the signal from O_3^- in NaO_3 and NH_3 (g_{av} = 2.012) [4.103]. In CeX zeolite the signal of O_3^- ion, produced at the interaction of gaseous O_2 with O^- on the surface, had the following parameters: g_1 = 2.024, g_2 = 2.018, and g_3 = 2.030 [4.104]. The production of O_3^- at the low-temperature adsorption of O_2 was also observed by other authors. The adsorption of O_2 at -195°C on the Cr/SiO_2 catalyst, which already contains O^- ions (g_{av} = 2.028), destroys the existing signal and gives rise to a new anisotropic signal with g_z = 2.008 and $g_{x,y}$ = 2.003. This signal is ascribed to the complex $Cr \ldots O^- \ldots O - O$ of an angular appearance, enabling the $O - O$ fragment to rotate about the $Cr-O^-$ .. O bond [4.105]. The anion radicals O_3^- on the surface of V_2O_5/SiO_2 have a T-shaped structure with two of the three oxygen atoms being equivalent to each other [4.106]. Upon heating to room temperature O_3^- disintegrates into O^- and O_2 [4.101,106,107]. Sometimes the structures formed are even more sophisticated.

Adsorption of O_2 on TiO_2 and exposure to UV radiation give rise to an EPR signal (apart from the signal of O_2^-) with g_z = 2.008 and $g_{x,y}$ = 2.001, indicating high disparity of oxygen atoms. This signal was attributed in [4.77] to the O_3^{3-} radical formed in the reaction

$$O^{2-}(latt.) + O_2(gas) \xrightarrow{h\nu} O_3^{3-} + \oplus$$

this surmise, however, is challenged in [4.85].

At very low (liquid helium) temperatures the signals from pairs $O_2^- \ldots O_2^-$ were observed; depending on the separation between ion radicals and their respective orientation they form both the triplet and the singlet states [4.108]. On the surfaces of Y_2O_3, La_2O_3, and other rare-earth oxides the forms of O_2^- with Wg = 0.053 and 0.119 persisted to -195°C; they are assumed to belong to the radical pairs $O_2^- \ldots O_2^-$.

A weak form of adsorption of oxygen of kind $O_2^- \ldots O_2$ was observed in [4.109]. The formation of a weak complex of neutral molecular oxygen with the ion radical O_2^- was found to destroy the signal from O_2^-. Raising the temperature from -196°C to normal destroyed the complex; neutral oxygen was desorbed and the signal from O_2^- restored. The strength of bonding between the neutral molecule and the ion radical, as estimated, did not exceed 8 - 9 kJ/mol.

Of interest is the question regarding the relative stability of the main ion-radical forms of adsorbed oxygen. The adsorption of oxygen on oxide surfaces at room temperature results in ion radicals O_2^- and a certain quantity of O^-. One observes a variety of forms having different stability. The molecular forms of adsorbed oxygen usually do not stand the rise in temperature above 200°C. One can see that the maximum activation energy of desorption of oxygen from the state O_2^- [4.64,110,111] (or, according to some authors, the energy of transition from O_2^- to O^-, which results in the disappearance of the signal from O_2^-) corresponds to the level 0.8 - 1.1 eV in the forbidden band of the oxide. The activation energy of desorption of O_2^- from the (1010) face of a ZnO single crystal, as estimated from the measurements of electroconductivity, equals 1.13 eV [4.62].

As reported in [4.112], the signal from O_2^- on ZnO with g_1 = 2.049, g_2 = 2.008, and g_3 = 2.002 arises upon adsorption of oxygen at 25°C. This signal disappears entirely at 170°C, replaced by a singlet with g = 2.002, which was attributed to the O^- ion (adsorbed). It ought to be observed that the parameters of this signal do not coincide with those known to be typical of O^-. In our view, this signal should rather be attributed to some lattice defect, which becomes manifest on adsorption of oxygen. Thus the disappearance of the only form

190

of oxygen (O_2^-) on ZnO at 170 0C is accompanied by the desorption of oxygen; in other words, the equilibrium O_2^- ET $O_2 + e^-$ is maintained at least to some extent.

The EPR spectra of radical forms of oxygen, produced on the adsorption of O_2 on partly reduced MoO_3/AL_2O_3 catalysts, were obtained in [4.66]. The total number of particles detected by EPR constituted only a small fraction of the total amount of the oxygen adsorbed. The signal, attributed to O_2^- stabilized on an anion vacancy (g = 2.039), appears at low-temperature chemisorption and persists to 150 0C. The EPR signal from Mo^{5+} ions remained unchanged at low-temperature adsorption and decreased at the high-temperature adsorption.

It must be noted that it was the equilibrium adsorption that was studied in all the temperature ranges. It is assumed in [4.66] that the low-temperature adsorption of oxygen involves the oxidation of Mo^{4+} ions, whose signals suffer full broadening, while the high-temperature adsorption is associated with the oxidation of Mo^{5+} ions. The observed regularities relate to the high values of coverage (20 to 80% of a monolayer) of Al_2O_3 by MoO_3. At low temperatures the adsorption of oxygen does not take place on the Mo^{5+} ions. At high temperatures (above 300 0C) the dissociative adsorption is reversible:

$$O_2(gas) + 4e^-(2e^-) \rightleftarrows 2O^{2-}(2O^-)(ads.).$$

This evidently agrees with all the above-cited works. As to the low-temperature chemisorption, the assumption regarding the production of O^- made in [4.66] seems to have little credibility. In our view, this process can be represented as

$$O_2 + 2Mo^{4+} \longrightarrow [O_2]^{2-}Mo_2^{5+} \rightleftarrows O_2^-Mo^{5+} + Mo^{4+} ,$$

$$O_2^-Mo^{5+} \longrightarrow O_2 + Mo^{4+} ,$$

$$
\begin{array}{ll}
\begin{array}{c}
O-O \\
| \quad | \\
Mo-Mo
\end{array} \text{(a)} &
\begin{array}{c}
O\cdot \ \ O\cdot \\
| \quad | \\
Mo-Mo
\end{array} \text{(b)} .
\end{array}
\qquad (4.14)
$$

The first reaction apparently results in a complex of type (4.14a) which is more advantageous energetically than the complex of type (4.14b), since the transition from (a) and (b) requires rupturing the O - O bond in the oxygen molecule, and the one-electron transitions resulting in O_2^- and desorption of oxygen are more favorable than the two-electron transition

$$[O_2]^{2-}-Mo_2^{5+} \rightarrow Mo_2^{4+} + O_2(gas) .$$

The creation of neutral molecular forms of O_2 upon adsorption was assumed on the basis of IR spectroscopic data. For instance, the adsorption of O_2 on TiO_2 specimens heat-treated in vacuum at 300°C gave rise to absorption bands at 1630, 1650, and 1680 cm^{-1}, attributed in [4.113] to the vibrations of the O = O bond in neutral molecular oxygen; these bands disappeared at 100°C. It was also reported in [4.113] that the band at 1180 cm^{-1}, characteristic of O_2^-, does not appear in the IR spectra of highly reduced specimens, and neither does the EPR signal from O_2^-. The absorption ν_{0-0} band is known to fall within the range 800 - 900 cm^{-1} for peroxide radicals of type O_2^{2-}, 1580 cm^{-1} for molecular oxygen O_2, and 1880 cm^{-1} for the ion radical O_2^-. The comparison leads the author of [4.113] to the conclusion that the oxygen on TiO_2 either occurs in a neutral molecular form or (less probably) has a weak positive charge.

The adsorption of O_2 at -195°C on the Cr/SiO_2 catalyst gave rise to a new signal - against the background of the EPR spectrum of Cr^{5+} ions in tetrahedral coordination - with $g_z = 1.95$, $g_{x,y} = 1.92$, and $B(^{53}Cr) = 25$ Oe. This was attributed in [4.114] to the complex of low-coordination surface Cr^{5+} ion with the molecular oxygen. The unpaired electron here is localized chiefly on the Cr^{5+} ion, while the O_2 molecule occurs in the complex in a diamagnetic (singlet) state.

A new rapidly developing branch of investigation is the study of the production and reactions of excited singlet molecular oxygen $^1\Delta_g$ on the surface of oxides. The authors of [4.115] observed a striking dissimilarity in the adsorptive properties of the common (triplet) and the singlet $^1\Delta_g$ states of oxygen. The former is adsorbed by MgO with a high value of activation energy, whereas the latter has an almost-zero activation energy. The production of singlet oxygen on the surface with subsequent desorption into the gaseous phase was observed with Li-Sn-P oxide catalyst [4.116].

A novel phenomenon - the electron transfer to the O_2 molecule from a remote electron-donor center in partly reduced oxide catalyst - was observed by us in [4.51,64,117-129]. We have studied the catalysts MoO_3, WO_3, and V_2O_5 supported on Al_2O_3, MgO, and BeO, and the solid solutions $Co_xMg_{1-x}O$ and $Ni_xMg_{1-x}O$. The data on the charge and the coordination of the reduced forms in these systems are given in Sect.4.1. The results concerning the EPR spectra of the ion radical O_2^- and the transition ions of Mo, V, and W supported on Al_2O_3, MgO, and BeO, and the Co ions in $Co_xMg_{1-x}O$ at the adsorption of oxygen at -196 and +25°C are compiled in [4.64]. In brief, the results lead to the conclusions in the following paragraphs.

The oxygen is adsorbed in the form O_2^- at the expense of the electron transfer from a donor ion (V^{4+}, Mo^{5+}, W^{5+}, Co^{2+}). However, the adsorption of oxygen by all partly reduced catalysts studied (except W/Al_2O_3), does not result in the stabilization of the ion radical O_2^- on the donor ion. The anisotropy of the g factor of signals from O_2^- in this case is exactly the same as Δg_z of signals from O_2^- on pure radiation-treated carriers ($\Delta g_z = g_z - g_e$); $g_z = 2.070$ for MgO supported systems, $g_z = 2.039$ for Al_2O_3-supported systems, and $g_z = 2.038$ for BeO-supported systems. Hence it follows that in almost all systems in the range of temperatures from -196 to +25°C the electron is transferred from the donor ion (transition-metal ion M^{n+}) to the oxygen being adsorbed, the latter subsequently stabilizing in the form O_2^- on the cation of the carrier lattice

$$M^{n+} + O_2 + \text{cation} \longrightarrow M^{(n+1)+} + O_2^-/\text{cation} ,$$

and some of this oxygen persists to $+25^{\circ}C$. The low-temperature adsorption results in a weakly bound form O_2^-, stabilized on the donor ion in the systems V/BeO, Mo/BeO, Mo/Al$_2$O$_3$, Co$_x$Mg$_{1-x}$O, and Mo/Al$_2$O$_3$ (g_z = 2.019 for Mo^{6+} and W^{6+}, g_z = 2.050 for Co^{3+}; the signal persists at $-196^{\circ}C$).

The square-pyramidal Mo^{5+} in Mo/MgO gives an electron to oxygen in a reversible way, while the electron transfer from Mo^{5+} in Mo/Al$_2$O$_3$ is irreversible; the ion radical O_2^- is stabilized on Mo^{6+} at $-196^{\circ}C$ and on Al^{3+} at $+25^{\circ}C$; the Mo^{5+} ion in Mo/BeO donates an electron to oxygen at $25^{\circ}C$ but not at $-196^{\circ}C$. Square-pyramidal V^{4+} (vanadyl VO^{2+}) in V/MgO and V/Al$_2$O$_3$ does not donate an electron to oxygen at $-196^{\circ}C$ while the same in V/BeO donates irreversibly. Tetrahedral Mo^{5+} in Mo/MgO and Mo/Al$_2$O$_3$ transfers an electron to oxygen at $-196^{\circ}C$ but not at $25^{\circ}C$; tetrahedral V^{4+} in V/MgO and V/Al$_2$O$_3$ donates an electron irreversibly at $-196^{\circ}C$ and $+25^{\circ}C$; and V^{4+} in V/BeO also gives an electron to oxygen irreversibly.

We see that on the surface of the three carriers studied the isolated d^1 ions (Mo^{5+}, V^{4+}, W^{5+}) in square-pyramidal coordination do not act as centers of strong adsorption and in most cases do not donate electrons to adsorbate oxygen (VO^{2+} in V/MgO and V/Al$_2$O$_3$); in other words, their adsorptive power is low.

The appearance of the signal from O_2^- upon adsorption of oxygen on Mo/Al$_2$O$_3$ at $25^{\circ}C$ and the increase in the signal from Mo^{5+} are interrelated. We have assumed [4.118] that on the surface of Mo/MgO and Mo/Al$_2$O$_3$ there exists a reduced form of molybdenum ions not detected by EPR; namely, the Mo^{4+} ions which donate electrons to oxygen molecules and oxidize to Mo^{5+}; at the same time, the O_2^- ion radical at room temperature cannot be stabilized on either Mo^{5+} or Mo^{6+} and goes to a remote lattice cation. (The diffuse reflection spectra of Mo/MgO do actually exhibit a band at 13.600 cm^{-1}, which may be attributed to Mo$_{T_d}^{4+}$).

There also exist forms of Mo^{5+} not detected by EPR because of exchange interactions: $Mo^{5+}-O-Mo^{5+}$, $Mo^{5+}-O-e^-$, as well as other antiferromagnetic or paramagnetic pairs, whose EPR signal is fully broadened already at $-196°C$. The authors of [4.66] point to an increase in the signal from Mo^{5+} together with the appearance of the signal from O_2^- ($g_z = 2.035$, $g_{x,y} = 2.010$) at the adsorption of oxygen on Mo/Al_2O_3, as well as an increase in the signal from Mo^{5+} with the rise in the temperature from -170 to $-150°C$; these features were associated with the Mo^{4+} ion:

$$Mo^{4+} + \square \longrightarrow Mo^{5+} + [e^-] \quad , \quad O_2 + [e^-] \longrightarrow O_2^- \square .$$

However, the ion Mo^{5+}, directly associated with the vacancy which has captured an electron, may not show up in the EPR spectra, since both these particles are paramagnetic. This empirical finding could have been explained by the existence of a pair $Mo^{5+} - O - Mo^{5+}$ next to a vacancy. Then the transition of an electron to the vacancy might expose one of the two Mo ions in the EPR spectrum:

$$Mo^{5+}-O-Mo^{5+} + \square \longrightarrow Mo^{5+}-O-Mo^{6+} + [e^-].$$

According to our XPS data [4.120], the action of oxygen on a Mo/Al_2O_3 catalyst at room temperature results chiefly in the transition $Mo^{5+} \to Mo^{6+}$, while the intensity of the Mo^{4+} signal remains about the same.

Study of the behavior of Mo and V ions on the surface of reduced catalysts and the interconversions of various forms of adsorbed oxygen allows one to assume that even at small surface concentrations of Mo and V the latter species form clusters (Sect.3.1), which are responsible for the pecularities in the adsorption of oxygen in various radical states. Clustering is also observed with the solid solutions $Co_xMg_{1-x}O$ and $Ni_xMg_{1-x}O$.

We have estimated the size of clusters formed on V and Mo catalysts, supported on Al_2O_3 and MgO, having studied the

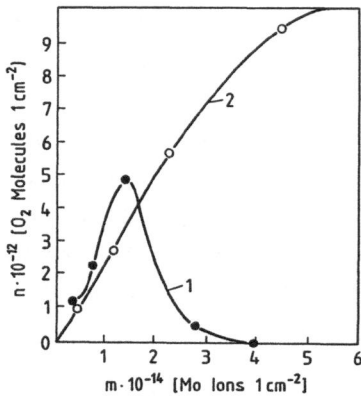

Fig.4.20. The quantity of adsorbed oxygen versus concentration of MoO_3 on MgO (1) and Al_2O_3 (2) [4.127]. The sample was heat-treated at 500°C in oxygen and in vacuum

adsorption of oxygen using the techniques of adsorptive micro-calorimetry and EPR [4.127].

Figure 4.20 shows the quantity of oxygen adsorbed at 25°C as a function of concentration of supported partly reduced MoO_3. The adsorption of O_2 on MoO_3/MgO (Curve 1) is greatest at a the concentration of $1.5 \cdot 10^{13}$ cm^{-2} (6% MoO_3 by mass), while at the concentration of $4 \cdot 10^{14}$ cm^{-2} the adsorption of oxygen is almost zero. The adsorption of oxygen on MoO_3/Al_2O_3 (Curve 2) goes up steadily with the increase in the concentration. The peculiar shape of Curve 1 is due to the fact that at low concentrations the Mo ions tend to cluster, while at high concentrations a regular molybdate is formed, which resists reduction and is incapable of adsorbing oxygen. By contrast, the aluminum molybdate (the creation of which is confirmed by the diffuse reflection spectra, Sect.3.1.6) is reduced easily, after which the Mo/Al_2O_3 becomes capable of adsorbing large amounts of oxygen.

The plot in Fig.4.21a shows the experimental values of the heat of adsorption of oxygen Q as a function of surface coverage by adsorbate for MoO_3/MgO specimens with varying concentrations of MoO_3. A similar plot for MoO_3/Al_2O_3 specimens is presented in Fig.4.21b. The heat of adsorption decreases as the coverage increases, and is the greater the higher the concentration of MoO_3 on Al_2O_3. The curve for MgO as a carrier displays a maximum. At the start of adsorption the values of the

196

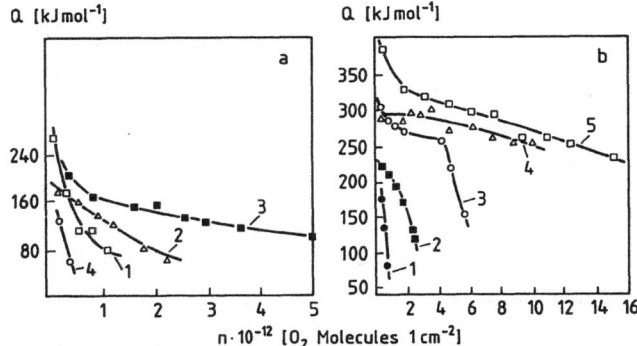

Fig.4.21. The heats of adsorption of O_2 [4.127]: (a) MoO_3/MgO, (b) MoO_3/Al_2O_3; mass percentage of MoO_3: (1) 1%, (2) 3%, (3) 6%, (4) 10%, (5) 15%

heat of adsorption are high. As the coverage of MoO_3/MgO by oxygen increases (at Q = 140 –170 kJ/mol), the EPR spectrum starts to display an anisotropic signal with g_z = 2.039 and $g_{x,y}$ = 2.009, pertaining to the ion radical O^- and the surface anion vacancy (V center). This ion radical reacts with CO. A further increase in coverage (at Q = 130 kJ/mol) gives rise to the ion radical O_2^-, stabilized on Mg^{2+} (g_z = 2.094 and 2.068, $g_{x,y}$ = 2.009) [4.130]. With MoO_3/Al_2O_3 at low concentrations of MoO_3 the signal comes from the ion radical stabilized on Al^{3+} (g_z = 2.038, $g_{x,y}$ = 2.008), while with 10 to 15% of MoO_3 the signal comes from O_2^- ... Mo^{6+} (g_z = 2.023, $g_{x,y}$ = 2.008) [4.84].

After adsorption of O_2 the intensity of the EPR signal from Mo^{5+} decreases by 15 to 20%. This reduction corresponds to about 20% of the total number of O_2^- ions produced at the adsorption of oxygen. Apparently, most of the molybdenum ions which take part in the chemisorption do not show up in the EPR spectrum, while some of the Mo^{5+} ions, detected by EPR, take no part in chemisorption.

The amount of oxygen, adsorbed after the appearance of the EPR signal from O_2^-, is similar – and in the case of 10% MoO_3/MgO very similar – to the total amount of O_2^- as calculated from the EPR spectrum. The mean heat of adsorption of

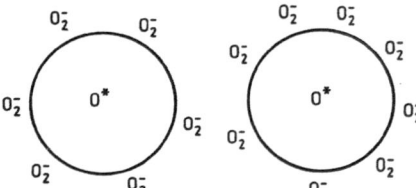

Fig.4.22. Adsorption of oxygen by a cluster on supported catalysts: O^* is oxygen adsorbed inside the cluster, not manifest in the EPR spectrum

oxygen on such a specimen (i.e., adsorption in the form O_2^-) is about 85 kJ/mol. Similar values were obtained for the heat of adsorption of oxygen in the form O_2^- by other supported catalysts, as well as by the solid solutions $Co_xMg_{1-x}O$ [4.124,125]. The corresponding value for O^- is above 140 kJ/mol. The values of 180 –200 kJ/mol and higher pertain to the strong nonparamagnetic forms of adsorption (O^{2-}, O_2^{2-}).

As we have pointed out in Sect.3.1, the creation of clusters on supported catalysts can be deduced from the existence of the lower concentration threshold in the reduction of MoO_3 and V_2O_5, and for vanadium ions also from HFS in the EPR spectra due to the interaction between the ions VO^{2+} ... V^{5+}. The production of O_2^-, detected in the EPR spectrum and localized on the ions of the carrier (Mg^{2+}, Al^{3+}), evidently may take place only on the border of the cluster at the expense of electron transfer from peripheral donor ions (Mo^{5+}, V^{4+}) to the O_2 molecule (Fig.4.22). The oxygen adsorbed within a cluster gives no EPR signal, since the concentration of paramagnetic particles inside the cluster is high, and therefore the exchange interaction is high, too.

Assuming that the oxygen not detected by EPR is adsorbed within the cluster, whereas the ions O_2^- ... Mg^{2+} and O_2^- ... Al^{3+} are found on the cluster border, we can obtain estimates for the size of the cluster. If the cluster is circular and the sites of adsortion of O_2 are distributed evenly in the cluster and on its border, we get

$$\frac{S}{\ell} = \frac{(x-y)a^2}{ay} = a\,\frac{x-y}{y} = \frac{\pi d^2}{4\pi d} = \frac{d}{4} = \frac{an}{4} , \qquad (4.15)$$

198

where x is the total amount of oxygen adsorbed, y is the amount of O_2^-, S is the area of the cluster, ℓ is its circumference, d is the cluster diameter, a is the separation between adsorptive sites, and n is the number of adsorptive sites across the diameter of the cluster.

Hence we can estimate the relative size of the cluster: $n = 4(x-y)/y$. The size of cluster n on MoO_3/MgO is 7 to 16 atoms of Mo across, and remains about the same as the concentration of MoO_3 increases. In the case of MoO_3/Al_2O_3 the increase in the concentration of MoO_3 results in an increase in not only the number, but also the size of clusters (n goes up from 3 to 106 Mo atoms), which – with the concentrations over 6% by mass – leads to the formation of aluminum molybdate $Al_2(MoO_4)_3$. Then there are no free lattice cations on the surface, and the adsorption of oxygen in the form O_2^- ... Al^{3+} does not take place.

The mean size of crystals, according to X-ray diffraction measurements, is 24±7 nm for the MoO_3/MgO and V_2O_5/MgO specimens studied and 26±8 nm for MoO_3/Al_2O_3 [4.127]. Assuming the distance between the adsorptive sites inside the cluster to equal 1 nm (which is the separation between Mo atoms in molybdates), the maximum diameter of a cluster on MoO_3/Al_2O_3 (6%) and V_2O_5/MgO (19%) is 16 nm, and 110 nm for MoO_3/Al_2O_3 (6%). Apparently the growth of the nucleus of the surface phase is restricted by the size of microcrystals of MgO. As to MoO_3/Al_2O_3, the surface clusters seem to be unrestricted by the dimensions of microcrystals of the carrier and grow beyond their limits.

Very informative is the low-temperature adsorption of oxygen by supported catalysts at –196ºC in the form O_2^- with g_z = 2.026. We presume this form of oxygen to be an intermediary in the electron transfer from the ions of Mo or V to the molecular oxygen O_2^-/Mg^{2+} (g_z = 2.090). This is suggested by the change in the strength of signals with g_z = 2.026 and 2.090 with successive admissions of oxygen at –196ºC.

The invariability of the signal with g_z = 2.026 for W/MgO, V/MgO, and Mo/MgO implies that this signal comes from the adsorbed oxygen rather than from the vacancies, because the signals from vacancies in MgO are usually very sensitive to changes in the structure and charge of the foreign ion, which is responsible for these vacancies [4.131]. Furthermore, the form of oxygen from which this signal comes does not interact with hydrogen at -196°C. Since it is known that the atomic ion radical O^- becomes easily engaged in the interaction with hydrogen already at -196°C, the signal with g_z = 2.026 certainly cannot be ascribed to O^-(ads.) [4.132]. We presume this signal to be produced by the O_2^- radical, stabilized on the lattice oxygen ions near the ion of the transition metal. It is known that the irradiation of molybdates and tungstates usually gives rise to signals from stabilized holes (V centers), which are smeared over the oxygen orbitals of the polyanion $[Mo_2O_8]^{3-}$ or $[W_2O_8]^{3-}$. Such anions near defects do actually exhibit donor properties and are capable of giving away an electron (capturing a hole), becoming V centers. The low-temperature adsorption of oxygen resulting in O_2^- ions stabilized on the vanadyl anions was also reported in [4.133].

The activation energy of desorption of weakly bound O_2^- (g_z = 2.026) is 6 and 8 kJ/mol for Mo/MgO and V/MgO, respectively. The form of O_2^- with g_z = 2.026 is intermediate in the sequence leading to stabilization on the cation (Z is a donor center):

$$O_2 + Z \xrightarrow{-196^{\circ}C} O_2^- \ (g_{\parallel} = 2.026)Z^+ \xrightarrow{O_2;\ -196^{\circ}C} Z^+ \quad O_2^-/Mg^{2+}.$$

Appropriate calculations point to the possibility of the Coulomb interaction of the ion radical O_2^- with the polyanion $[M_n^{z+}O_m^{2-}]^{(2m-nz)-}$ over oxygen, provided the charge Z^+ of the metallic ion in the polyanion is large enough (>5). The parameters of EPR spectra of such forms in some cases are the same as the parameters of the spectra of strongly bound forms. This indicates that the strength of the bonding between O_2^- and the adsorptive site has little effect on the EPR parameters of this radical. The weakly bound form of O_2^- may occur at the intermediate stages of deep catalyzed oxidation of hydrocarbons and CO. We shall raise this matter again in Sect.4.7.

In addition, let us consider the signals from O_2^- on Mg^{2+}. The signals from Mo/MgO, V/MgO, and W/MgO have g_z = 2.070, 2.090, and 2.080; the values of g_1, g_2, and g_3 of these forms are very close. The signal with g_z = 2.070 is observed at both -196^oC and $+25^oC$. The signals with g_z = 2.080 and 2.090 arise after adsorption at -196^oC and are much stronger than the signal observed upon adsorption at room temperature. This may be due to the rearrangement of the adsorptive site, O_2^- with g_z = 2.026 serving as an intermediate. As with Mo/Al_2O_3, this rearrangement takes place at low temperatures, when the lifetime of O_2^- with g_z = 2.026 is sufficiently long. Even more symptomatic is that these signals (2.090 and to a lesser extent, 2.080) arise only after reduction of the specimen by hydrogen.

The change in g_z we relate to the change in the effective charge of Mg^{2+} caused by reduction in hydrogen. In the course of the reduction some of the hydrogen adsorbed at 500^oC forms hydroxyl groups near Mg^{2+} ions of the complex adsorptive site; thereupon both the coordination and the effective charge of Mg^{2+} ions suffer a certain change:

$$H\diagdown O \diagdown Mg^{2+} \cdots O^{2-}.$$

Essential for the understanding of the phenomena associated with the catalytic action is the fact that the adsorbed oxygen affects the adsorptive properties of the adjacent donor sites. For instance, the Mo/MgO specimen showed an increased adsorptive capacity at 25^oC after adsorption at -196^oC, as compared with the fresh catalyst. After adsorption of oxygen at 25^oC, the form with g_z = 2.026 fails to appear on Mo/MgO, V/MgO, and W/MgO at -196^oC, which suggests the complete reconstruction of

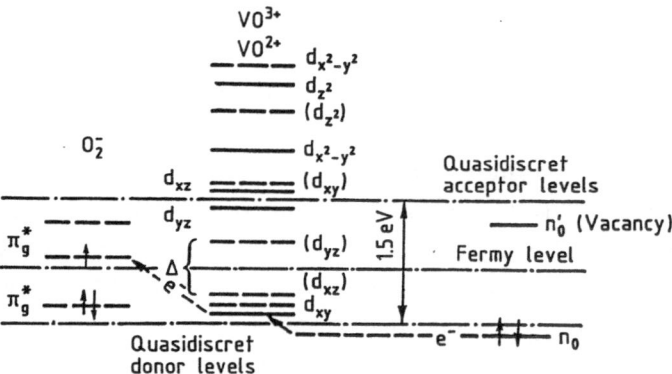

Fig.4.23. The band model of a portion of MgO surface containing a donor site V^{4+} showing adsorption of oxygen. Bracketing refers to the position of levels after oxygen adsorption. The energy in eV counted from the conductivity zone edge is drawn along the y-axis

the surface and the reduction in the number of nonbonding orbitals on the surface available for the adsorption of O_2.

This adsorptive "hysteresis" can possibly be explained by the change in the cluster size caused by the change in the temperature. This change may be retarded by adsorption. A similar phenomenon is discussed in Sect.4.6 with reference to the adsorption of CO by diluted solid solutions.

The energy-band scheme allowing one to explain the processes of adsorption of oxygen on V/MgO and Mo/MgO catalysts was constructed in [4.64] (Fig.4.23). This scheme is based on the band model of the MgO surface with a donor site VO^{2+} ... V^{5+} (Sect.3.5.2, Fig.3.22) and the experimental findings, which indicate that oxygen adsorbed on oxides as O_2^- creates a local level 0.7 to 1.0 eV above the bottom of the conduction band. As already indicated, this value is rather close to the tentative energy of activation of the desorption of oxygen adsorbed in molecular form (90 to 170 kJ/mol or 0.9 to 1.1 eV).

The O_2^- ion becomes polarized by the field of the adsorptive site, and the π^*_g level (0.9 eV below the narrow band of quasidiscrete acceptor levels) is split into $\pi^*_{g_x}$ and $\pi^*_{g_y}$. The magnitude of splitting Δ is assumed to be 0.5 – 1.0 eV, the same as the polarization shift of the π orbital in the square-pyramidial V^{5+} ion, produced from VO^{2+} after the capture of an electron by oxygen. We presume also that the site has not had time to rearrange itself, and the set of vanadyl levels persits, now the d_{xy} orbital is vacant. The inclusion on O_2^- into the coordination sphere of vanadium may change the order of the unoccupied t_{2g} orbitals (d_{xy} above d_{xz} and d_{yz}); d_{xz} and d_{yz} are also split by Δ. Thus we get a set of levels shown in the diagram by dashed lines. This set is unstable, because the unpaired electron of O_2^- occupies the level above the vacant levels of V^{5+} (VO^{3+}), which leads to desorption of oxygen.

Should the d_{xz} level (Fig.4.23) be close enough to the narrow band of quasi-discrete donor levels (which here is a counterpart of the valence band), it may accept an electron from this band even at low temperatures (dashed arrow in the diagram). This will make the adsorptive system (oxygen + the adsorptive site) more stable, since the vacant level d_{yz} now lies below the oxygen's level. The O_2^- ion, however, will not show up in the EPR spectrum (a pair of ions O_2^- – V^{4+}); what we observe is the signal from a V center, since now there exists an unpaired electron on the level n_0 of the nonbonding oxygen orbitals in the polyanion. This bonding is also rather loose, because the oxygen level lies close to the vacant d_{yz} level of the vanadyl ion and the electron may go into the upper "band" of the local levels, which will result in desorption of oxygen. Having got into the upper set of levels, the electron may jump from one V ion to another, and be eventually captured by a vacant trap (n_0^+) or by a neutral molecule. The latter will then stabilize in the form of O_2^- on any available site, which may be removed from the donor center (e.g., on a vacancy $\square_{Mg}^{2+}$).

In all probability, the stabilization of O_2^- on the donor ion d^1 itself is possible only if, during the lifetime of oxygen on this ion (or directly at the instant of adsorption), the site's levels are reconstructed in such a way that the vacant ones are much above the oxygen level, and the nonbonding oxygen levels of the same site are occupied. Such rearrangements involve the displacement of centers of atoms which form the defect, so they must involve the lattice phonons. Apparently, the phonons must be of appropriate symmetry and energy (Sect.4.2). The required energy comes from the adsorbed oxygen; in most cases (for donor–type centers) this is only possible at low temperatures.

All this can be visualized in the following simplified way: at the instant of adsorption the system O_2^- + adsorptive site is in a excited state; its stabilization requires dissipating the energy of the excitation into the solid or elsewhere. This being impossible, the oxygen is desorbed (which is similar to the reactions in gaseous phase):

$$A + B + M \rightarrow [AB]^* + M \rightarrow AB + M^* .$$

Adsorption takes place if

$$O_2 + Z \rightarrow [O_2^-Z^+]^* \quad \hbar\omega' \quad [O_2^-Z^+] + \hbar\omega'' ,$$

where ω is the phonon frequency ($\omega' < \omega''$), $\hbar\omega$ being the phonon energy.

If the phonon spectrum of the catalyst is capable of supplying phonons with frequencies $\omega_1 \neq \omega'$, whose symmetry facilitates the electron transitions from O_2^- to the quasi-discrete acceptor levels rather than the reconstruction of the site, then the oxygen molecule is desorbed.

Detailed reviews on the generation and structure of atomic and molecular oxygen ions on oxide surfaces were published recently by CHE and TENCH [4.134].

4.5 Adsorption of Olefins on Oxides

Detailed knowledge about the mechanism of adsorption of oxygen was gained with the aid of the EPR technique which presently is the most sensitive and informative method of studying surface compounds.

Now we shall take another example: the adsorption of olefins by oxides. The main information about the mechanism of this kind of adsorption was supplied by IR spectroscopy. The theory of IR spectroscopy of adsorbed molecules is less advanced than the theory of the EPR, which leads to certain difficulties in interpreting the IR spectra. The IR spectra of adsorbed molecules are interpreted mostly by analogy with the spectra of homogeneous compounds and complexes of similar composition. For this reason the proposed mechanisms of adsorption of olefins are more controversial than the mechanisms of adsorption of oxygen, derived from EPR spectroscopy.

4.5.1 Ethylene

Of all the olefins, ethylene forms the weakest bond at the low-temperature adsorption by oxides. Only the initial portions of adsorbed ethylene exhibit elevated values of the heat of adsorption (sometimes to 100 kJ/mol), which is usually due to side reactions (oxidation or polymerization of C_2H_4). In general, however, the adsorption of ethylene at room temperature is rapid and reversible. The heat of adsorption of ethylene Q is 35 kJ/mol for NiO [4.135], about the same for the Al-Cr-K-O catalyst [4.136], and somewhat higher for ZnO [4.137] and certain zeolites [4.138].

Considering the low values of the heat of adsorption, it was presumed [4.136] that at low temperatures ethylene is adsorbed physically. However, comparison of the adsorption of ethylene (bp-104°C) and ethane (bp-89°C) on ZnO indicates that at 0°C and 100 Pa the adsorption of ethylene is by two orders of magnitude greater than that of ethane [4.139], whereas quite the opposite ought to be expected from physical adsorption. The heat of adsorption of ethylene is anyway much higher than the heat of condensation (15 kJ/mol).

Evidently the adsorption of ethylene is chemical, but the bonding is weak. The appropriate kind of bonding is the π bond of the metallic atom on the surface with the olefin:

$$M \cdots \begin{matrix} C \\ \| \\ C \end{matrix}$$

(a π complex).

The formation of π complexes in homogeneous systems, in homogeneous catalysis in particular, has been studied quite well (Sect.4.1, Fig.4.5). The stability of a π complex relies on the donor-acceptor bonds (interaction of the unoccupied p_z and d_{z^2} orbitals of metal with the occupied orbitals of the ligand) and back donation (interaction of the unoccupied antibonding π^* orbitals of metal with the occupied orbitals of the olefin ligand). Back donation (dative bonds) leads to the production of stronger complexes.

The literature contains an abundance of conflicting opinions regarding the relative strength of bonding in π complexes. There is no way of giving an unambiguous explanation for the metal's tendency to form back donation and donor-acceptor bonds. Apart from depending on the number of d electrons it would depend on the number of ligands, as well as on their particular configuration and electron structure (this also applies to ligands not taking part in π interaction), on the configurational symmetry, etc. In general, the element's capability for back donation is the greater, the lower its degree of oxidation; for isoelectronic complexes (within the same period) this capability increases with the rise in the negative charge of the complex.

The olefins form π complexes with a large number of metals, the π bonding being especially stable in planar square complexes having $4d^8$ and $5d^8$ electron configuration (Rh^+, Ir^+, Pd^{2+}, Pt^{2+}). The complexes with metals of the first long period are generally less stable [4.140].

The IR spectra of olefinic complexes point to a considerable lowering of the multiplicity of the carbon–carbon bond. In gaseous ethylene the stretching vibrations of the C = C bond produce the band ν_2 = 1621 cm^{-1}, forbidden in the IR spectrum and active in the Raman spectrum. In the spectra of ethylene-metal complexes this band is sometimes shifted by 100 cm^{-1} and becomes active in the IR spectrum. In the spectrum of the Zeise salt KPtCl$_3$(C$_2$H$_4$) these vibrations produce the band at 1518 cm^{-1}. The internuclear distance C – C in a complex is much greater (0.150 nm in the Zeise salt) than in a free ethylene molecule (0.134 nm). The lowering of the multiplicity of the bond is caused both by the decrease in the electron density on the bonding π orbtial of the olefin (donor–acceptor interaction) and by the transfer of electrons from the orbitals of the metal to the antibonding π^* orbital of the olefin (back donation).

The creation of π complexes upon adsorption of ethylene on solid catalysts was reported in a number of works dealing with IR spectroscopy of adsorbed molecules.

Analysis of the IR spectra and the heat of chemisorption Q for the adsorption of ethylene on zeolites AgX, CdX, LiX, NaX, CaX, KX, and BaX reveals the existence of a correlation between the heat of chemisorption Q and the shift of the ν_2 band (stretching vibrations of the C–C bond) with respect to the frequency of C–C vibrations in gaseous ethylene [4.138]. The shifting in ν_2 is much smaller than the shifting in the spectra of homogeneous π complexes; it is the greatest with the AgX zeolite (52 cm^{-1}) and decreases in the decreasing order of values of the heat of chemisorption. The authors of [4.138] relate the difference between CdX and AgX to the fact that Cd^{2+} has a smaller ionic radius (0.098 nm) than Ag$^+$ (0.126 nm). Upon adsorption on CdX the occupied π orbitals of olefin overlap with the 5s orbitals of Cd (and possibly also with the 5sp orbitals). In the case of AgX a back donation d-π^* bond is also created at the expense of the overlap of the unoccupied π^* orbitals of olefin with the 4d orbitals of the metal.

The correlation between the frequency shift $\Delta\nu_2$ of C = C vibrations in a π complex of ethylene and the charge of the cation was reported in [4.141]: $\Delta\nu_2 = 20 - 30$ cm^{-1} for doubly charged cations (Zn^{2+}, Cd^{2+}), $\Delta\nu_2 = 50 - 70$ cm^{-1} for singly charged cations (Cu^+, Ag^+), and $\Delta\nu_2 = 100 - 150$ cm^{-1} for "zero-valence" metals (Cu, Ag, Au, Ni, Pd). The greater the contribution of back donation, the greater the frequency shift $\Delta\nu_2$. This correlation, however, contradicts the findings of [4.142]. SELEZNEV and KADUSHIN [4.142] studied the adsorption of ethylene by zeolites CuY, CoY, FeY, AgY, NiY, and HY at room temperature. The spectra of all the zeolites studied display absorption bands at 1400 - 1500 cm^{-1}, typical of bending vibrations of CH_2 groups, observed also with C_2H_4 in the gaseous phase (1450 cm^{-1}). The band at 1450 cm^{-1}, associated with physically adsorbed ethylene, is supplemented by the band at about 1420 cm^{-1}, which pertains to the same vibrations in the ethylene molecule bound to the transition-metal ion. It must be observed that, unlike NaY, other zeolites adsorb ethylene rather strongly (the band at 1420 cm^{-1}), so that ethylene is desorbed from the surface only at 150°C. Heat treatment of specimens in ethylene results in an increase in the intensity of the 1420 cm^{-1} band and a decrease in the intensity of the band of weakly bound ethylene (1450 cm^{-1}).

Especially thorough IR spectroscopy investigations were carried out by KOKES and DENT [4.137,139,143-145] for adsorption of olefins on ZnO. Their results are widely employed for interpreting the spectra of olefins adsorbed on the oxides of transition metals. Let us consider more closely the experimental results of Kokes and Dent for the adsorption of C_2H_4 by ZnO at room temperature and a pressure of 13.3 kPa.

No bands arise in the region of Zn-H and O-H valence vibrations; this points to nondissociative adsorption of ethylene. A strong band at 2984 cm^{-1} and weaker bands at 2993, 3055, and

Table 4.4. Comparison of IR spectra of gaseous, chemisorbed, and π-complexed ethylene [cm^{-1}] [4.139]

Type of vibration	C_2H_4 (gas)	C_2H_4 (ads.on ZnO)	π complex $KPtCl_3(C_2H_4)$
ν_9	3105[a]	3140–3122[c]	3094
ν_5	3044[b]	3055–3078[c]	3079
ν_1	3008[b]	2993–3000[c]	3013
ν_{11}	2980[a]	2984	2988
ν_2	1621[b]	1600	1518
ν_{12}	1455[a]	1458	1426
ν_3	1342[b]	1451	1419

[a] Infrared spectrum
[b] Raman spectrum
[c] The range of variations in the absorption maximum with the pressure of C_2H_4 changing from 3 to 32 kPa.

3125 cm^{-1} appear in the region of C-H stretching vibrations. The band at 1600 cm^{-1} is attributed to the stretching vibrations of the double C=C bond in the chemisorbed ethylene, and the bands at 1451 and 1438 cm^{-1} to the deformation vibrations of the C-H bond. This was verified by comparing the observed bands with the spectrum of adsorbed C_2D_4. The intensity of the bands at 3125 and 3055 cm^{-1} is roughly proportional to the pressure; the intensity of the bands at 1451, 1438, 1600, and 2984 cm^{-1} shows no dependence on pressure. Accordingly, the first two bands were attributed to physically adsorbed ethylene, and the rest to the chemisorbed ethylene.

According to DENT and KOKES [4.137,139], the set of bands of the chemisorbed ethylene points to the creation of a π complex. Their results are presented in Table 4.4 along with the data on the IR spectra of gaseous ethylene and ethylene in a π complex (Zeise salt $KPtCl_3(C_2H_4)$). Certain IR bands, forbidden in the spectrum of gaseous ethylene, do show up on adsorption.

Comparison of the peaks of complexed, gaseous, and chemisorbed ethylene reveals that the shifting of the ν_2 and ν_3 bands for C_2H_4 chemisorbed on ZnO is much smaller than the shift observed in the IR spectrum of the Zeise salt and other homogeneous π complexes. Apparently, the bonding in the chemi-sorptive π complex is weaker.

As demonstrated in [4.146], the intensity of the IR spectra of ethylene adsorbed on solid solutions $Ni_xMg_{1-x}O$ at room tem-perature is very low, but is greatly improved when pure ethy-lene is replaced by a mixture of ethylene with oxygen. This can be explained first by the fact that the adsorption of C_2H_4 takes place chiefly on the Ni^{3+} sites, and secondly by the partial oxidizing of C_2H_4 on the surface. However, there are no bands in the region 3300 - 3800 cm^{-1} characteristic of the vibrations of surface OH groups, pointing to nondissociative adsorption. The strong band at 1600 cm^{-1}, observed at the adsorption of C_2H_4 on $Ni_xMg_{1-x}O$ [4.146], was identified with a π complex; in contrast to [4.139], however, this band shifts only slightly upon adsorption of deuterized ethylene. This allows this band to be attributed to dimerization of ethylene, leading to π complexing of butylene. The strong bands at 1660 and 1310 cm^{-1} fall within the absorption region of the surface carbonate-carboxylate compounds and are identified with the symmetric and asymmetric vibrations of the carboxylate ion

$$\left[R-C\begin{smallmatrix} \nearrow O \\ \searrow O \end{smallmatrix} \right]^-.$$

The oxidative conversions of ethylene adsorbed at low tem-perature on the surface of transition metal oxides were reported by a number of researchers. The greater part of the ethylene, adsorbed on NiO at room temperature, is desorbed at above 150°C in the form of hydrocarbons and oxygenated com-pounds, and only a smaller part is released intact at 110 - 130°C from decomposing surface π complexes [4.135]. The pro-portion of irreversibly adsorbed ethylene (i.e., ethylene adsorbed in a chemically different form) increases with the rise in the temperature of adsorption [4.136].

4.5.2 Propylene

Propylene can be adsorbed by transition-metal oxides both reversibly and irreversibly [4.135,147,148]. The measurements of conductivity and work function of all the oxide semiconductors investigated point to the positive charging of the surface with adsorption of propylene, thus revealing propylene's electron-donor properties.

The reversible adsorption has zero activation energy and results in the creation of π complexes ($Q = 30 - 60$ kJ/mol). The irreversible adsorption takes place to some extent already at room temperature, but normally occurs at $150 - 200^{\circ}C$ with nonzero activation energy. The irreversible adsorption proceeds chiefly at the expense of the interaction of propylene with lattice oxygen, resulting in the oxidation of propylene.

As compared with the adsorption of ethylene, the adsorption of propylene is distinguished by the possible splitting off of hydrogen from the methyl group, facilitating the creation of the so-called π-allyl complexes:

(4.16)

Analysis of the NMR spectra, X-ray diffraction, and to a somewhat lesser extent, the IR spectra revealed the essential information regarding the structure of the homogeneous π-allyl complexes of transition metals [4.149-154]. The NMR spectra point to the identity of the two protons H_a and the two protons H_b. The different chemical shifts indicate that H_a are located closer to the metal than H_b and H_c. In the complex $(\pi-C_3H_5)_2Pd_2Cl_2$ the angle between the plane of the π-allyl group and the plane of the rhomb $Pd\begin{smallmatrix}Cl\\Cl\end{smallmatrix}Pd$ is 112°. The central carbon atom C_2 is somewhat closer to the metal than C_1 and C_3.

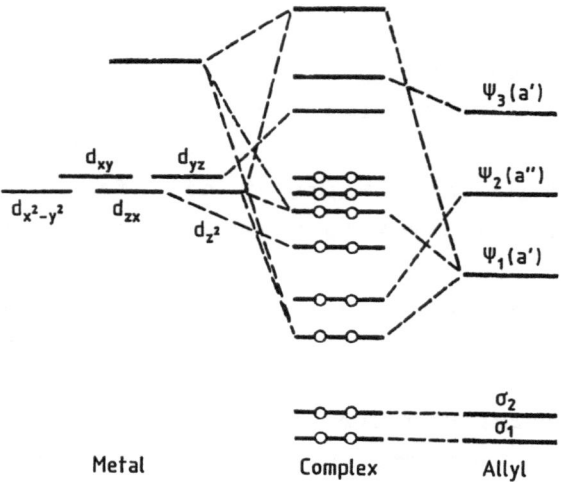

Fig.4.24. Molecular orbitals in an allyl complex of a transition element

The interaction between the orbitals of the metal (Pd) and the allyl group is shown in Fig.4.24 [4.150]. The bonding molecular orbital of the allyl group, containing a pair of electrons ($\psi_1 = 1/2(\phi_1+\phi_2+(2)^{1/2}\phi_3$, where ϕ_1 is the atomic orbital of the allyl's carbons), may combine with the $4d_{z^2}$, 5s, and 5p orbitals of the metal, permitting allyl-to-metal electron transitions. A still a more important contribution to complex formations comes from the interaction between the $4d_{yz}$ orbital and the "nonbonding" allyl orbital $\psi_2 = 1/2(\phi_1-\phi_3)$, which contains one electron (Fig.4.24). The central atom C_2 takes no part in this interaction. The interaction of the third kind, consisting in the back donation of the electrons from the metal to the allyl group, is possible between the d_{xz} orbital and the antibonding allyl orbital $\psi_3 = 1/2(\phi_1+\phi_2-(2)^{1/2}\phi_2)$. Back donation works against the interaction of the first kind. We see that the back donation (dative bonding) takes place not only in the simple π complexes, but also in the π-allyl complexes.

The actual geometry of the complex depends on the particular type of metal and ligands X. The charge of the allyl group in the complex depends on the relative contributions of the back donation and the donor-acceptor interactions and varies for different metals and ligands. Usually the allyl group in the complex has a negative charge and acts as an anion. Ligands X with strong electron-acceptor properties have a stabilizing action on the π-allyl complex. The stability of π-allyl complexes generally increases with the increase in the element's number in a group: Ni < Pd < Pt, Cr < Mo < W. Insofar as the stability of the allyl complex depends critically on the interaction of the unpaired electron in the nonbonding ψ_2 orbital, the presence of an unpaired electron on the metal's d orbital is considered [4.155] essential for the complex to be formed.

Further details on the structure of surface π complexes were found by the FTIR method [4.147]. In homogeneous allyl complexes the σ and π complexes may exist in equilibrium, which is easily detected by IR spectroscopy and NMR [4.148,149]. There also exists a "dynamic" σ allyl which exhibits rapid exchange

$$X_nMCH_2-CH\overset{*}{=}CH_2 \rightleftharpoons X_nM\overset{*}{C}H_2-CH=CH_2 .$$

This type of complex may be distinguished from σ and π-allyl complexes with the aid of NMR spectroscopy. The NMR spectrum of the π allyl displays three signals from protons, whose intensities are in the ratio 1 : 2 : 2; the spectrum of the σ allyl exhibits four signals and that of the dynamic σ allyl two. According to NMR spectroscopy [4.154], in the complex $C_3H_5PtCl \cdot PPh_3$ at low temperatures (-44^oC) there exists only the π allyl, while at elevated temperatures the two σ forms and one π form exist in equilibrium

$$(4.17)$$

213

After CLARKE [4.154], a complex with π and σ bondings is termed a μ-allyl complex. For example, in the complex (4.18)

$$
\begin{array}{c}
\text{CH}_2 \quad\quad \text{PPh}_3 \\
\text{HC} \quad\quad \text{Pd} \\
\text{CH}_2 \quad\quad \text{PPh}_3 \;,
\end{array}
\tag{4.18}
$$

the two C–C bonds have different lengths (0.140 and 0.147 nm). The μ form is possibly the first step in the transition from π to σ allyl.

From a comparison of the IR spectra of allyl complexes and gaseous propylene it becomes clear that the lowering of the frequency of stretching vibrations $\Delta\nu$ has a typical value for each type of complex, thus allowing preliminary conclusions about the type of bonding in the complex: for the symmetric π-allyl complexes of Pd, Ni, and Mn with the uniform distribution of charge, $\Delta\nu = 120 - 160$ cm^{-1}; for σ-allyl complexes, $\Delta\nu = 20 - 30$ cm^{-1}; and for μ-allyl complexes, $\Delta\nu$ is somewhere in between.

The idea that the adsorption of propylene on oxides gives rise to π-allyl complexes was put forward – on the basis of the catalytic data – long before the physical measurements could supply the necessary data on the structure of the complexes.

ADAMS and JENNINGS [4.156] and SACHTLER [4.157] have studied the oxidation of propylene, deuterium-labeled at one end, on Cu_2O and bismuth molybdate, and succeeded in demonstrating that in the product of oxidation (acrolein) the deuterium label is uniformly distributed between the extreme atoms. This points to the symmetric nature of the intermediate complex:

$$
CH_3CHCD_2 \xrightarrow{O_2} CH_2CHCD_2
\begin{cases}
CHCHCD_2 \longrightarrow CD_2CHCHO \\
CH_2CHCD \longrightarrow CH_2CHCDO \;.
\end{cases}
$$

The replacement of all three of the propylene's carbons by ^{14}C also indicates that C_1 and C_3 atoms have similar properties and show the same rate of oxydation in the intermediate complex on bismuth molybdate [4.158].

These results lead to the conclusion that the adsorption of propylene on the oxide surface gives rise to a π–allyl complex with the splitting off of hydrogen. The use of labeled molecules in studying the mechanism of hydrogenation of propylene and isomerization of locally deuterated propylene on oxides also points to the participation of surface allyl π complexes in these reactions.

The IR spectroscopic data indicating the formation of π–allyl complexes on the surface of oxides were first obtained by DENT and KOKES [4.144]. The IR spectrum of propylene adsorbed on the surface of ZnO in the range above 3000 cm^{-1} is distinguished from the spectrum of ethylene by the presence of the band at 3593 cm^{-1}, which stem from OH vibrations. No new bands arise in the region of Zn-H vibrations. Obviously, the adsorption of propylene is dissociative, the C_3H_5 group being bound to the Zn atom:

$$C_3H_6 + O-Zn-O \longrightarrow \overset{\displaystyle C_3H_5 \quad H}{O-Zn-O}.$$

Five bands (3055, 2970, 2947, 2915, and 2868 cm^{-1}) fall within the region of stretching C-H vibrations. The bands at 1443, 1390, 1288, and 1203 cm^{-1} correspond to bending vibrations of the C-H bond. Close to these is the band at 1545 cm^{-1}, attributed to the C-C vibrations.

In the spectrum of adsorbed C_3D_6 the band at 3593 cm^{-1} is shifted to 2600 cm^{-1}, and the band at 1545 cm^{-1} to 1473 cm^{-1}. Analysis of the spectrum of adsorbed molecules of $CH_3-CH=CD_2$, $CH_3-CD=CH_2$, and $CD_3-CH=CH_2$ reveals that hydrogen at adsorption is removed from the propylene molecule and forms symmetric allyl complexes.

A large shift in the frequency of C-C vibrations (107 cm^{-1}) in relation to the band of $\overset{\diagdown}{\diagup}C=C\overset{\diagup}{\diagdown}$ vibrations in gaseous propylene (1652 cm^{-1}) indicates that the reduction in the multiplicity of the bond is greater than in simple π complexes, and bears close resemblance to the IR spectra of π–allyl complexes of transi-

215

tion metals. On the basis of the theory of intramolecular vibrations, DENT and KOKES [4.144] determined all types of vibrations in the $CH_2-CH-CH_2$ fragment of C_{2v} symmetry; 16 out of 18 ought to be active in the IR spectrum, and the surface effects remove the forbiddenness of the remaining two. The authors managed to detect 11 absorption bands; the other 7 fall below 1200 cm^{-1}. The spectra of adsorbed isotope-labeled molecules of propylene agreed quite well with the spectra calculated on the basis of the known allyl structure and the data on the adsorption of C_3H_6. Thus the creation of π-allyl complexes upon adsorption of propylene on ZnO can be taken for a solid fact. According to [4.144], propylene on ZnO is stabilized as an anion $C_3H_5^-$.

The removal of hydrogen from propylene (giving rise to allyl) occurs more easily at oxidative conversions. The production of π-allyl complexes was observed in [4.128,159] for adsorption of $C_3H_6+O_2$ on the solid solutions $Ni_xMg_{1-x}O$ and $Co_xMg_{1-x}O$. The separate adsorption of C_3H_6, although partly irreversible, did not give rise to the pertinent IR bands (at room temperature).

This point is exemplified in Fig.4.25 by the IR spectra obtained after the adsorption of propylene-oxygen mixtures on the solid solution $Co_{0.06}Mg_{0.94}O$ [4.128]. A strong and complex absorption band with peaks at 1660, 1640, 1600, and 1580 cm^{-1}, together with some bands in the range 1270-1500 cm^{-1}, is observed between 1200 and 1800 cm^{-1}. The introduction of deuterium into the propylene molecule almost totally removes the bands at 1365, 1580, and 1640 cm^{-1}, while the bands at 1505 and 1430 cm^{-1} are shifted down by 30 - 40 cm^{-1}. Simultaneously, the spectrum contains the bands typical of the stretching C-H vibrations (3100, 3045, 3015, 2820, and 2930 cm^{-1}) and the absorption band at 3665 cm^{-1} related to the creation of surface O-H groups.

It would be natural to attribute the bands at 1200-1500 cm^{-1}, which are killed by deuteration, to bending C-H vibra-

216

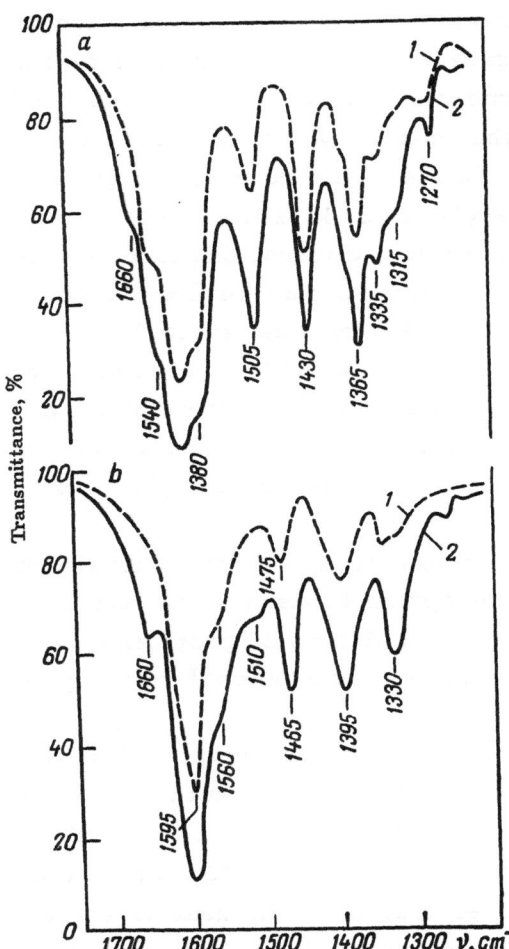

Fig.4.25. The spectra after simultaneous adsorption of oxygen and (a) propylene and (b) deuterated propylene on $Co_{0.06}Mg_{0.94}O$ [4.128]: (1) at $P_{in} = 1300$ Pa, (2) after evacuation, P_{res} 10^{-2} Pa

tions, and the shifting band at 1505 cm^{-1} to the creation of a π-allyl complex. The latter is corroborated by the dissociative nature of the adsorption of propylene, giving rise to O-H groups.

Very appealing is the large shifting of the said bands with respect to the frequencies of stretching C=C vibrations in free propylene (1552 cm^{-1}) and the π-allyl complex on ZnO (1545 cm^{-1}). At the same time propylene is bound to the surface much more strongly than ethylene, and is desorbed from the surface at 200–300°C. Calorimetric investigations of the adsorption of propylene on $Co_xMg_{1-x}O$ indicate that at room temperature the major proportion of propylene is adsorbed with the heat of adsorption of 85 kJ/mol, the initial value of the heat of adsorption being 220 kJ/mol [4.124]. The bonding energy in a similar complex (palladium–allyl) is 230 kJ/mol [4.160].

In a simple π complex the band of C–C vibrations is typically shifted by just several tens of inverse centimeters. It would be reasonable therefore to attribute the band at 1505 cm^{-1} to the allyl complex. The band at 1600 cm^{-1}, pertaining to strongly bound propylene, is unaffected by deuterization and evidently corresponds to the vibrations of the double bond in strongly adsorbed compounds, e.g., polymer propylene or μ complex (4.18).

The band at 1640 cm^{-1} is ascribed to the bending vibrations of the surface water. Weak absorption bands at 1660 and 1315 cm^{-1} may be attributed to the oxidized form of propylene, containing a carbonyl group, and the band at 1430 cm^{-1} to the products of oxidative decomposition.

As the coverage increases, the heat of adsorption of propylene goes down from 120 to 50 kJ/mol for bismuth molybdate [4.161], and from 170 to 40 kJ/mol for MoO_3/Al_2O_3. Presumably, the upper values are close to the heat of formation of surface π–allyl complexes, and the lower values to the heat of formation of simple π complexes. The study of thermal desorption of propylene from solid solutions $Co_xMg_{1-x}O$ indicates that the activation energy of desorption of propylene in nonmodified form is about 40 kJ/mol. This is, apparently, the heat of formation of π complexes [4.162].

The IR spectra of propylene adsorbed on Cu_2O and CuO were studied in [4.163]. The major part of propylene is adsorbed on Cu_2O reversibly, in particular, as a weakly bound π complex with a frequency of C=C vibration of 1626 cm^{-1}. The major part of propylene on CuO takes the form of highly oxidized carbonate–carbonyl structures. The authors do not exclude the possible production of π-allyl complexes, although their concentration on the surface is presumably small and they are quickly oxidized further. The band at 1440 cm^{-1}, observed at the chemisorption of C_3H_6 on CuO and Cu_2O, is identified in [4.163] with the asymmetric vibrations in the methyl group of the surface acetate:

$$Cu \overset{O}{\underset{O}{\diagdown}} C - CH_3 .$$

Other surface carboxylates can also be formed: two formate ions HCOO$^-$ out of the ethylene molecule, acetate CH_3COO^- and formate from propylene, and propionate $C_2H_5COO^-$ and formate from butylene-1.

MIKHAL'CHENKO with co-workers [4.164] challenge the conclusions made in [4.163] by assuming that the band at 1400 cm^{-1} pertains to the π-allyl complex of propylene. Deuteration caused shifting of this band by 100 cm^{-1}. The isotopic shift due to deuteration is 70 - 90 cm^{-1} in π complexes. The band at 1400 cm^{-1}, arising at the adsorption of propylene on gallium molybdate, was ascribed to the π-allyl complex in both [4.163] and [4.164]. The band at 1600 cm^{-1} was attributed to the σ-allyl complex, which on the oxide catalysts may be fixed on the surface via oxygen: $M - O - CH_2 - CH = CH_2$.

The IR spectra of propylene adsorbed on CuO and Cr_2O_3, and on solid solutions $Cu_xMg_{1-x}O$, also detect π complexes, π-allyl complexes of propylene, and various oxidized structures [4.165-167], whose proportion increases with the rise in the temperature of adsorption. Contrary to the findings of [4.128,159], the π-allyl complex appears to be relatively unstable: desorption from $Cu_{0.1}Mg_{0.9}O$ at 150°C results in a considerable lowering of the strength of the band at 1440 cm^{-1} which, besides, undergoes splitting into 1450 and 1430 cm^{-1} bands. The intensity of the band at 1510 cm^{-1}, attributed to the π complex, also starts to decrease at 150°C and disappears completely at 250–350°C; deuteration shifts this band down to 1410 cm^{-1}.

The adsorption of allyl bromide CH_2=$CHCH_2Br$ on the solid solution $Cu_xMg_{1-x}O$ gave rise to the same band (1430 cm^{-1}) [4.166,167]. Since the allyl bromide decomposes according to the reaction $C_3H_5Br \rightarrow C_3H_5 + Br$, this has been taken for another piece of evidence in favor of attributing the band at 1430 cm^{-1} to the surface allyl complex [4.166,167]. The difference between the observed value (1430 cm^{-1}) and the value quoted by DENT and KOKES [4.144] (1545 cm^{-1}) is assumed to arise from the fact that allyl is cationic $(C_3H_5)^+$ on CuO/MgO, and anionic $(C_3H_5)^-$ on ZnO. The recharging of adsorbed particles on the surface of transition oxides takes place easily; however, the positive charging could hardly produce such a large frequency shift. More likely, the production of the band at 1430 cm^{-1} by adsorption of C_3H_6 and C_3H_5Br is associated with oxidized structures.

The band at 1580–1590 cm^{-1}, arising from the adsorption of C_3H_6 on $Cu_xMg_{1-x}O$ and other catalysts, is attributed in [4.164–167] to the vibrations of the double C=C bond in a relatively loosely bound "σ complex" of propylene:

$$CH_2=CH-\overset{H}{\underset{H}{C}}-H$$

where the double bond takes almost no part in the interaction with the surface. However, despite the allegedly weak bonding of this complex with the surface, the band at 1590 cm^{-1} does not disappear from the IR spectrum completely even at 300°C, and has little sensitivity to deuteration ($\Delta\nu$ =20 – 25 cm^{-1}). This band, therefore, may be related to strongly bound polymer complexes.

Apart from carboxylate complexes, the IR spectra allowed detection of other products of propylene oxidation on the oxide catalysts.

The higher symmetry of the crystalline field on the surface of solid catalysts, as compared with homogeneous complexes, ought to increase the probability of production of σ complexes and dynamic σ complexes upon the adsorption of olefins on the surface.

Very promising are the methods of microwave spectroscopy. Japanese researchers [4.168] have proposed a novel technique for identifying the surface compounds of propylene, which possibly are intermediates in various catalytic reactions on the oxide surface. The existence of various surface propylene complexes was proved by studying the H-D exchange and the isomerization of deuterated isomers of propylene via monitoring the distribution of label in the propylene molecule. It turned out that π allyl is formed on ZnO; σ allyl at low temperatures and dynamic σ allyl at high temperatures (300°C) are formed on bismuth molybdate; 1-propyl-CH_2-CHD-CH_3 and 1-propenyl-CH=CH-CH_3 arise on the charge transfer complex potassium + nickel phthalocyanine; and 2-propyl CH_3-CH-CH_3 is formed in solutions of strong acids.

It has been assumed [4.169] that σ-allyl complexes are predominantly formed on acid surfaces. The allyl groups then are neutral or carry a small positive charge. The dynamic σ-allyl complexes are more likely to form on basic surfaces and are of an anionic or a radical nature.

Negatively charged complexes are often formed on oxide catalysts [4.170]. It is possible that after the adsorption of propylene the allyl complex becomes negatively charged at the expense of the charge transfer from catalyst to allyl due to the interaction of hydrogen atoms in the π-allyl complex with lattice oxygen. Judging by the IR spectra, the π-allyl complexes are easily formed on basic catalysts (Bi_2O_3, CoO, MgO). The C-C bonds in such complexes must be weakened due to the redistribution of the excess charge via the antibonding orbital (hence the high reactivity):

$$C_3H_6 + O^{2-} \longrightarrow C_3H_5^- + OH^-.$$

This step may be followed by the interaction of $C_3H_5^-$ with the metal in the upper oxidized state:

$$C_3H_5^- + M^{(n+1)+} \longrightarrow C_3H_5 \cdot M^{n+}.$$

In this complex the neutral allyl group has the properties of a free radical and exhibits high reactivity. The increased reactivity of allyl complexes facilitates the transition from the π-allyl state to the σ-allyl state. The potential barrier for this transition in homogeneous complexes is low (50 – 65 kJ/mol). This results from the ability of the allyl ligand to continuously change the energy of bonding with the metal depending on its orientation in the complex. The activation energy of this transition on the surface must be about the same. The desorption of allyl radical in the oxidation of propylene on manganese dioxide was detected by mass spectrometry [4.171].

The surface complexes formed in the adsorption of propylene and other olefins were also studied with the aid of diffuse reflection spectroscopy in the visible and UV ranges. As a rule, the sensitivity of this technique is higher as compared with IR spectroscopy. Of special interest is the detection of carbonium ions containing positively charged carbon.

In the case of hydrocarbons, the carbonium ion can be formed due to the heterolytic breakage of the covalent C–H bond with the splitting off of a hydride ion (4.19a) or via the attachment of a proton to the multiple bond (4.19b):

$$\text{>C:H} \longrightarrow \text{>C}^+ + \text{H}^- . \tag{4.19a}$$

$$\text{>C=C<} + \text{H}^+ \longrightarrow \text{>C}^+ - \text{C-H<}. \tag{4.19b}$$

On the oxide surface the first process may take place on the Lewis centers, and the second on the Brönsted centers. The specificity of the production of carbonium ions on the surface of transition-metal oxides appears to be entirely due to the acidic properties of the surface.

For example, the UV absorption bands at 354, 330, and 265 –
275 nm were observed after the adsorption of propylene on
γ-Al$_2$O$_3$ and subsequent heat treatment at 150 – 300°C [4.130].
The intensity of these bands increased considerably after
doping the γ-Al$_2$O$_3$ surface with Ni^{2+} or Co^{2+}. However, as indicated by special research, the role of dopants consists solely
in raising the acidity of the surface. The band at 354 nm was
attributed to the linear carbonium ion (L standing for the
Lewis center):

$$2L + 2C_3H_6 \longrightarrow 2(L\ C_3H_6(ads.)) \longrightarrow C=C-\underset{L^-}{C^+}-C=C-C \longrightarrow L + H_2 ,$$

and the band at 330 nm to the cyclic carbonium ion:

$$C=C-\underset{L^-}{C^+}-C=C-C\ +\ C=C-C \longrightarrow \langle \ \rangle^+ L^-\ +\ \text{gas products (CH}_4 \text{ etc.)}.$$

GORDYMOVA and DAVYDOV [4.172] also observed the UV bands at
about 260 nm after the adsorption of propylene on γ-Al$_2$O$_3$. They
attributed this band to π allyl because the oxidized structures
are unlikely to produce absorption bands in this range.
However, the strong dependence of this band on the valence
state of the transition cation and the results of the study of
thermal desorption allow one to assume that the band at
265-275 nm belongs to the conjugated hydrocarbon with terminal
carbonyl groups: ... – HC=CH–CH=CH–CH=O [4.130]. Reviews
dealing with the production of carbonium ions on acid catalysts
can be found in [4.173,174].

The attempts to detect allyl groups on the surface by NMR
did not succeed. The first work to employ NMR for studying the
mechanism of the chemisorption of olefins by oxides was apparently that of WHITNEY and GAY [4.175], who investigated the
adsorption of paraffins and olefins on ZnO at room temperature
with coverage below 0.5 of a monolayer. The lines obtained in
the NMR spectrum (100 MHz) are broad (with a halfwidth of 30 to
120 Hz) because of the averaged dipole–dipole interaction.

The experiments indicate that physical adsorption does not give rise to any appreciable chemical shifts with respect to the nonadsorbed hydrocarbon. Adsorbed ethylene produces one highly biased band, explained naturally by the creation of a π complex, in accordance with the findings of Dent and Kokes. Structures like CH_2–CH_3 and CH_2–CH_2 are precluded.

The situation with propylene is more complicated. About one half of the C_3H_6 adsorbed on ZnO is removed by pumping out at room temperature. The remaining (strongly bound) propylene (according to Dent and Kokes, π allyl) does not show up in the NMR spectrum. The NMR spectrum of weakly bound propylene displays three lines, identified with the protons of the methyl and olefinic groups. The chemical shift is much greater for olefinic protons than for the methyl protons. This fact is well accounted for by the creation of π complexes of C_3H_6 on ZnO.

Another version of NMR was employed for studying the surface complexes of olefins with the transition ions by KAZANSKY and BOROVKOV [4.176]. The main idea is similar to the use of NMR for studying the complexes of the transition metals in solutions. The creation of complexes can be detected by the high paramagnetic shifts of the lines in the NMR spectra of ligands (as high as tens of thousands hertz). Although the sensitivity of NMR is inferior to, e.g., EPR, under conditions of rapid exchange between the ligands in the complex and the appropriate molecules in the solution the NMR spectrum easily detects the creation of a complex by the large paramagnetic shifts:

$$\delta = N_K \delta_K , \qquad\qquad (4.20)$$

where N_K is the molar part of molecules bound in the complex and δ_K is the NMR line shift in the complex molecule. A similar exchange may take place between the chemisorbed molecules and the molecules in solution or in the state of physical adsorption.

Study of the adsorption of ethylene on silicagel doped with 0.5 - 1 of Co^{2+} and Ni^{2+} at -50°C indicates that the NMR spectrum consists of one line, shifted with respect to the NMR line of physisorbed molecules [4.176]. The dependence of δ on the magnitude of adsorption points to the firm nature of adsorption.

The adsorption of propylene on Ni^{2+}/SiO_2 at -50°C gave rise to a more complicated NMR spectrum, consisting of four lines. Three of these, shifted toward strong fields, correspond to the protons at the double bond, and the fourth is shifted toward weak fields and corresponds to the protons in the CH_3 group. The protons at the double bond are thus nonidentical, indicating to the strong distortion of the propylene molecule at chemisorption. It has been assumed [4.176] that the chemisorption of olefins under these conditions results in the production of π complexes of olefins with Co and Ni ions. The NMR spectrum of ethylene adsorbed on Ni^{2+}/SiO_2 at room temperature displays new lines indicating the onset of oligomerization.

4.5.3 Butylene

Butylene exists in three isomeric forms: α butylene (a), cis-butylene (b), and trans-butylene (c); the adsorption of butylene by oxides may be accompanied by isomerization:

$$CH_2{=}CH{-}CH{-}CH_3 \qquad \begin{matrix} CH{=}CH \\ CH_3 \quad CH_3 \end{matrix} \qquad \begin{matrix} & CH_3 \\ CH{=}CH \\ CH_3 \end{matrix} \qquad (4.21)$$

$$\text{(a)} \qquad\qquad\qquad \text{(b)} \qquad\qquad \text{(c)}$$

At low temperatures butylenes, like the lower olefins, form π complexes with the ions of transition metals. The heat of adsorption for π complexes varies in line with the number of carbon atoms in the moelcule. For instance, the values of the heat of low-temperature adsorption on CoO and NiO constitute (in kJ/mol) [4.136,177]:

Ethylene ... 25   Pentenes ... $35\frac{1}{N}40$

Propylene ... 30   Hexenes ... $40\frac{1}{N}50$

Butylenes .. $30\frac{1}{N}35$,

The adsorption of α butylene and other α olefines proceeds in a manner generally similar to the adsorption of propylene: weak reversible adsorption with π complexing at room temperature, very low adsorption at 100–200°C, and irreversible adsorption accompanied by oxidation and polymerization at higher temperatures.

Upon the adsorption of butylenes the π–allyl structures may assume either of the two isomeric forms:

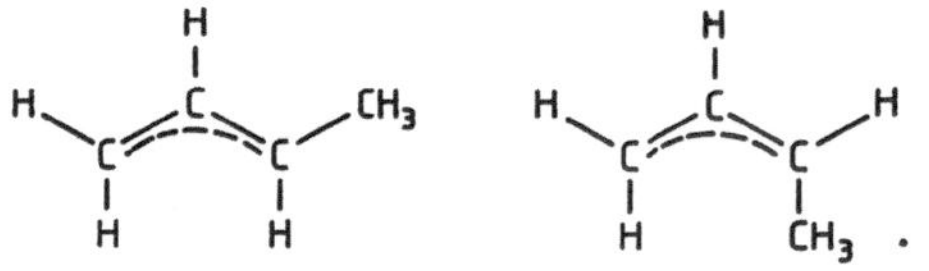

$$(4.22)$$

(a)   (b)

Cis-butene initially gives rise to the anti-form (4.22b), while the trans-butene produces the syn-form (4.22a), usually more stable. In the homogeneous π–allyl complexes of the transition metals the direct syn-anti conversion of the π–allyl ligands apparently does not take place. The isomerization of olefins is facilitated by the dynamism of the π–allyl complexes, π allyl converting into σ allyl [4.137,146]:

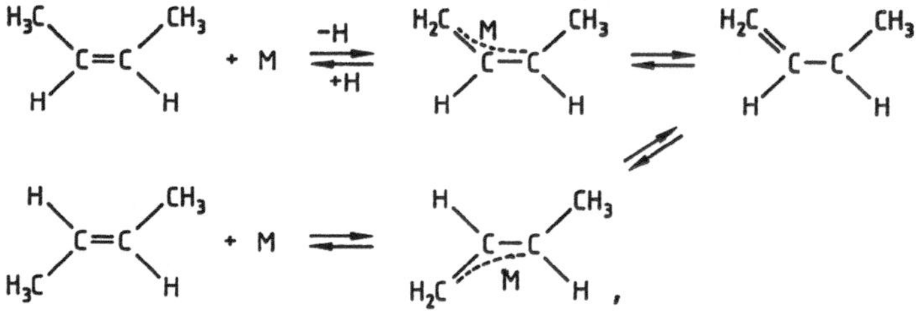

where M is the central metallic ion. The isomerization of butylenes on oxide catalysts apparently follows the sequence

cis-butene ⇌ butene-1 ⇌ trans-butene .

226

The data obtained in studying the olefinic complexes with Ni, Pd, and Fe at -195°C allow one to assume that the isomerization of olefins with respect to the double bond may also take place without allyl complexing. The activation energy of isomerization of π-bound olefins in such complexes is about 2 kJ/mol [4.178].

DENT and KOKES [4.137,139,145] have identified the π-allyl complexes in the IR spectra taken after the adsorption of butenes on ZnO. The adsorption of butene-1 at room temperatures on ZnO gives rise to a new band at 3587 cm^{-1}, pertaining to the stretching vibrations of the OH group and indicating the removal of a hydrogen atom from butylene. In this case, in contrast to the adsorption of propylene, the appearance of the OH band is extended in time, which points to its activated nature. Pumping out destroys this band even at room temperature, which also is different from the adsorption of C_3H_6 on ZnO.

The most characteristic IR bands of π complexes and π allyl complexes correspond to the stretching C-C vibrations and fall in the range between 1500 and 1700 cm^{-1}. For the adsorption of butene-1 on ZnO the band at 1650 cm^{-1} pertains to the gaseous 1-C_4H_8 (1645 cm^{-1}) and quickly disappears with time. The band at 1610 cm^{-1}, also disappearing, was ascribed to the π complex. The shift of the band with respect to gaseous 1-C_4H_8 constitutes 35 cm^{-1} and is greater than with the π complexes of C_2H_4 (23 cm^{-1}) or C_3H_6 (30 cm^{-1}) on ZnO, which illustrates the greater strength of the π complex of butylene. The intensity of the band at 1630 cm^{-1} grows with time. Apparently the isomerization of butene-1 results in π complexing of cis- and trans-isomers. Indeed, the adsorption of cis-butene initially gives rise to the 1625 cm^{-1} band of gaseous cis-butene), while at the adsorption of trans-butene the 1640 cm^{-1} band ($\Delta \nu$ =36 cm^{-1}) appears first. The band at 1625 cm^{-1} (π complex of cis-butene) is by far the strongest of the three. Obviously, the equilibrium proportion of π complexes on a surface is essentially different from the proportion of isomers in gaseous phase (77% trans-, 20% cis-, 3% 1-butene).

The band at 1572 cm^{-1} was attributed to the production of anti-π allyl (4.22b) [4.179]. The shift with respect to gaseous butylene is 73 cm^{-1}, which is much lower than that associated with the creation of propylene π-allyl complexes ($\Delta\nu$ = 107 cm^{-1}). The adsorption of cis-butene gives rise to the same band, whose intensity does not show any appreciable variations with time. The adsorption of trans-butene-2 results in the appearance of the band at 1582 cm^{-1}, ascribed to syn-π allyl (4.22a). This band is gradually replaced by the band at 1572 cm^{-1}, i.e., the syn-form slowly converts into the anti-form. The existence of several forms of π and π-allyl complexes is also indicated by the appearance of more than one band in the region of C–H vibrations.

In [4.179] these results were explained on the basis of the following scheme:

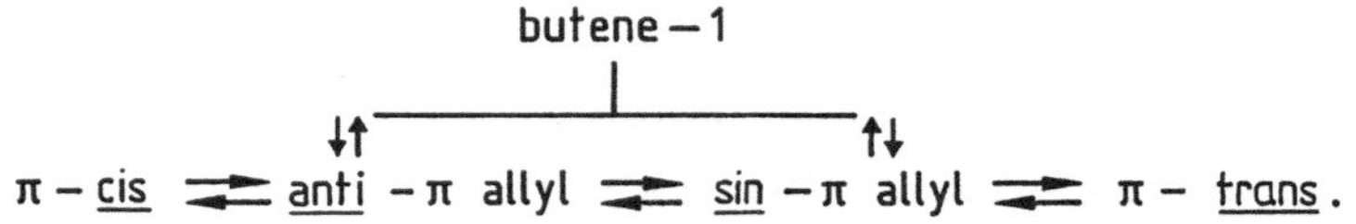

$$
\begin{array}{c}
\text{butene} - 1 \\
\downarrow\uparrow \overline{\hspace{4cm}} \uparrow\downarrow \\
\pi - \underline{cis} \rightleftharpoons \underline{anti} - \pi \text{ allyl} \rightleftharpoons \underline{sin} - \pi \text{ allyl} \rightleftharpoons \pi - \underline{trans}.
\end{array} \tag{4.23}
$$

The slowest stage here is represented by the interconversion of anti-π and syn-π allyls. This interconversion has been assumed to proceed via the intermediate σ-allyl complexes, similar to the scheme proposed for homogeneous complexes. The IR spectra gave no indication of π-allyl complexes in the adsorption of butylenes on ZnO.

In homogeneous complexes (e.g., with Co) the order of relative stability of the allyl complexes is reverse of that observed for the adsorption on ZnO: the syn-form is more stable than the anti-form. This might be attributed to the fact that the adsorbed π allyls carry a large negative charge and behave like surface-stabilized anions.

In [4.146] it was demonstrated that the adsorption of $C_4H_8 + O_2$ mixture on the solid solutions $Ni_xMg_{1-x}O$ gives rise to

IR bands at 1670, 1600, 1510, 1430, 1385, 1360, and 1310 cm^{-1} in the 1200 – 1700 cm^{-1} spectral range. Of these, the bands at 1670 and 1310 cm^{-1} may be attributed to the oxidized form of butylene containing a carboxyl ion, the band at 1600 cm^{-1} to the firmly bound complex (4.23a) or to the symmetric π complex (4.23b) bound with the surface ions of Ni, and the band at 1510 cm^{-1} to the π-allyl complex:

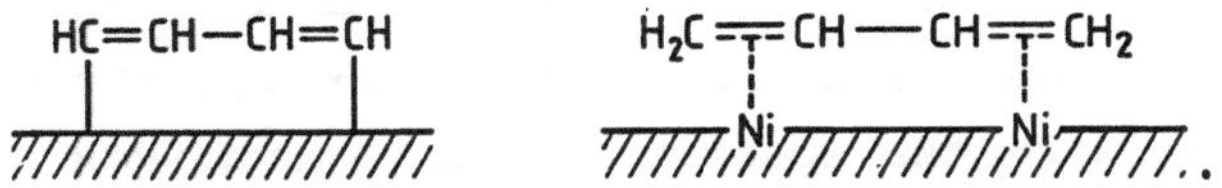

$$HC{=}CH{-}CH{=}CH$$

(a)

$$H_2C{=\!\!=\!\!}CH{-}CH{=\!\!=\!\!}CH_2$$

Ni   Ni

(b)

According to [4.180], the adsorption of butene-1 mixed with oxygen on oxidative catalysts gives rise to π complexes with a relatively small shift of the C=C band: $\Delta\nu$ =20 cm^{-1} for Cu_2O, $\Delta\nu$ =42 cm^{-1} for V_2O_5-P_2O_5. Further on, these π complexes convert into π-allyl complexes, which, however, do not show up in the IR spectrum.

From the above discussion it follows that the IR spectra of adsorbed olefins have not yet received unambiguous interpretation. The advances in this directions will depend not only on widening the scope of the experimental research, but also on the employment of other investigative techniques, providing the basis for comparison.

4.6 The Adsorption of Carbon Monoxide by Oxides

The adsorption of carbon monoxide can be reversible and irreversible. The reversible adsorption is usually explained by the creation of carbonyl structures. According to the electrophysical data, they carry a small positive charge. The irreversible adsorption on the oxide surface gives rise to oxidized structures. Our knowledge of the structure of the surface compounds comes chiefly from the data supplied by IR spectroscopy, treated by analogy with the spectra of the homogeneous complexes. The strong IR bands in the range 1800 to 2200 cm^{-1} pertain to carbonyls. The adsorption of CO on the oxides of transition metals gives rise also to a number of bands outside this region [4.36,121,181,182]. The band at 2350 cm^{-1} belongs to the physisorbed or highly polarized CO_2, resulting from the oxidation of CO. Very strong bands appear after the adsorption of CO on transition metal oxides in the 1200 – 1800 cm^{-1} range; they stem from the oxidized structures (carbonates, carboxylates, formates).

4.6.1 Carbonyl Structures

The absorption band of the C-O stretch vibrations in nickel tetracarbonyl $Ni(CO)_4$ is observed at 2057 cm^{-1}, and in chromium hexacarbonyl $Cr(CO)_6$ at 1968 cm^{-1}. According to X-ray analysis, all carbonyls in these compounds are attached to the metallic ion by a single carbon-metal bond (structure 4.24a). At the same time in the iron carbonyl $Fe_2(CO)_9$ the absorption bands at 2082 and 2019 cm^{-1}, related to the "linear" or "terminal" carbonyl groups, appear together with the lower-frequency bands at 1829 cm^{-1}, attributed to the "bridge" carbonyl groups (structure 4.24b) [4.181]:

$$\overset{\diagdown}{\underset{\diagup}{-}}M-C\!\equiv\!O \qquad \overset{>Fe\diagdown}{\underset{>Fe\diagup}{}}C\!=\!O\,. \qquad\qquad (4.24)$$

(a) (b)

The juxtaposition of a large number of IR spectra of carbonyls with the X-ray data allows one to assume that the absorption bands of the terminal carbonyl groups lie above 2000 cm^{-1}, and those of the bridge carbonyl groups lie below 2000 cm^{-1}.

EISCHENS and PLISKIN [4.182] who were the first to study the IR spectra of carbon monoxide adsorbed on a large number of transition metals and their oxides, took advantage of this correlation for identifying the absorption bands in the region of C-O stretch vibrations (1800 - 2200 cm^{-1}). In their view, the absorption bands between 1800 and 1900 cm^{-1} pertain to the bridge carbonyl structures (4.24b), and the bands between 2030 and 2120 cm^{-1} belong to the linear carbonyl structures of type (4.24a). As the coverage increases, the latter gain in intensity more quickly than the former. According to [4.182] the adsorption of CO on NiO gives rise to yet another band at 2143 cm^{-1}, which was attributed to CO adsorbed on the oxygen ion resulting in a structure of type (4.25c) (essentially the structure of carbon oxide):

$$\text{Ni} - \text{C} \equiv \text{O} \qquad \begin{matrix} \text{Ni} \\ \\ \text{Ni} \end{matrix} \!\! \Big\rangle \text{C} = \text{O} \qquad \text{Ni} = \text{O} = \text{C} = \text{O} \qquad (4.25)$$

$$\quad \text{(a)} \qquad\qquad \text{(b)} \qquad\qquad \text{(c)}$$

Study of the structure of carbonyls of transition metals indicates, however, that this criterion for distinguishing between the bridge and the linear carbonyls is not always correct. While the IR bands above 2000 cm^{-1} do really belong to the linear groups, the bands below 2000 cm^{-1} may pertain to both bridge and linear carbonyls. The shifting of the absorption band of the CO group in linear complexes may be due to any of several factors:

1) <u>Negative charging</u> of the carbonyl. For example, $Fe(CO)_5$ absorbs at 2028 and 1994 cm^{-1}, while $Fe(CO)^{2-}$ absorbs at 1780 cm^{-1}.

2) <u>Electron-donor ligands</u> in the carbonyl complexes. For instance, pyridine and ethylenediamine shift the bands of linear

carbonyls toward 1800–1850 cm^{-1}. Converely, an electronegative ligand or a π–complexing ligand shifts the absorption band to above 2000 cm^{-1}.

3) Other basic vibrations, whose intensity may be reinforced at the expense of the resonance with the basic vibrations of carbonyl [4.181].

BLYHOLDER [4.183] demonstrated the frequency-based criterion is incapable of furnishing a sound basis for calling the structure either linear or bridge. The bridge compounds must exhibit bands in the range 700–1000 cm^{-1}, corresponding to the asymmetric M–C vibration. At the same time, the linear compounds should not have bands in this range, although the absorption band of bending vibrations of the M–C–O structure must be manifest in the 400–500 cm^{-1} range. The adsorption of Co and Ni and on oxidized Ni gave rise to the band at 1940 cm^{-1} in the range of stretching C–O vibrations, and the low-frequency band at 435 cm^{-1}, similar to the 422 cm^{-1} band in the spectrum of the linear carbonyl. This implies that the band at 1940 cm^{-1} pertains to the linear rather than to the bridge structure.

In [4.184] it was observed that the dilution of solid solutions $Ni_x Mg_{1-x}O$ (reduction in x) does not affect the ratio between the mean integral intensities of the bands above and below 2000 cm^{-1}, pertaining respectively to the bridge and the linear structures. This is taken as an indication that all the bands observed belong to the linear complexes Ni–C–O. However, the ratio between the bands pertaining to the bridge and the linear structures might be expected to remain unaffected by the dilution under the assumption of the cluster structure of active centers (Sect.3.1.2).

The experiments on desorption reveal that the molecules first removed from the surface of the catalyst are those responsible for the high-frequency bands. Thus, the strength of M–C bonding is inversely related to the frequency of vibration. This, however, disagrees with the results of [4.185], where the

232

frequency of C–O vibration was observed to go in line with the value of heat of adsorption of CO on oxide catalysts in the 2160–2240 cm^{-1} range.

BLYHOLDER [4.183] gave a qualitative description of the chemical bonding in the adsorbed CO from the standpoint of the theory of molecular orbitals. His calculations, together with the later findings of other authors [4.186–188] explain some particular features of the IR spectra of adsorbed CO.

The diagram of the molecular orbitals of CO and of the adsorptive complex of CO with a transition metal is shown in Fig.4.26, along with the scheme of overlapping molecular orbitals.

In an isolated molecule of CO two electrons occupy the 1σ orbitals, and another two occupy the 2σ molecular orbitals, formed from the 1s orbital of oxygen and 1s orbital of carbon. Molecular orbitals 3σ, 4σ, $1\pi_x$, and $1\pi_y$, next in the energy,

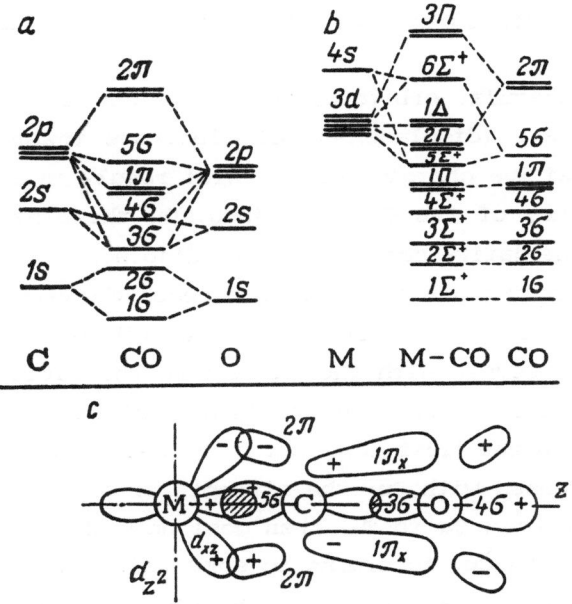

Fig.4.26. Molecular orbitals of carbon monoxide (a), carbonyl complex (b), and the scheme of overlapping of molecular orbitals in the carbonyl complex (c)

derive from the atomic orbitals 2s, $2p_x$, and $2p_y$, and are also occupied by two electrons each. In all cases the maximum electron density is biased towards the oxygen.

Two orbitals in CO are antibonding: the 2π orbital, arising mainly from the $2p_z$ orbitals of oxygen, and the 5σ orbital, produced by the $2p_z$ orbitals of carbon. The former corresponds to a pair of electrons near the oxygen atom, and the latter corresponds to a free electron pair near the carbon atom and is directed outward along the axis of the molecule. This free electron pair counterbalances the shifting of the remaining orbitals toward oxygen, making the net dipole moment of the CO molecule very small.

When the CO molecule forms a complex with a metallic ion, the antibonding 5σ orbital overlaps with the unoccupied d_{z^2} orbital of metal, thus producing the donor-acceptor bond between CO and metal and giving rise to a $5\Sigma^+$ orbital. The back donation of the electron from the occupied d orbital of metal (d_{yz}, d_{xy}) to the unoccupied 2π orbital of CO produces what is called a dative bond.

In this scheme of molecular orbitals, the formation of the donor-acceptor bond M ← C (displacement of electrons toward metal and hence the positive charging of CO) will result in an increase in the frequency of the CO vibrations. On the other hand, the creation of the dative bond M → C (the displacment of d electron to the antibonding orbital of CO and hence the negative charging of CO) will lower the frequency of CO vibrations. The bonding of the CO molecule with the surface in the latter case is weaker than in the former. The strength of the donor-acceptor bonding M-C-O in the first long row increases steadily from Ca to Ni and decreases with Cu; the strength of the back donation bond increases from Ca to Ti, then goes down from Ti to Ni, and up a little with Cu. According to some calculations [4.187], the low-frequency bands may be explained by the multisite adsorption of CO: one molecule may be attached to three or even four atoms of metal.

Still controversial is the question regarding the origin of the highest-frequency bands in the IR spectrum of adsorbed CO in the range between 2150 and 2200 cm^{-1}. Some authors challenge the conclusions of Eischens and Plishin and ascribe these bands to CO_2 rather than to CO. According to GARDNER [4.189], the high-frequency bands in the IR spectrum arise from adsorption-ionized CO molecules. It has also been assumed [4.181], that the high-frequency absorption bands of adsorbed CO pertains to the compounds attached to the surface only by a δ bond at the expense of the unshared pair of carbon electrons.

Semiempirical calculations of Ni-CO complexes, performed using the extended Hückel method [4.188] indicate that both the σ complexing and the charge transfer are responsible for the high-frequency bands. According to these calculations, the complex M-C-O is from all aspects (overlapping of orbitals, net energy of orbitals) more stable than the M-O-C complex.

The interaction of Ni^+, Ni^{2+}, and Ni^{3+} with CO results in much greater overlap in the carbon-oxygen bond in a complex than in a free molecule of CO, which leads to strengthening of the bond. The reverse is true for Ni^0, and the bond is weakened. The calculations of the electron density in the Ni-C bond indicate the transfer of a large amount of charge from CO to Ni: +0.02 for Ni^+...CO, +0.33 for Ni^{2+}...CO, and +0.59 for Ni^{3+}...CO. In the complex with neutral nickel Ni^0...CO the charge is transferred in the opposite direction, from Ni to CO. The charge transfer is due to the displacement of σ electrons from CO to Ni; as to π electrons, they are shifted in the opposite direction, although to a smaller extent.

These calculations generally agree with the findings of [4.190], which indicate that CO adsorbed on the $Ni^{II}Y$ zeolite gives rise to a much higher-frequency band in the IR spectrum (2217cm^{-1}) than does CO and $Ni^I Y$ zeolite (2188 cm^{-1}).

The calculations in [4.188] seem to confirm the interpretation of the high-frequency bands of adsorbed CO, based on the concept of σ complexing [4.181]. The interaction M...CO is much stronger than just the ion–dipole interaction; at the same time, the transfer of charge is not so large as to justify ions CO^+ and CO^{2+}.

According to the calculations using the extended Hückel technique [4.186], in the adsorptive complexes of CO the electron is transferred from CO to metal on the oxides of trivalent metals (V_2O_3, Cr_2O_3, Fe_2O_3, Co_2O_3), and from metal to CO on the oxides of bivalent metals (MnO, FeO, CoO, NiO).

Numerous experiments with the adsorption of CO on the transition-metal oxides [4.190-197] confirm in general the scheme discussed above. Other conditions being equal, the increase in the cation's charge strengthens the bonding of the adsorbed CO molecule and shifts the absorption band of the C-O stretch vibration towards higher frequencies. While for the adsorption on the pure transition metals the value of ν_{CO} is between 1900 and 2100 cm^{-1} and close to the corresponding values for the homogeneous carbonyl complexes, the adsorption on transition-metal oxides gives rise to much higher frequency bands (2100-2200 cm^{-1}), for the higher oxides CrO_3, TiO_2, ZrO_2, and V_2O_5 the value of ν_{CO} being close to the upper limit of this range.

As indicated earlier, the increase in back donation (π bonding) shifts ν_{CO} toward lower frequencies. This should give rise to characteristic shifts in frequency as we go from one transition metal to another. However, the analysis of a large bulk of data fails to detect any systematic changes in ν_{CO}, e.g., for the fourth-row transition-metal oxides. The large variance caused by different coordinations of the adsorptive complexes and the effects of foreign ligands effectively disguises those changes which might be associated with the transition from one metal to another. For the same reason there is little hope in using the correlation between the frequency of the C-O stretch vibration and the cation charge for determining the charge of the cation, as proposed in [4.197]. The results differ even for the equicharged cations of the same metal in

236

different matrices. For instance, the appearance of ν_{CO} bands in the 1800-2080 cm^{-1} range for the adsorption of CO on NiO was attributed in [4.197] to the presence of metallic Ni^0 on the surface. However, the UV-visible spectroscopic data indicate the production of oxidized structures of Ni, while metallic Ni is not detected either by the magnetic methods or by any other techniques. Most likely, these bands are associated with the multisite adsorption of CO.

The adsorption of CO by zeolites containing transition metals always produced only the high-frequency bands [4.189, 197-200] (2200 cm^{-1} for zeolites CoY, MnY, FeY, CuY). The band at 2212 cm^{-1}, arising after the adsorption of CO by NiY, may be due to the interaction between CO and the Ni cation near oxygen: $-O-Ni^{2+}...[C \equiv O]$ [4.200].

Since in zeolites the transition-metal ions are more isolated than in oxides, the correlation between ν_{CO} and the cation charge [4.197] may be more likely to be valid for zeolites than it is for the oxides. At least, the IR spectra of CO adsorbed by zeolites conform with the general principle: the more isolated and more highly charged transition ions M form a weaker bond with CO, but strengthen the intramolecular C-O bond and shift the frequency of its vibration toward higher frequencies.

In Sect.3.1.2 we have already mentioned the cluster structure of the surface of solid solutions $Co_xMg_{1-x}O$ and $Ni_xMg_{1-x}O$. The clustering of the interacting ions of the transition metals may be detected by the IR spectra of adsorbed CO.

The IR spectra of CO adsorbed on the solid solutions $Co_xMg_{1-x}O$ are reproduced in Fig.4.27 [4.125,201]. The region of 1800-2100 cm^{-1} shows a number of absorption bands, which must be identified with the bridge and the multisite forms of adsorption of CO [4.187]. The bands may be divided into two groups. From Fig.4.27 it is clear that the intensity of the bands at 1875, and 1930 cm^{-1} (Set I) decreases with the increase in the concentration of CoO, while the intensity of the

237

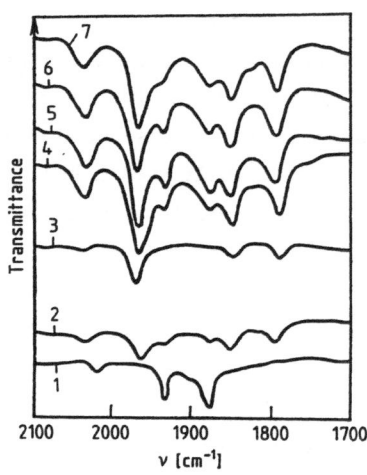

Fig.4.27. The spectra of CO adsorbed on solid solutions $Co_xMg_{1-x}O$ at room temperature. Molar percentage of CoO: (1) 1.8, (2) 4.3, (3) 8.0, (4) 9.0, (5) 9.2, (6) 12.0, (7) 14.4

bands at 1805, 1845, and 1965 cm^{-1} (Set II) increases. It was assumed [4.201] that Set I pertains to the adsorption on the isolated octahedral Co^{2+} ions, and Set II to the adsorption of CO on the interacting Co^{2+} ions (Fig.3.2). The strength of the bands comprising Set II depends on the concentration of Co^{2+} in the solid solution, but not on the frequency. This implies that the nature of bonding of CO adsorbed in clusters does not depend on the concentration of Co^{2+}.

In Sect.3.1.2 we have indicated the thermodynamic possibility for the cluster of transition ions in the oxide solid solutions to change its size according to the temperature. Larger clusters are favored at low temperature (T_2) and smaller ones at high temperature (T_1). These transitions are possibly responsible for the "temperature hysteresis" at the adsorption of CO on $Co_xMg_{1-x}O$ and $Ni_xMg_{1-x}O$ [4.36,201]. When the adsorption of CO took place at 20°C, with the subsequent cooling of the specimen to -195°C, the IR spectrum of the adsorbed CO remained unchanged. Quite a different picture is observed when the adsorption takes place at -195°C, which corresponds to the adsorption by large clusters. The IR spectrum of CO adsorbed by $Co_xMg_{1-x}O$ at -195°C becomes similar to the room-temperature spectrum ("small clusters") only after heating the specimen to room temperature. With $Ni_xMg_{1-x}O$, the low-temperature IR spectrum turns abruptly into the high-temperature spectrum after heating to 160°C. Thus, if CO has been adsorbed on the ions prior to cooling, the diffusion of Co and Ni ions on the surface is prevented. The adsorption of CO preserved the status of the surface of the solid solution, thus being responsible for the peculiar phenomenon of adsorptive hysteresis: the status of the surface phase at room temperature depends on the side from which it is approached. Large clusters persist on approaching from the lower temperatures; small clusters persist on approach from the higher temperatures. The hysteresis in the adsorption of CO on $Co_xMg_{1-x}O$ and $Ni_xMg_{1-x}O$ is detected also by the measurements of magnetic susceptibility. This process may be described by the scheme [4.26] ($T_1 > T_2$):

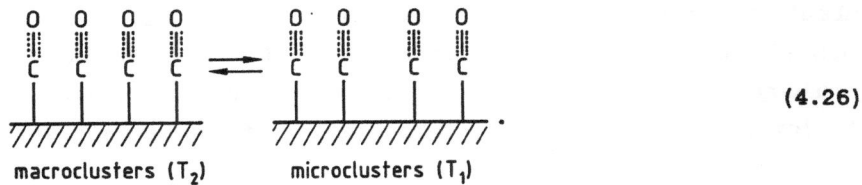

$$(4.26)$$

macroclusters (T_2)   microclusters (T_1)

The formation of negatively charged clusters $(CO)_6^{3-}$, $(CO)_5^{3-}$, etc. in CO adsorption on $Co_xMg_{1-x}O$ and $Ni_xMg_{1-x}O$ is discussed in [4.202]. The study of the adsorption of CO on the supported V_2O_5/SiO_2 catalysts indicates that CO is adsorbed only after the reduction of V_2O_5, which gives rise to V^{4+} ions. Prior to the adsorption of CO the EPR spectrum corresponds to the tetrahedral V^{4+} ions; after the adsorption the latter assume the coordination of the square pyramid [4.203]. The band at 2185 cm^{-1} appears in the IR spectrum. It may be safely concluded that CO enters the coordination sphere of vanadium and forms a σ complex with V^{4+}. The EPR signal, pertaining directly to the adsorbed CO, was detected in studying the adsorption of CO on ThO_2 at room temperature [4.198]. The signal with $g_z = 1.981$ and $g_{x,y} = 1.998$ corresponds to the paramagnetic particle concentration of $1.7 \cdot 10^{12}$ cm^{-2}, which is about 0.1 of the total number of the adsorbed CO molecules. The IR spectrum failed to produce bands indicating M–CO bonding, which might be attributed to the lower sensitivity of IR spectroscopy compared with EPR. According to [4.204], upon adsorption the two σ electrons of carbon are displaced toward the catalyst with simultaneous back donation of an electron to the antibonding π orbital. This results in a positively charged CO^+ particle. In this case treatment of the adsorption of CO based on the EPR spectra is essentially similar to the picture derived from the IR spectra and discussed in detail above.

4.6.2 Carbonates and Carboxylates

The bands in the IR spectrum between 1200 and 1800 cm^{-1}, pertaining to the oxidized structures, often occur in pairs: one at about 1600 and the other at 1300 cm^{-1}. A large bulk of experimental material has been accumulated [4.204–206] which, by analogy with the corresponding bulk structures, leads to the following enumeration and interpretation of the bands observed:

6. $M\left[C\diagdown_O^O -\right]$ $\underset{a}{C\diagdown_O^O}$ $\underset{b}{C\diagdown_O^O}$

7. $M\left[-\diagdown_O^O C-H\right]$ $\underset{a}{H-C\diagdown_O^O}$ $\underset{b}{H-C\diagdown_O^O}$.   (4.27)

a) The <u>noncoordinated carbonate</u> ion CO_3^{2-} (ionic form, structure 4.27a); asymmetric stretching vibration at 1415–1470 cm^{-1}. It usually is firmly attached to the surface and decomposes at high temperatures.

b) The <u>"monodentate" carbonate</u> (structure 4.27b): a doublet of asymmetric stretching vibration (a) at 1420 – 1540 cm^{-1} and symmetric stretching vibration (b) at 1330 – 1390 cm^{-1}, and an additional band ν_{CO} at 980 – 1050 cm^{-1}. Thermal stability is high, close to the stability of the noncoordinated carbonate.

c) The <u>"bidentate" carbonate</u> (structure 4.27c): a doublet of stretching vibration (a) at 1600 – 1670 cm^{-1} and asymmetric stretching vibration (b) at 1280 – 1310 cm^{-1}, and a band of symmetric stretching vibration ν_{COO^-} at 980 – 1050 cm^{-1} and a further band at 830 cm^{-1}. The thermal stability of the structure is reportedly lower than that of the preceding two.

d) The <u>bridge carbonate</u> (structure 4.27d): a wide doublet of stretching vibration (a) at 1780 – 1840 cm^{-1} and asymmetric stretching vibration (b) at 1250 – 1280 cm^{-1}, and a band of symmetric stretching vibration ν_{COO^-} at about 1000 cm^{-1}. This structure is not so common as the first three.

e) The <u>bicarbonate ion</u> CO_3H^- (structure 4.27e): a doublet of asymmetric stretching vibration (a) at 1615 – 1630 cm^{-1} and symmetric stretching vibration (b) at 1400 – 1500 cm^{-1}, and bands of O–H vibrations $\nu_{OH} \simeq 3600$ cm^{-1} and $\delta_{OH} \simeq 1225$ cm^{-1}. This structure involves the surface OH groups and oxygen; its thermal stability is low and decomposition occurs at a little above room temperature.

f) The <u>carboxylate ion</u> COO$^-$ (structure 4.27f): a doublet of asymmetric stretching vibration (a) at 1570 – 1630 cm^{-1} and symmetric stretching vibration (b) at 1350 – 1390 cm^{-1}, with no bands at 1000 cm^{-1}. Thermal stability is somewhat lower than that of the carbonates.

g) The <u>formate ion</u> HCOO$^-$ (structure 4.27g): a doublet of asymmetric stretching vibration (a) at 1580 – 1620 cm^{-1} and symmetric stretching vibration (b) at 1340 – 1390 cm^{-1}, and a band of the valence C–H vibration at 2740–2850 cm^{-1}. This structure is formed by the reaction

$$CO(ads.) + OH^- \rightarrow HCOO^-(ads.) \; ;$$

its thermal stability is higher than that of the bicarbonate ion.

On decomposition, all these structures produce CO_2. Consequently they may be viewed as being the intermediate stages in the oxidation of CO on the oxide surface. The adsorption of CO at above room temperature on the oxides of transition metals results usually in structures of types 4.27a–g only, and does not lead to the production of the weakly bound carbonyl forms, whose absorption bands in the IR spectrum are in the 1800 – 2200 cm^{-1} range.

Structures 4.27a–e are formed at the interaction of CO with two atoms of oxygen or with the surface OH groups; structures 4.27f and 4.27g are produced at the interaction with one oxygen atom or OH group.

Observe that similar carbonate and carboxylate structures are often detected in the adsorption of various organic substances on the oxide surface, including the adsorption of hydrocarbons, as already mentioned in Sect.4.5. For this reason the importance of such structures is appreciated in many schemes proposed for the catalytic oxidation of hydrocarbons. When considering these mechanisms, however, one must be constantly aware that the coefficients of extinction of the carbonyl and carboxyl groups in the IR spectrum are one or two orders of magnitude higher than those of the low-polar C-H and C-C groups. It is therefore possible that the observed carbonates and carboxylates on the surface actually represent a side route in the conversions of the organic substances and CO, taking place on the surface.

Calorimetric measurements performed for the adsorption of CO on the oxides of transition metals allow one to estimate the strength of binding of different forms of CO with the surface [4.36, 121, 125, 127, 207-210].

For the adsorption of CO on the oxidized forms of the transition oxides (the ions Ni^{3+}, Ce^{4+} in the oxides), the initial values of the heat of adsorption reach 150 - 250 kJ/mol. As indicated by IR spectroscopy, only the oxidized carbonate and carboxylate structures 4.27a-g are formed (usually with zero activation energy). On some oxides CO was adsorbed with a heat of adsorption of about 120 kJ/mol and nonzero activation energy. Apparently, this adsorption also resulted in the oxidized forms.

The adsorption of CO, which gives rise to the high-frequency IR bands, is characterized by low values of the heat of adsorption. For instance, for the adsorption of CO on the solid solutions $Co_xMg_{1-x}O$ at room temperature the heat of adsorption does not depend on the coverage and equals 130 kJ/mol [4.121]. It can be assumed that the different bands of CO correspond to the CO molecules adsorbed on the Co^{2+} ions in different coordination. Treatment of the surface with oxygen gives rise to the bands of the carbonate structures in the IR spectrum and results in an increase in the heat of adsorption to 400 kJ/mol.

The final portions of the adsorbate on oxides correspond to the weak forms of adsorption (the heat of adsorption of 30 -60 kJ/mol). In the IR spectra they produce extreme high-frequency bands at 2100 - 2200 cm^{-1}. For the adsorption of CO by zeolites the heat of adsorption is also not high [4.209]: 60 - 65 kJ/mol for zeolites MnY, NiY, and FeY. At the same time high-frequency bands appear in the IR spectrum. The heat of adsorption for the

CuY zeolite (where the oxidized forms were detected) was a little higher (80 kJ/mol).

4.7 Adsorption and Catalysis on the Diluted and the Supported Oxide Systems

In Sect.3.1 we discussed the properties of the surface of diluted solid solutions of the transition-metal ions in the matrix of a dielectric (a nontransition-metal oxide). As regards adsorption and catalysis, such systems prove to help in elucidating the most important features of the mechanism of these processes: 1) the role of the local and the collective interactions, studied by comparing the catalytic and chemisorptive activity per ion of transition metal at different dilutions in one and the same matrix; 2) the role of the crystal field, studied by comparing the activity of an ion of one and the same transition metal in various inert matrices having different symmetry of the cation sites (e.g., T_d and O_h); 3) the role of the d-electron configuration, studied by comparing the activity of different transition ions (d^n and d^m) in the same matrix at the same concentration; and 4) the role of the oxidation state of the transition metal, studied by implanting into the matrix foreign inert ions whose valence differs from the valence of the native lattice cations (the principle of controlled valence, Sect.3.1).

A large number of works by STONE [4.211-215], CIMINO [4.216-220], and other researchers [4.221-223] have dealt with the decomposition of N_2O on the diluted solid solutions $Ni_xMg_{1-x}O$, $Co_xMg_{1-x}O$, and $\alpha-Cr_{2-x}Al_xO_3$. To a degree the matrices themselves (MgO, $\alpha-Al_2O_3$) exhibit catalytic activity, which, however, is greatly enhanced by introducing the transition-metal ions. The catalytic activity per square meter of surface area of diluted specimens (one transition ion to 100 - 1000 lattice cations) is but a little lower than the activity of

244

the high-concentration specimens. The catalytic activity turns out not to depend on the concentration of the transition-metal ions, or even to be higher with lower concentrations. In some cases the activation energy E was lower with the diluted specimens. For example, with $Co_xMg_{1-x}O$, E = 70 kJ/mol when the (atomic) concentration of Co is 0.05%, and E = 125 kJ/mol for the concentration of 50% (the same as for pure CoO) [4.217]. With the system $Ni_xMg_{1-x}O$ the activation energy increases from 75 kJ/mol for 1% Ni to 145 kJ/mol for pure NiO, while the reaction rate constant per Ni atom increases by two or three orders of magnitude [4.221]. The rise in the activity per Cr atom is observed also with the increased dilution of the Cr_2O_3-Al_2O_3 system, although the overall range of the variations of catalytic activity is smaller [4.224].

Studies of the adsorption of oxygen on diluted solid solutions indicate quite the opposite: chemisorption is the strongest and activation energy the smallest with the pure oxides CoO, NiO, and Cr_2O_3; dilution weakens the bonding of oxygen.

POMONIS and VICKERMAN [4.222] paid attention to the fact that for a large number of various solid solutions the catalytic activity in the reactions of N_2O decomposition, H_2-D_2 exchange, and decomposition of isopropyl alcohol first increases with an increase in x, has a maximum at x = 0.1 - 0.15, then goes to a minimum at x = 0.3 - 0.5, and further increases with the increase in the concentration of the transition ion. This behavior was explained on the basis of the polaron theory of jumps of d electrons toward the adsorbed molecule. In this case the electron transport from the bulk to the active site is an adverse phenomenon for catalysis. With low values of x the ions of transition metals are separated by a large number of nontransition ions (Al, Mg, O) and the jump has a high activation energy. At x = 0.08 for Ni^{2+} and Co^{2+} in MgO and x = 0.14 for Cr^{3+} in Al_2O_3, each transition ion has at least one close neighbor of the same kind, which makes the jump easy, and the electron goes from the localized to the delocalized state.

However, the ease of electron transport to the adsorbate molecule makes the bonding stronger and thus hampers the release of the product into the gaseous phase. The catalytic activity has a minimum. At high concentrations the catalytic activity increases because of the increase in the total number of active sites.

The role of the matrix and the symmetry of the cation configuration was studied for the decomposition of N_2O on the solid solutions $Ni_xMg_{1-x}O$ and $Ni_xZn_{1-x}O$ [4.217,219]. The catalytic activity of both matrices (MgO and ZnO) is low and more or less equal. With the same concentration (1% Ni), the activation energy of the reaction for Ni^{2+} in tetrahedral coordination (ZnO) is twice as high as for Ni^{2+} in octahedral coordination (MgO). Accordingly, the reaction rate for $Ni_xMg_{1-x}O$ is much higher. The activity of Ni ions in the $MgAl_2O_4$ matrix has an intermediate value. Similar results were obtained for CoO in MgO and ZnO. The reverse was found once again to be true for the adsorption of oxygen: oxygen is held more firmly when the cation is in the tetrahedral coordination. This may be related to the fact that the Ni-O bond in the tetrahedron is shorter and more covalent in nature [4.219].

The effects of the cation charge on the catalytic activity were investigated by introducing Li_2O into the MgO or Al_2O_3 lattice [4.218-220,224]. In accordance with the principle of controlled valence a number of Ni^{2+} ions (in proportion to the number of Li^+ ions) assume the Ni^{3+} form, or, similarly, Co^{2+} ions become Co^{3+}, Mn^{2+} become Mn^{3+}, etc. In general, the catalytic activity per transition ion falls with the increase in the concentration of lithium cations, the activation energy of the decomposition of N_2O increases, and the strength of the bonding of oxygen and the coverage of the surface by oxygen also increase. In other words, the higher-oxidized transition-metal ions Co^{3+}, Ni^{3+}, and Mn^{3+} are less active in the decomposition of N_2O than the ions in the lower oxidized states (Co^{2+}, Ni^{2+}, Mn^{2+}).

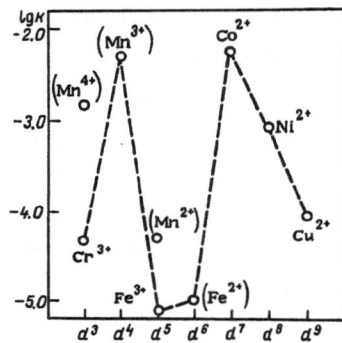

Fig.4.28. The catalytic activity of transition-metal ions in the MgO matrix in the decomposition of N_2O: atomic concentration 1% [4.217)

The comparison of the catalytic activity of different transition-metal ions brings us to the two-peak pattern, quite familiar to us from the discussion of the role of the crystal field in catalysis (Sect.4.1). Figure 4.28 gives an example of the variations in the catalytic activity of different oxides of transition metals, contained in the MgO matrix at 1% concentration (atomic) [4.217].

These results indicate that the charge transfer in the catalytic act does not involve the electrons (and holes) from the conduction band. The following scheme was proposed for the decomposition of N_2O [4.218-222]:

$$
\begin{array}{ccccccc}
\square_S & & O_S & & \square & & \\
| & & | & & | & & \\
-M^{n+}-O_S^{2-} & \xrightarrow[\text{(a)}]{N_2O} & -M^{n+}-O_S^{2-}+N_2 & \xrightarrow[\text{(b)}]{} & -M^{n+}-O_2^{2-} & \xrightarrow[\text{(c)}]{N_2O} & \\
/|\backslash & & /|\backslash & & /|\backslash & &
\end{array}
$$

$$
\begin{array}{ccc}
O_S^- & & \square_S \\
| & & | \\
\longrightarrow -M^{n+}-e\square_S+N_2+O_2 & \xrightarrow[\text{(d)}]{} & -M^{n+} \\
/|\backslash & & /|\backslash ,
\end{array}
\qquad (4.28)
$$

where $[\,]_S$ is the surface oxygen vacancy, e^- is the electron, n is the valence of the cation, and subscript S denotes the surface.

The entire process may be described without the cation M^{n+} changing its charge, although the first stage (the rupturing of the N-O bond) involves an electron from the cation of the transition metal. The reaction takes place hardly at all on the pure MgO or ZnO. The complex produced at the stage (4.28a) may be described as $M^{(n+1)+}-O_S^-$. It is essential that stage (4.28a) involves the transfer of no more than one electron and the production of weakly bound oxygen. At stage (4.28b) this weakly bound oxygen combines with the lattice oxygen to form the peroxide ion O_2^{2-}, again without electron transfer. Later on, the peroxide ion may either migrate in the lattice

$$O_2^{2-} + 2O^{2-} \rightarrow 2O^{2-} + O_2^{2-} \ ,$$

interact with N_2O (stage 4.28c), or disproportionate with the production of gaseous oxygen:

$$O_2^{2-} + O_2^{2-} \rightarrow 2O^{2-} + O_2 \ .$$

None of these processes requires the transport of electrons from the bulk of the oxide.

If the interaction between the ions of the transition metal gives rise to continuous levels or bands, the easy transfer of electrons to the weakly bound oxygen strengthens the bonding

$$M^{n+}-O + 2e^- \rightarrow M^{n+}-O^{2-} \ ,$$

and blocks the catalytic action of the site.

Similar results were obtained in studying the effects of dilution on the activity in the reactions of hydrogenation of ethylene [4.225] and H_2-D_2 exchange [4.214,226,227]. In these cases too the catalytic activity was found to be constant or even to increase with dilution.

If the catalyzed reaction depends on the supply of electrons from the catalyst, the situation is different: the catalytic activity per transition ion falls with the decrease in the concentration. With high dilutions the catalytic activity in the oxidation of CO and hydrocarbons is low, although the catalytic action does not depend on the creation of the continuous band structure. The pairwise or cluster-type centers are sufficient

248

to facilitate transport of the required number of electrons to the active center or to the adsorptive site [4.228].

BORESKOV [4.229] has studied the effects of dilution on the catalytic activity of the Cu^{2+} and Co^{2+} ions in an MgO matrix in the oxidation of H_2 and the conversion of CO by water vapor. The catalytic activity of $Cu_xMg_{1-x}O$ in the oxidation of H_2 and in the conversion of CO by water vapor, related to one Cu atom, increases rapidly with the increase in x. If we relate the catalytic activity not to the isolated ions of Cu, but rather to the atoms that have close neighbors of the same kind, then the activity will appear to be the same whatever the concentration of Cu in the solid solution. This means that the active center of the conversion of CO must comprise at least two ions of transition metal capable of interacting with one another. The constant activity per pair of adjacent Co atoms in the oxidation of CO on the solid solutions $Co_xMg_{1-x}O$, heat-treated at 500°C, in the oxidation of CO on the solid solutions $Co_xMg_{1-x}O$ was reported in [4.230].

MATYSHAK and co-workers [4.127,201,231] studied the catalytic activity of the solid solutions $Co_xMg_{1-x}O$, heat-treated at 500°C in the oxidation of CO and hydrocarbons. As indicated in Sect.3.1.2, the structure of the active sites depends on the value of x (Fig.3.2). The oxidation of the adsorbed CO was studied in the temperature range from -10° to 150°C by variations in the optical density of the IR band of CO, in the course of the reaction between CO and gaseous oxygen. The comparison was based on the two strongest bands of CO: 1875 cm^{-1} ("Set I", Sect.4.6.1, Fig.4.27) and 1965 cm^{-1} ("Set II"). From Fig.4.29 it is clear that the activation energy E of the oxidation of the first of CO exhibits no dependence on the concentration of Co^{2+} in MgO (Curve 1). This agrees well with the assumption that this set of bands in the IR spectrum relates to the adsorption of CO by isolated Co^{2+} ions. The activation energy of the oxidation of the second form of CO (adsorbed on the interacting ions of Co^{2+}) decreases from 25 kJ/mol to 5 kJ/mol as the concen-

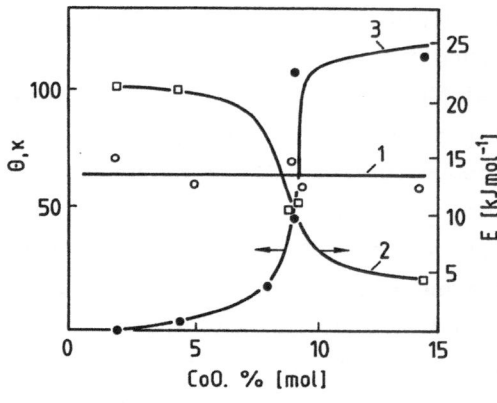

Fig.4.29. Activation energy E of oxidation of surface forms of CO and $Co_xMg_{1-x}O$ versus concentration [4.125] (1) form 1 (1875 cm^{-1} band), (2) form 2 (1965 cm^{-1} band), (3) Weiss constant θ versus x in $Co_xMg_{1-x}O$

tration of Co^{2+} increases (Curve 2) and correlates with the change in the Weiss constant (Curve 3), which characterizes the interionic interaction. The greatest change in E is observed for the second set of bands in the region of the steep increase of the Weiss constant θ.

To a first approximation the ions of Co^{2+} in clusters may be divided into two groups: the outer (boundary) ions and the inside ions. Then, assuming that the activation energy of the oxidation of CO is greater for the outer ions than for the inside ions of Co^{2+} in the clusters, the behavior of E(x) may be explained in the following way (Curve 2 in Fig.4.29): at low values of x the majority of Co occurs in the form of isolated ions or outer ions of small clusters; hence E is high. With the increase in x the average size of clusters increases, the inside ions begin to outnumber the outer and the isolated ions, and the value of E decreases.

Rough estimates of the cluster size may be derived from the dependence of E on x [4.37]. Assuming $E \simeq 0$ for the inner ions in a cluster, we get $d_2/d_1 \propto E_1/E_2$, d being the diameter of the cluster. According to the experimental data, the activation energy of small clusters is 3 – 6 times greater than for large clusters. If d_1 corresponds to two or three atoms, then the large clusters of Co^{2+} will be 10 – 15 atoms (4.5 – 7.0 nm) across, and will comprise 100 – 200 atoms. The size of the cluster is easily proved to be smaller than the size of the microcrystal of MgO ($\simeq$ 20 nm).

As pointed out in Sect.4.6.1, the nature of bonding of CO, adsorbed in a cluster, does not change with the increase in the concentration of Co^{2+}. It may therefore be assumed that the lowering of the activation energy with the increase in the cluster size derives from the fact that oxygen in a cluster becomes activated more easily. The following mechanism has been proposed in [4.125] for the low-temperature interaction between oxygen and the adsorbed CO:

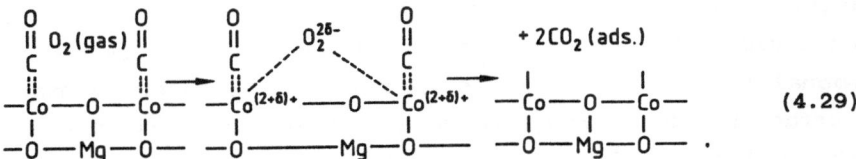

$$(4.29)$$

Scheme (4.29) may be modified by assuming that the adsorption and the activation of O_2 also require an ion of Co^{2+} (three-site adsorption). The required number of active sites (three Co atoms) may be readily available inside the cluster, while on the boundary of the cluster they become available only after activated structural rearrangement. The oxidation of CO on an isolated Co^{2+} ion takes a different route, possibly via the CO_2^- ion. The value of E then shows no dependence on x (Curve 1, Fig.4.29).

The redox catalyzed reactions seem to occur more readily on the clusters of several transition atoms than on isolated atoms, because (1) the multisite adsorption facilitates the redox process, (2) the redox reaction usually requires a number of electrons for the elementary act, which are more readily supplied from a number of metallic atoms of variable valence, and (3) a considerable number of closely spaced electron levels become available in a cluster, raising the probability of finding the levels having the appropriate energy for the catalytic act.

The diluted solid solutions and the supported systems provide a convenient object for studying the mechanism of complex catalyzed reactions of the oxydation of hydrocarbons, which involve the transfer of a large number of electrons.

As already indicated in Sect.4.4, in the dilute supported oxide systems the electron is transferred from the donor ion (Mo, V, Co) to the oxygen molecule, but the latter is stabilized in the form of O_2^- on the lattice ion (Al, Mg). The adsorption of the mixture $C_3H_6+O_2$ on MoO_3/MgO at 25°C gave rise to a very strong EPR signal from O_2^- on Mg with $g_z = 2.090$, and a second signal from O_2^- with $g_z = 2.070$, identical with the signal observed after the adsorption of pure oxygen [4.230]. The signal with $g_z = 2.090$ was ascribed to O_2 adsorbed near the surface hydroxyl group. We have surmised that the creation of the surface complex of olefin takes place with the complete or partial removal of hydrogen from the olefin and the electron transfer from the olefin to oxygen:

$$
\begin{array}{ccc}
O_2^- & H...C_3H_5^+ & \\
| & | \quad | & \\
Mg...O...Mo^{5+}. & &
\end{array}
$$

The charged allyl then apparently migrates over the surface toward the oxygen and combines with it (the purely ionic reaction). Having reacted with oxygen, the complex departs from the site, the catalytic act completed.

The adsorption of the mixture $C_3H_6+O_2$ on $Co_xMg_{1-x}O$ gives rise to an EPR signal from O_2^- similar in parameters and intensity to the signal observed after the adsorption of pure oxygen: $g_z = 2.080$ and 2.070 [4.231]. After the adsorption of the mixture $C_3H_6+O_2$ at 25°C the signal with $g_z = 2.080$ rapidly loses its intensity. In this case the process may involve a proton, split off from the olefin:

$$
\begin{array}{ccccc}
C_3H_5 & H^+ & O_2^- & & C_3H_5 \quad HO_2 \\
| & | & | & \longrightarrow & | \qquad | \\
Co^{2+}\!-\!O\!-\!Mg^{2+} & & & & Co^{2+}\!-\!O\!-\!Mg-.
\end{array}
$$

The radical HO_2, produced in the reaction between O_2^- and H^+, readily reacts with C_3H_5. A similar activating action of protons was observed in the reactions of complexes of Co^{2+} with O_2^- in the liquid phase [4.232].

The mutual enhancement of the adsorption of O_2 and hydrocarbons at temperatures far below the temperature of the catalysis was observed earlier by the adsorptive techniques [4.233].

The products of the interaction of anion radicals O_2^- with olefins on diluted catalysts (Ti, Nb, V, Mo oxides supported on SiO_2) were analyzed in [4.234]. The adsorption of $C_3H_6+O_2$ at 20 - 100°C was found to give rise to formaldehyde, acetaldehyde, and propionaldehyde in quantities comparable to the quantity of O_2^-, which on these catalysts is stabilized on the ions of the transition metals M^{n+}. The following intermediate structure has been deduced [4.234] to exist

$$\left[\begin{array}{c} H\diagdown \quad \diagup H \\ \quad C{-}C \\ H \diagup \; | \quad | \; \diagdown CH_3 \\ \quad O \vdots O \\ \quad M^{n+} \end{array} \right]^- \diagdown\!\!\!< \begin{array}{l} HCHO + CH_3CHO + M^{(n-1)+} \\ \\ CH_3CH_2CHO + M^{n+}O^- \end{array} \quad .$$

The interaction of O_2^- with ethylene on these catalysts results mainly in deeply oxidized products ($CO+CO_2$).

The anion radicals O^- on the surface are far more reactive than O_2^-. The interaction of O^- with ethylene on Mo/SiO_2 even at -183°C gives rise to the radical $H_2C{-}CH_2O^-$, which becomes the radical $CH_2{=}CH$ as the temperature is raised. The interaction of CO with O^- on the surface of Mo/SiO_2 and Cr/SiO_2 at -195°C gives rise to the radical CO_2^- [4.235].

The free radical reactions on the dilute oxide system, reported in [4.230,231,234,235] may be the initial stages of the complicated oxidative catalyzed reactions. However, in the diluted systems (below 1% of the transition metal), where the radical products are conveniently measured by their EPR spectra, the IR detection of the molecular products is impeded

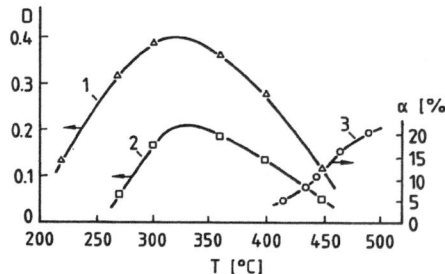

Fig.4.30. The temperature dependence of absorption bands of surface compounds, produced at the oxidation of propylene on Bi_2O_3 · MgO catalyst: (1) carboxylate complex (1430 cm^{-1}), (2) π-allyl complex (1510 cm^{-1}), (3) gaseous CO_2 yield

by the inferior sensitivity of IR spectroscopy. For this reason the conclusions regarding the mechanism of catalysis in [4.230,231] were based on the comparison with the spectra of the products of conversion of olefins on more concentrated systems (Fig.4.25].

The role of the allyl complexes in the deep and partial oxidation of olefins is established today beyond doubt. In any case, the isotopic data [4.156,157] irrefutably testify in favor of the creation of a symmetric complex in the course of the catalyzed oxidation of propylene. However, there are no indications that the allyl complexes (detected by the IR spectra) participate in the steady-state process of oxidation (Sect.4.5.2).

The kinetics of the production and decomposition of a number of surface compounds in the course of the catalyzed reaction of oxidation of propylene was studied by means of IR spectroscopy on the supported 5 Bi_2O_3 · 5 MoO_3 · MgO and 5 Bi_2O_3 · MgO catalysts [4.236]. Figure 4.30 depicts the temperature dependence of the concentration of a number of surface compounds: π allyl (the band at 1510 cm^{-1}) and carboxylate (1430 cm^{-1}). It has been assumed that all these surface structures arise from the oxidation of propylene adsorbed on the surface. The release of the oxidized products (CO_2 for the Bi_2O_3/MgO catalyst and CO_2 + acrolein for the Bi-Mo/MgO catalyst) starts at 320 - 340°C. The similar rates of the catalytic reaction and the consumption of the allyl complex allow one to assume that the latter is the intermediate product in catalysis.

254

The investigations of the chemisorptive and the catalytic activity of the ions of transition metals were performed mainly with dielectric matrices (SiO_2, Al_2O_3, MgO). The same applies to the molecular-layered structures [4.237,238]. In all these cases the electron exchange between the surface atoms of transition metal and the carrier is of little importance. On the other hand, many studies of the effects of the electronic subsystem of the solid on chemisorption and catalysis were based on model semiconductors (Ge, Si, GaAs, ZnO) whose surface atoms exhibit a rather low chemisorptive activity [4.239]. Much is to be expected from the creation of dilute systems of active atoms on the surface of semiconductors, in which the surface atoms of the transition metals would establish good electronic contact with the allowed bands of the semiconductor. Such systems can be produced by the molecular layering technique or by implanting the transition ions into the semiconductor.

This idea is directly supported by the studies of the photolysis of water molecules adsorbed on the semiconductor surface [4.240]. The quantum yield was shown to increase by nearly an order of magnitude after deposition of two layers of Cr_2O_3 (molecular layering) on the real surface of silicon. The deposition of Cr_2O_3 also resulted in a shift of the mean quantum yield $\eta(H_2)$ toward lower energies. According to our electron-vibration model of the capture of the semiconductor charge carriers by the slow adsorptive states [Ref.4.38,Chap.8] this shift arises from the difference in the maximum vibrational modes in the complexes of water with silicon and chrome. The energy released at the recombinations of the photoproduced electrons and holes goes mainly to the excitation of these vibrational modes, thus stimulating the heterolysis of the adsorbed molecules [4.38]. The coating of silicon by Cr_2O_3 resulted also in a substantial increase in the activity of the surface in the reactions of photocatalyzed decomposition of formic acid (dehydration and dehydrogenation). An illustration is given in Fig.4.31, which shows the change in the catalytic

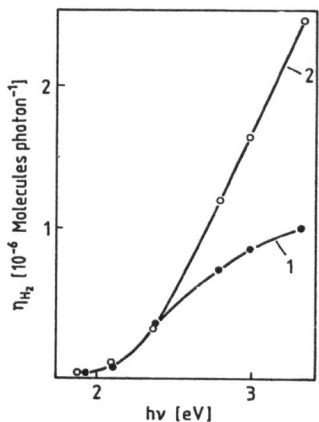

Fig.4.31. The yield of decomposition products of formic acid on Si (1) and on Si covered by one monolayer of Cr_2O_3 versus energy of photoirradiation

activity of silicon in the reaction of dehydrogenation of COOH, as a monolayer of Cr_2O_3 is being built on the surface.

4.8 Catalysis by Zeolites Containing Ions of Transition Elements

Another example of catalytic systems that have been studied to elucidate the relative importance of the collective and the local interactions in catalysis is furnished by zeolites containing atoms of transition elements. The structure of these systems was discussed in Sect.3.2. In particular, it was indicated that at the low degrees of exchange of the ions of Na^+ or H^+ by the transition ions, the latter tend to occur singly, whereas at the high degrees of exchange they occur predominantly in associations $M^{n+} \ldots O^{2-}) \ldots M^{n+}$, where O^{2-} is an oxygen ion not belonging to the aluminosilicate skeleton of the zeolite.

Thorough investigations revealed the possibility of oxidation of a number of simple molecules (CO, NH_3, H_2, C_2H_4) by NaY zeolites, in which Na^+ was partly (50-70%) exchanged for the transition-metal ions (Ag^+, Cu^{2+}, Fe^{3+}, Co^{2+}, Cr^{3+}, Ni^{2+}, Mn^{2+}) [4.241-244]. Heat treatment at up to 600°C does not destroy the zeolite structure (with the exception of CrY zeolite). The experiments indicate that the catalytic activity of zeolites varies within wide limits: from highly active CuY and AgY to practically inactive NaY and MnY. According to their activity in the oxidative reactions, the zeolites may be ordered as follows (for comparison, we also reproduce the activity series of the oxides of transition metals):

Oxidation of hydrogen:

$AgY > CuY > FeY > CoY > (CrY, NiY) > MnY$

$Co_3O_4 > CuO > MnO_2 > NiO > Fe_2O_3 > Cr_2O_3$;

Oxidation of carbon monoxide:

$CoY > AgY > CuY > CrY > NiY > FeY > MnY$

$MnO_2 > Co_3O_4 > MnO > CuO > NiO > Cu_2O > Co_2O_3 > Fe_2O_3 > Cr_2O_3$;

Oxidation of ethylene:

$CuY > CrY > AgY > FeY > CoY > (NiY, MnY)$

$Co_3O_4 > Cr_2O_3 > Ag_2O > Mn_2O_3 > CuO > NiO > Fe_2O_3$;

Oxidation of ammonia:

$CuY > CrY > AgY > FeY > CoY > MnY > NiY$

$Cu_2O > Co_3O_4 > MnO_2 > Cr_2O_3 > NiO > Fe_2O_3$.

Despite certain parallels in the activity series of the oxides and the zeolites, there are some distinctions. The oxides of cobalt and nickel zeolites are relatively inactive (with the exception of oxidation of CO by CoY). In almost all reactions the most active are the zeolites of copper and silver. The activity of the zeolite MnY is very low, almost the same as that of the initial zeolite NaY. This observation agrees with the predictions of the crystal-field theory, according to which the ion Mn^{2+} (whose crystal-field stabilization energy is zero) must exhibit minimum catalytic activity (Sect.4.1). The picture in general, however, is rather at variance with the theory. The experimental results indicate that the zeolites with cations in higher oxidized states (Cu^{2+}, Fe^{3+}, Cr^{3+}) are more active than the zeolites with the cations in the lower oxidized states (Co^{2+}, Ni^{2+}, Mn^{2+}). This conclusion is different from the results of experiments on the decomposition of N_2O on diluted solid solutions (Sect.4.7). This is an argument in favor of the assumption that the limiting stage in the conjugated process of oxidation-reduction is the reduction of the catalyst.

Some of the zeolites showed capability for stepwise catalyzing action. For this purpose the catalyst was treated with excess oxygen, followed by pulsewise admission of H_2, CO, NH_3, and C_2H_4 in succession, until all the active oxygen was exhausted. It was found that the greatest capacity for storing the active oxygen is observed with CuY; the oxygen in FeY is capable of oxidizing H_2 and CO, but not NH_3 and C_2H_4; and the active oxygen in CrY interacts in equal measure with all reductants. In contrast to these zeolites, MnY and CoY were found incapable of stepwise oxidation, while the same process for NiY was possible only above 500°C.

The low catalytic activity of zeolites containing the ions of transition elements as compared with the respective oxides (per ion of transition metal) can be attributed to the absence in the zeolite skeleton of the mobile oxygen, which could be conveyed to the active site. This surmise is confirmed by the IR spectra of the surface carbonate structures [4.199]. In the course of adsorption on CoY, CuY, and FeY, the absorption bands of CO_3^{2-} ions arise only for the oxidation of CO, but fail to appear after the admission of CO_2. The bands of carbonates on oxides also arise due to the interaction of CO_2 with the atomic oxygen of the oxide lattice.

The data on the oxidation rate of CO at 300°C on CoY, the degree of oxidation of C_2H_4 at 350°C, and the mobile oxygen content in type-Y zeolite are plotted in Fig.4.32 as functions of the percentage of the exchange of Na^+ for Co^{2+} or Cu^{2+} [4.244]. In this case the catalytic activity and the adsorptivi-

258

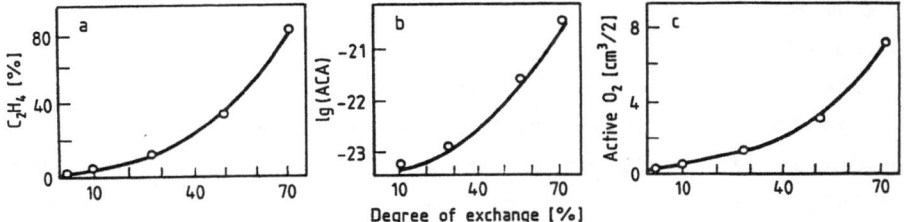

Fig.4.32. The degree of conversion of C_2H_4 at 350°C on CuY zeolite (a), the rate of oxidation of CO at 300°C on CoY, (b) the active oxygen content in CuY (c) versus the degree of exchange for the transition-metal ions

ton capacity exhibit a nonlinear dependence on the degree of exchange. The activation energy of adsorption of ethylene decreases from 155 to 100 kJ/mol, as the degree of exchange of Na^+ by Cu^{2+} in type-Y zeolite increases [4.244]. A similar strictly nonlinear increase in the activity was observed for the adsorption of CO by type-Y zeolite, as the percentage of exchange of Na^+ for mulitply charged ions increased [4.245]. This nonlinearity is explained in [4.245] by the fact that the multiply charged cations show a tendency to take the sites S_1 (Fig.3.9), on which the adsorption of CO is impossible. This tendency decreases in the order Ce^{3+} > Ca^{2+} > Ni^{2+} > Mn^{2+}, UO_2^{2+} > Cu^{2+} > Zn^{2+}. However, even with the ZnY zeolite, which in this series is the most accessible for the adsorption of CO, no more than 10% of the Zn^{2+} ions participate in the adsorption of CO. At the same time, out of 56 cation sites in the elementary cell of type-Y zeolite, 16 are represented by the least-accessible S_1 sites, one-third of the total number available.

The nonlinear behavior seems to arise from two causes: the cations' preference to occupy the unfavorable sites S_1, and the clustering of ions in the accessible zeolite cages at high concentrations of cations. The data on the stepwise oxidation of CO and C_2H_4 on CuY [4.245] indicate that for every two atoms of copper there is one atom of oxygen capable of participating in the catalytic process. The same proportion is indicated by the Mössbauer spectra for the FeY zeolite [4.246], where the bridges $Fe^{3+}-O-Fe^{3+}$ are presumably formed.

The catalytic activity of type-Y zeolites as a function of the exchange of native cations for the transition ions in the oxidation of simple molecules was also studied by BORESKOV [4.247]. The catalytic activity per atom of transition element (the atomic catalytic activty or ACA) increases with the increase in the degree of exchange for NiY, CoY, and CuY, and remains the same for FeY. The pertinent data for CoY is given in Fig.4.32b. Apparently, in the FeY zeolites the paired and clusterwise centers, containing several ions of iron and interstitial oxygen, form even at the low degree of exchange of Na^+ for Fe^{3+}. In this case, already in the course of preparing the catalyst, iron in the solution occurs as $Fe(OH)^{2+}$ rather than in the form of Fe^{3+} ions.

4.9 Adsorption and Catalysis on Magnetic and Ferroelectric Oxides

The peculiar magnetic and electronic properties of certain transition-metal oxides were already mentioned in Sect.3.4.3. A large number of works have dealt with the variations of the adsorptive and catalytic properties at the points of magnetic phase transitions [4.250-253]. Many catalysts exhibit the so-called Hedvall effect: a sharp increase in the temperature coefficient of the reaction rate at the Curie point, when the ferromagnet becomes a paramagnet. According to [4.5], Cr_2O_3 starts to exhibit catalytic activity only above the antiferromagnetic phase.

NAGAEV and LAZAREV [4.254] were the first to give a theoretical treatment of the adsorption of a one-electron atom on the surface of an antiferromagnetic crystal. It is assumed that the adsorbed atom is localized near the lattice anion, and its electron is engaged in strong exchange interaction with the cation spins. In terms of the exchange energy, the most advantageous condition is the creation on the surface of a ferromagnetic microdomain (surface ferron). This tendency is opposed by the exchange interaction between the cations, which establishes the antiferromagnetic order in the crystal. Taking account of these two competing processes, the authors of [4.254] estimated the radius of the localized ferron, and plotted the number of adsorbed atoms as a function of temperature (the adsorption isobar). It was demonstrated that the adsorbed atoms may acquire a magnetic moment equal to the magnetic moment of the adjacent ferron. For instance, in the case of the strong s–d exchange the magnetic moment of the atom adsorbed on the compounds of rare-earth elements is approximately equal to 50 atomic moments $\mu_B S$ (μ_B being the Bohr magneton, and S being the spin of the magnetic atom).

According to the results obtained in [4.254], the ferrons, localized at the adjacent adsorbed atoms, tend to merge together. This results in a peculiar attraction between the adsorbed atoms, not associated with their exchange interactions. The electron spins of the attracting atoms become parallel. This implies the possible clustering of an arbitrary number of atoms, whereas the exchange interaction allows only the pairwise attraction of atoms (atoms with the antiparallel spins). For these reasons adsorption is especially advantageous – in terms of energy – on surface defects, on which the ferrons are already localized. The experimental verification of these ideas is of considerable interest for surface science.

Catalytically important compounds of transition elements are often found among the binary oxide systems displaying the ferromagnetic ↔ paramagnetic, or ferroelectric ↔ paraelectric transitions. They crystallize into the structures of perovskite, garnet, spinels, scheelite. Currently under active investigation are perovskites, which turn out to be good catalysts of deep oxidation, capable of replacing platinum in exhaust converters [4.255]. The activity series of perovskites

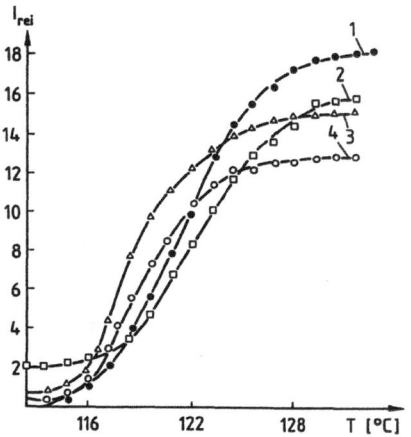

Fig.4.33. The temperature dependence of the intensity of the EPR spectrum (signal of Fe^{3+}) for adsorption of different gases on $BaTiO_3$: (1) CO, (2) pyridine, (3) C_3H_6, (4) vacuum

($LaCoO_3$ > $LaMnO_3$ > $LaFeO_3$ > $LaCrO_3$) was correlated in [4.256] with the occupancy of d orbitals and the lattice parameter.

The classical ferroelectric – barium titanate – already mentioned in Sect3.4.3, also belongs to the perovskites. Its adsorptive and catalytic properties were investigated by ROZEN-TULLER and coworkers [4.257]. The adsorption was found to affect the Curie point T_C. The Curie point was determined by means of EPR spectroscopy: by monitoring the temperature dependence of the intensity of EPR lines from the paramagnetic ions Mn^{2+} and Fe^{3+}, naturally occurring in the matrix of $BaTiO_3$ (below 0.01% by mass). In the ferroelectric phase the EPR lines of Mn^{2+} and Fe^{3+} (configuration d^5) are broadened because of the local tetragonal distortions, and after the transition to the paraelectric phase their intensity increases by nearly an order of magnitude. The temperature dependence of the intensity of the EPR lines of Fe^{3+} after the adsorption of various gases is plotted in Fig.4.33. It is clear that the adsorption of CO and C_3H_6 shifts T_C toward lower temperatures, while the adsorption of O_2 and NO causes an upward shift with respect to the Curie point for the pumped out $BaTiO_3$ sample. We know that CO and propylene are adsorbed by the donor-type mechanism, and the adsorption of NO and O_2 follows the acceptor-type mechanism. Thus the direction of the shift gives us an indication of

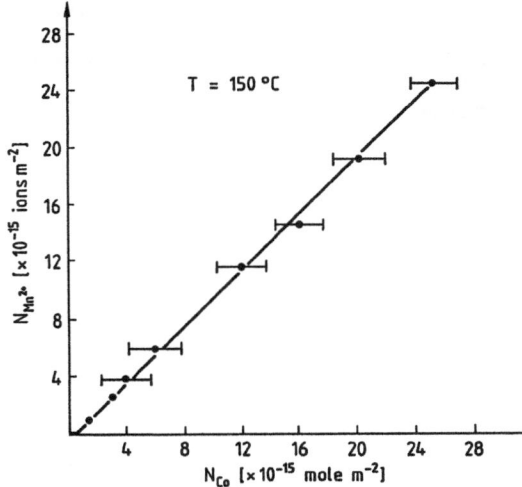

Fig.4.34. The concentration of Mn^{2+} ions in $BaTiO_3$ versus the concentration of adsorbed oxygen at 150ºC at different initial concentrations

whether a given molecule is adsorbed by the donor-type or the acceptor-type mechanism. This method of determining the sign of charging of the adsorbed molecule eliminates the need for contact electrodes, inevitable in measuring the electroconductivity, work function, and other electrophysical parameters.

As the coverage of the surface by the adsorbed CO molecules increases, the shift of the Curie point ΔT_C becomes greater, and reaches its maximum value (2.4º) at the coverage of $2.7 \cdot 10^{16}$ m^{-2}.

Simultaneous study of adsorption and EPR spectra indicates that the properties of the paraelectric phase of $BaTiO_3$ at 130 – 170ºC are entirely determined by the impurity ions Mn^{2+}. The adsorption of CO leads to the reduction of the bulk ions of Mn^{4+}

$$Mn^{4+} + CO + O^{2-} \rightarrow Mn^{2+} + CO_2 \ .$$

From Fig.4.34 it becomes clear that in the range of CO coverages from $0.3 \cdot 10^{15}$ to $2.6 \cdot 10^{16}$ m^{-2}, the number of Mn^{2+} ions emerging in the sample exhibits linear dependence on the number of the adsorbed molecules of CO: one ion of Mn^{2+} per molecule of CO. All the ions detected by EPR are located in the bulk of the catalyst, in deep layers, which do not experience any surface distortions.

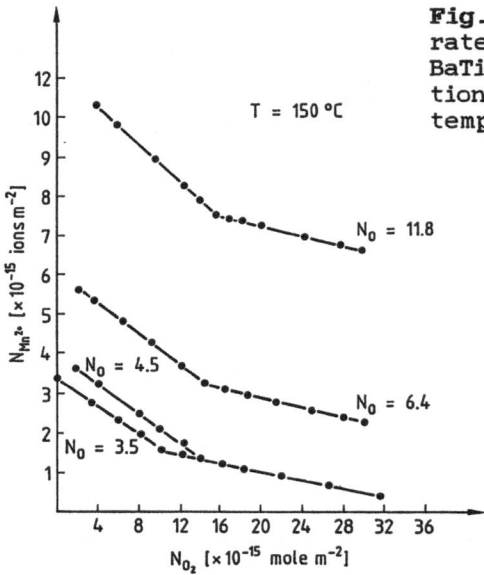

Fig.4.35. The initial logarithmic rate of (1) oxidation of CO on BaTiO$_3$, (2) superadditive adsorption of CO+O$_2$ versus reciprocal temperature

The adsorption of O$_2$ on BaTiO$_3$ gives rise to the opposite effect: the elimination of Mn^{2+} ions. The process of adsorption was found to occur in two steps (Fig.4.35), each having a different stoichiometry:

$$2Mn^{2+} + O_2 \rightarrow 2Mn^{4+} + 2O^{2-} \ ,$$

$$Mn^{2+} + 2O_2 \rightarrow Mn^{4+} + 2O_2^- \ .$$

The admission of the mixture 2CO + O$_2$ at these conditions gives rise to the catalyzed reaction of oxidation of CO, and the steady-state concentration of Mn^{2+} ions is set up, as indicated by EPR. Thus the active sites – the ions of Mn^{2+} – belong not to the surface, but rather to the bulk of BaTiO$_3$. The electron exchange between the active sites of the catalyst and the adsorbate takes place in a layer 10 – 20 nm thick, apparently by virtue of the transport of charge carriers over the polaronic band.

Figure 4.36 shows the logarithm of the initial oxidation rate of CO and BaTiO$_3$ plotted against reciprocal temperature. The catalytic properties suffer an abrupt change at the Curie

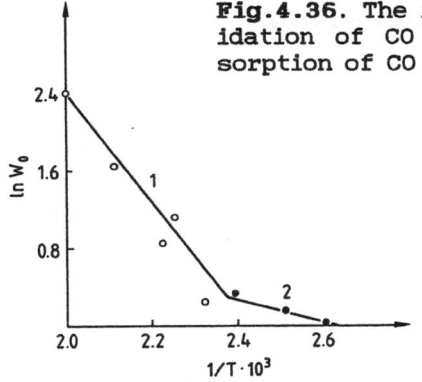

Fig.4.36. The initial logarithmic rate of (a) ox-idation of CO on $BaTiO_3$, (b) superadiative adsorption of $CO + O_2$ versus inverse temperature

point. The catalytic reaction producing gaseous CO_2 occurs only on the paraelectric phase. The separate adsorption of CO or O_2 is also observed only above the Curie point. On the ferroelectric phase – below the Curie point – there is neither catalysis nor separate adsorption of CO or O_2, although the superadditive adsorption of the mixture $CO+O_2$ does take place (Fig.4.36).

We see that the adsorptive and catalytic properties of $BaTiO_3$ exhibit a drastic change at the point of ferroelectric phase transition. The inactivity of the ferroelectric phase toward adsorption and catalysis is related apparently to the strong bending of energy bands on the surface of a spontaneously polarized crystal. This bending prevents the migration of charge carriers into the bulk of the crystal. In the paraelectric phase the spontaneous polarization disappears, and the electron transport is facilitated by the absence of the bending of energy bands.

5. The Surface of the Transition Metals

The main characteristic of a metal is the presence of free electrons (the so-called electron gas), which account for the high electroconductivity of metals. In the transition metals the d electrons are more localized than the sp electrons, the latter therefore being mainly responsible for the conduction of electricity. However, the extremely high differences in the electroconductivity (and its temperature dependence) of metals and oxides have little effect on the chemisorptive and catalytic properties of their surface. The catalytic and adsorptive properties of both the metals and the oxides depend mainly on the d electrons, which exhibit a high degree of localization.

5.1 The Band Structure of the Transition Metals

The transition metals are characterized by densely packed crystalline structures. The most commonly used catalysts γ-Fe, β-Co, Ni, Cu, Ag, Au, and all six platinum metals occur in the face-centered cubic lattice A1, where each atom has twelve closest neighbors at the vertices of a dodecahedron. A closely related coordination is assumed by the atoms in the A3 lattice (hexagonal close packing): Ti, Zn, α-Co, Mg, Cd, and some other metals. The metals V, Cr, δ-Mn, Nb, Ta, Mo, and W (and all alkali metals) crystallize in the body-centered-cubic lattice. These structures differ little in energy, and easily convert one into another.

266

The most important achievements in the study of the electron structure of metals are associated with the development of the theory of energy bands. The bands arise from the splitting of the energy levels of separate atoms when they are arranged in the crystal lattice [5.1,2]. The metals are characteristically distinguished from semiconductors and dielectrics by the over-lapping of the upper bands (e.g., the overlapping of 3p, 3d, and 4s bands in the first-period transition metals). The conductivity of metals, in contrast to the conductivity of semiconductors, decreases with the rise in the temperature because of the interactions of electrons with thermal vibrations of atoms in the lattice. The occupation of the energy band of metal by electrons is subject to the Fermi-Dirac statistics, and the borderline between the occupied and the unoccupied levels is called the Fermi level.

The special properties of the transition metals arise from the presence of d electrons. At the same time, the transition metals differ little in conductivity from the nontransition metals. In current views, the main cause of the high conductivity of metals is the existence of mobile electrons thanks to the partly occupied overlapping s and p bands. As to the d electrons, they occupy more narrow orbitals. Of these, the most localized are the 3d levels, while the 5d levels are of a more collective, bandwise nature.

It must be observed that the actual band structure of the transition metals is quite sophisticated. Considering the high concentration of electrons in metal, the calculations of its electric properties must account for the correlative effects, arising from the interelectron interactions, and the effects of screening. Strictly speaking, the problem of calculating the energy spectrum is a typical many-electron problem [5.1,3].

Simplified theories are usually used for describing the behavior of electrons in metals, theories which lay stress on either the collective or the local interactions. Most commonly used today are the models of s-d exchange, which to a certain extent combine both approaches, simultaneously accounting for the localized system of d (or f) electrons, and the collective system of conduction electrons [5.4]. The density of levels of d electrons is much greater than the density of levels of free s electrons in the same energy interval.

The qualitative energy-band model, based on the method of molecular orbitals employed by the ligand-field theory (Sect.2.1), is very perspicuous and favored by chemists. This theory is consistently presented and applied to studying the structure of metals and alloys, their magnetic properties, and the chemical bonding in the book by GOODENOUGH [5.5]. His approach subjects to analysis the direct overlapping (or nonoverlapping) of space-oriented atomic t_{2g} and e_g orbitals of the adjacent atoms of metal (rather than considering the overlapping between the metal and ligands, as done in the ligand-field theory). An essential part of these conceptions is the assumption that the atomic d orbitals retain their symmetry in the solid metal.

The configuration of d orbitals in the face-centered-cubic lattice Al is shown in Fig.5.1. The coordination number twelve corresponds to the overlapping of the twelve lobes of t_{2g} orbitals (four in each of the perpendicular planes d_{xy}, d_{xz}, d_{yz}). The remaining d electrons occupy the e_g orbitals, directed toward further-off atoms in the second coordination sphere

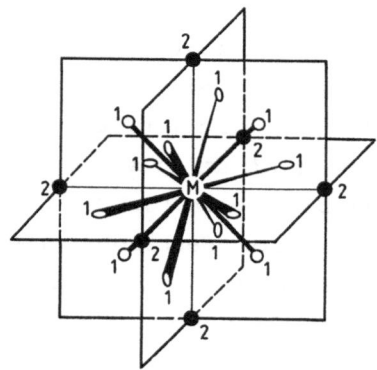

Fig.5.1. The directions of orbitals in the FCC lattice: (1) twelve closest neighbors of the central atom M, (2) six nest-nearest neighbors

along the axes x,y, and z though not associated with these atoms. The overlapping t_{2g} orbitals make up a band, and thus are responsible for the metallic bonding and electroconduction in what may be viewed as a single metallic molecule. The localized e_g orbitals do not overlap and form two levels due to Hund's splitting, or two very narrow bands.

The occupation of the metal's orbitals by electrons determines a number of properties, including the magnetic properties. For densely packed structures (A1 and A3) the criterion for ferromagnetism is the existence of partly occupied localized electron states. GOODENOUGH [5.5] quotes the empirical values of the critical localization radius R_{cr} for d electrons: R_{cr} = 0.306 nm for 3d metals, 0.394 nm for 4d metals, and 0.442 nm for 5d metals. Paramagnetism prevails above R_{cr}. The fifth- and sixth-row transition metals are paramagnetic. In the fourth row the paramagnets are the metals in the beginning of the period, in which the t_{2g} band is occupied by electrons, and the e_g levels remain free. The ordering of spins starts as the localized e_g levels begin to take up electrons.

The directionality of d orbitals and their occupancy by electrons determine the peculiarities of the crystalline structure of the transition metals. For instance, a hole in the t_{2g} orbital (in the t_{2g} band) gives rise to distortions bearing a cooperative nature. As a result, the lattice of γ-Mn (d^5s^2) becomes tetragonal rather than cubic, with axial ratio c/a = 0.93 < 1.

The heat of sublimation of the transition metals as a function of the elements' position in each period is plotted in Fig.5.2. The elements of the first and second (long) periods exhibit a two-peak pattern, similar to that observed with the oxides (Figs.2.5,6) and explained in terms of CFSE. The minimum falls on the metal of group VII [i.e., Mn(d^5s^2)]. The variations in the metallic radius of the transition metals of the first (long) row exhibit maxima in the beginning and at the end of the period, and for Mn [5.6].

Fig.5.2. The heat of sublimation of transition metals [5.6]: (1) first long period, (2) second long period, (3) third long period. The group number is plotted on the x axis

The study of GOODENOUGH [5.5] adduces a number of schemes of the energy structure of transition metals, based on the experimental findings and the ligand-field theory. These schemes were constructed under the assumption that the collectivized d electrons have the same symmetry with respect to the lattice as the individual d orbitals, which contribute to their creation. The symmetry of the latter was discussed in Sect.2.1 (Fig.2.1).

Figure 5.3a shows the most convenient arrangeement of coordiate axes in the body-centered-cubic cell of a metal. The electrons on the e_g levels are localized, and by the intraatomic fields these levels are split into two sublevels, each having the capacity for two electrons. The t_{2g} electrons form a band, although rather narrow. The relatively narrow d bands and

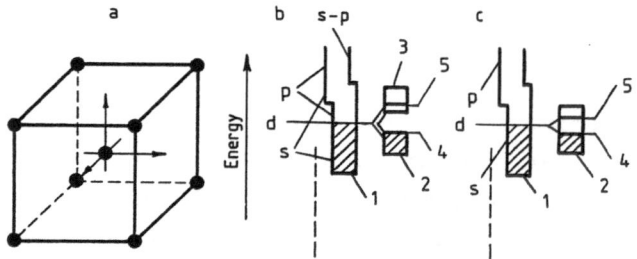

Fig.5.3. The band structure of the BCC lattice of transiiton metals [5.5]: (a) the arrangement of coordinate axes, (b) the t_{2g} orbitals less than half occupied ($n_{t_{2g}} \leq 3.0$), (c) the t_{2g} orbitals more than half occupied ($n_{t_{2g}} > 3.0$); (1) sp band, (2) the bonding t_{2g} orbital, (3) the antibonding t_{2g}^* orbital, (4) the localized $e_g\uparrow$ levels, (5) the localized $e_g\downarrow$ levels

levels may be overlapped by the wide sp bands. If the number of electrons on the t_{2g} level $n_{t_{2g}}$ is no more than three, and there are no electrons on the e_g levels ($n_{e_g} = 0$), then the transition metal with body-centered-cubic lattice is a paramagnetic (Sc, Ti, V). If $n_{t_{2g}} \leq 3$ and $n_{e_g} \neq 0$, the metal is an antiferromagnet, and a ferromagnet if $n_{t_{2g}} > 3$. The schematic energy diagrams are given in Fig.5.3b for an antiferromagnet, and in Fig.5.3c for a ferromagnet. In the former case the t_{2g} band is split into the bonding and the antibonding bands. In the latter case the splitting is impossible, but the bonding and the antibonding parts of the t_{2g} band must differ from one another. The e_g levels may overlap with the bonding t_{2g} band.

In order to understand the magnetic and other properties of the transition metals, it is also essential to know the extent to which the sp band overlaps with the d band. So far there are no reliable theoretical calculations of the relative energy of the levels and their occupancy by electrons. One has therefore to resort to the experimental data, drawing conclusions from the X-ray K spectra, UPS, synchrotron-radiation spectroscopy, X-ray scattering, and magnetic and neutron measurements.

According to [5.5], the most consistent treatment of the experimental data can be given on the assumption that $n_{sp} = 3$, n_{sp} being the number of sp electrons per atom.

A simplified empirical curve of density of states $N(E)$ for the metals Ti, V, Cr, and Mn with the body-centered cubic lattice, based on the measurements of neutron diffraction, is reproduced in Fig.5.4. The diagram shows the location of the Fermi level with respect to the bands for different metals having the same band structure, and the splitting of the t_{2g} band into the bonding and the antibonding components (Curves 1 and 2). In the case of ferromagnetic iron, the antibonding t_{2g} band forms a common sublevel with the lower e_g orbital (parallel spins), which contains 3.5 electrons per atom. According to different authorities, Fe has five or six 3d electrons. The bonding band is located in a manner similar to that of the anti-

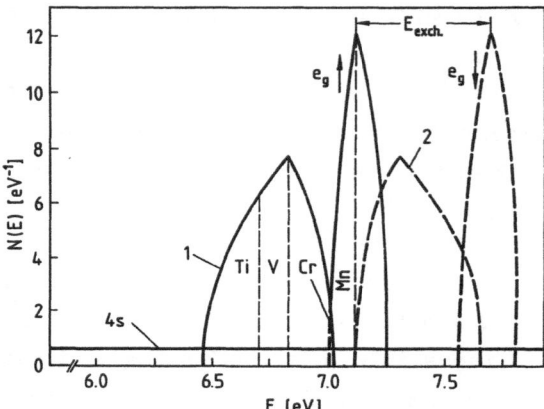

Fig.5.4. The density of states N(E) for the fourth-row metals with the BCC lattice [5.5]. Dotted straight lines indicate the location of the Fermi level for different metals. The values of energy are counted from the bottom of the sp band, 1-t_{2g} bonding band, 2-t_{2g}*-antibonding band

ferromagnetic metals, and the upper e_g^* level with antiparallel spins partly overlaps with the lower-lying $t_{2g}+e_g$ band.

The crystal structure and the relative energy of various levels and bands for the face-centered-cubic lattice of Al are shown in Fig.5.5 (indicating also the number of electrons in each state per atom). In the FCC lattice the e_g electrons are localized and split into two sublevels by the intraatomic fields according to Hund's rule. The collectivized t_{2g} electrons form a metallic band. The data on the magnetization of Ni-Cr alloys indicate that the upper e_g level lies below the upper edge of the t_{2g} band (Fig.5.5b). The stabilization of the p band in the Al structure, in contrast to the BCC lattice, is not possible, and the number of occupied sp states n_{sp} is about one, decreasing with the increase in the atomic number. In accordance with the experimental data, $n_{sp} = 0.55$ for Ni, 0.75 for Co, 0.95 for Fe, and 1.15 for Mn.

At $4 < n_{t_{2g}} < 5$ (Co) one hole falls into the d_{xy} orbital, giving rise to the bonding d_{xy} band, and at $3 < n_{t_{2g}} < 4$ there are two holes, and a d_{yz}, d_{zx} band. The concerted distortion of

272

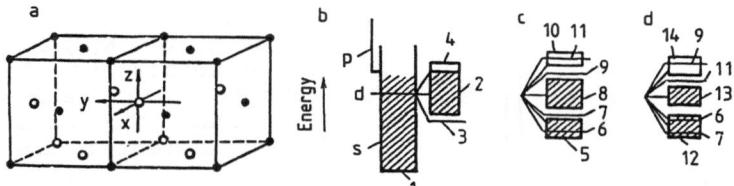

Fig.5.5. The band structure of fourth-row metals with the FCC lattice [5.5]: (a) the arrangement of coordinate axes, (b) the structure of the d band at $n_{t_{2g}} > 3$ (c/a = 1), (c) the d band at $4 < n_{t_{2g}} < 5$ (c/a > 1), (d) the d band at $3 < n_{t_{2g}} < 4$ (c/a < 1); (1) metallic s band (2 electrons), (2) metallic t_{2g} band (2 electrons), (3) localized $e_g\uparrow$ levels (2 electrons), (4) localized $e_g\downarrow$ levels (2 electrons) (5) the bonding d_{xy} band (1 electron), (6) localized $d_{z^2}\uparrow$ levels (1 electron), (7) localized $d_{x^2-y^2}$ levels (1 electron), (8) metallic d_{xy}, d_{xz} band (4 electrons), (9) localized $d_{z^2}\downarrow$ levels (1 electron), (10) the antibonding d_{xy}^* band (1 electron), (11) localized $d_{x^2-y^2}$ levels (1 electron), (12) the bonding d_{yz}, d_{xz} band, (13) metallic d_{xy} band (2 electrons), (14) d_{xz}, d_{yz} band (12 electrons per atom)

the lattice leads to tetragonal symmetry with c/a > 1 in the former case and c/a < 1 in the latter, ultimately resulting in the two-sublattice structure and antiferromagnetism at low temperatures. The corresponding band structures are shown in Fig.5.5c (c/a > 1) and Fig.5.5d (c/a < 1). At $5 < n_{t_{2g}} < 6$ the transition metals have the cubic structure down to the lowest temperatures.

The density of states N(E) for the metals with FFC lattice is reproduced in Fig.5.6. Nickel contains 10 outer electrons, and, according to [5.7], has one hole in the e_g band. According to GOODENOUGH [5.5], the number of holes in the d band of Ni is n_s = 0.55; this number is shared between the t_{2g} band (0.41) and the e_g band (0.14). The number of holes in the d band is $n_s + 1$ for Co, $n_s + 2$ for Fe, and $n_s + 3$ for Mn. The e_g levels of the metals in the beginning of the period are more localized than at the end of the period.

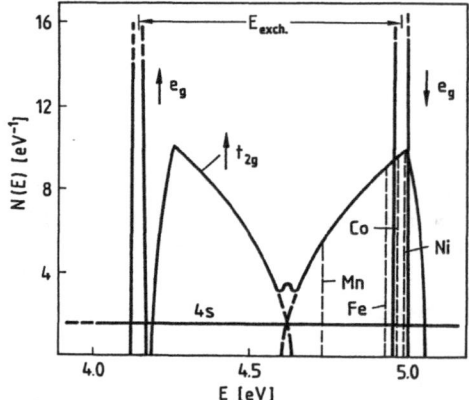

Fig.5.6. The curves of density of states N(E) for fourth-row metals with the FCC lattice [5.5]. Dashed lines indicat the Fermi level for different metals. The energy E is counted from the bottom of the sp band

Improved calculations of the electron structure of the transition metals appeared more recently in the wake of the study of Goodenough [5.8]. The most successful were the quantum-chemical calculations for clusters, comprising a small number of metallic atoms (Sect.5.6).

The wide employment of XPS and UPS techniques [5.9,10] made possible the experimental study of the structure of d bands of the transition metals. The width of the d band is about 3 eV for the first (long) period (lower in the beginning of the period), 3–5 eV for the second, and 6–8 eV for the third. The predicted structure of the d band was also detected. The metals in the second and third long periods display two maxima, characteristic of the delocalized d electrons. The occupancy of the d bands generally agrees with the schemes discussed above (Figs.5.4,6).

A similar treatment was given to other lattices including those of the heavier metals.

5.2 The Surface Orbitals in the Transition Metals

The models of the band structure of metals described above have been employed for describing the catalytic and adsorptive properties of the transition metals and alloys. BOND [5.11] used these models for producing a clear-cut description of the

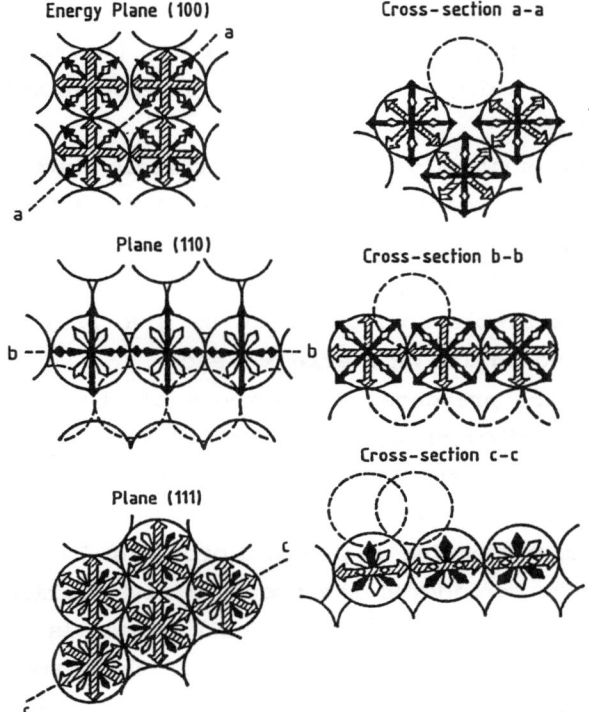

Fig.5.7. The arrangement of axes of d orbitals on the (100), (110), and (111) faces of metals with the FCC lattice [5.11]. Solid arrows denote e_g orbitals in the plane of the diagram for the (100) and (110) faces and at an angle to the plane of diagram for the (110) and (111) faces; hatched arrows denote t_{2g} orbitals in the plane of the diagram; plain arrows denote t_{2g} orbitals at an angle to the plane of the diagram

molecular orbitals, projecting outward from the surface of a transition metal, on the assumption of a simple termination of the lattice.

The scheme of the axes of the d orbitals of the surface atoms of a metal with the FCC lattice is shown in Fig.5.7. The axis of one e_g orbital (vertical) and four axes of t_{2g} orbitals (at 45°) project upward from each atom on the cubic (100) face. On the other hand, on the rhombic dodecahedral face (110) the t_{2g} orbital is vertical, and the e_g orbital projects at 45° to the surface. None of the orbitals is vertical on the octahedral

(111) face: both the t_{2g} and the $e_g{}'$ orbitals have three directions in which the electron density is the greatest (trigonal symmetry), the inclination being 54°44' for the t_{2g} orbitals and 35°16' for the e_g orbitals.

The symmetry of the outwardly projecting orbitals determines the type of possible bonding both between the surface atoms and with the adsorbed atoms and molecules. The interaction between the surface atoms may result in their rearrangements, leading to a more stable configuration, or give rise to surface conduction (the so-called skin effect), etc.

The bonding requires the presence of electrons of the projecting orbitals. The distribution of electrons between the s, p, and d orbitals determines the chemical properties of the surface. WEINBERG [5.12] has assumed equiprobable distributions of electrons between the projecting s, p, and d orbitals. In the preceding section we have considered the possible distribution of electrons between the s and d bands in the bulk of metals of the first long period. Accordng to MELIUS [5.13], if the structure of the ground state of the transition metal is s^2d^n (n = 6 for Fe, n = 8 for Ni), then the prevalent electron configuration of the surface atom, participating in the bonding, will be s^1d^{n+1} (s^1d^7 for Fe, s^1d^9 for Ni). The prevalent configuration of the surface atom of metal early in the period (Sc, Ti, V) is s^2d^n.

The main distinction between the structure of the surface orbitals of the transition metals and that of the oxides arises from the fact that in a metal the d orbitals of adjacent atoms are close to each other. Their overlapping (δ interaction) is responsible for the peculiar properties of the surface of transition metals.

According to [5.14], a consistent theoretical description of the surface phenomena of the transition metals must take into account both the localized and the free electrons. The role of the nearly free electrons consists in screening. The space around the two bound atoms of metal may be separated into the "bonding region", which comprises most of the electrons in-volved in establishing the bonding and counterbalancing the nuclear repulsion, and the two antibonding regions. The nonzero density of electrons in the antibonding regions inevitably weakens the bonding.

For a long time information regarding the electronic proper-ties of the surface of metals was derived chiefly form electro-physical measurements.

While in semiconductors the surface conductivity σ depends largely on the concentration of free charge carriers in the subsurface region of the crystal, in the metals, where the concentration of free electrons is quite high, the value of σ depends wholly on the mobility of the electrons. It has long been noted that the conductivity of thin metallic films is lower than that of the same metal in the bulky samples. J.J. Thomson was the first to ascribe this fact to the additional scattering of electrons on the film's surfaces, see [5.15]. These effects become noticeable when the thickness of the film becomes comparable with the mean free path of the electrons. The theory developed by BELETSKY and FUCHS [5.15] gives a fair description of the conductivity of thin films of simple sp metals, taking into account the scattering on the surfaces. It was found that, depending on the particular metal, the scatter-ing can be both specular and partly specular (changing the quasi-momentum of the electron reflecting from the surface).

It must be observed that the interpretation of the data ava-ilable on the conductivity of thin films is still ambiguous. Theoreticians are dealing with perfect monocrystalline films. The patchy structure of real films and the existence of steps, kinks, and dislocations may highly enhance the scattering of the charge carriers with respect to the same effects in the bulk.

Another electrophysical "surface" parameter is the elec-tronic work function. The work function ϕ, or the energy needed to remove an electron form the Fermi level of a metal to infinity, is, strictly speaking, a bulk property: it defines the location of the Fermi level in the bulk. Nevertheless, being affected by the asymmetry of the potential on the metal–vacuum boundary, it can be used for studying the electron structure of the surface [5.15,16].

Apart from measuring the thermoelectronic emission, the thermoelectronic work function ϕ_t can also be determined by measuring the contact potential difference ϕ_{cont} by the method of vibrating capacitor. In this case $\phi_t = \phi_{cont} - \phi_{el}$, ϕ_{el} being the work function of the material of the reference electrode. The drawback is that ϕ_{el} depends strongly on the state of the surface of the reference electrode. Finally, the work function for the individual faces of a single crystal can be determined by the method of electron emission (field–emission microscope FEM, Sect.2.2.4).

The numerous measurements of the work function for single crystals of metals positively indicate that ϕ_t is different for different faces of the same crystal [5.16]. For instance, $\phi_t = 4.58$ eV for the (100) face of W, 4.40 eV for the (111) face, and 4.30 eV for the (110) face. Similarly, $\phi = 4.40$ eV for the (100) face of Mo, 4.15 eV for the (111) face, and 5.10 eV for the (110) face. In most cases the values of ϕ_t quoted in literature are average; the real surface is patchy, each patch having a slightly different value of ϕ_t.

A thorough survey of theoretical calculations of the work function of metals and alloys, together with an analysis of the influence of adsorbed atoms on the work function, can be found in [5.17], where the authors make extensive use of the density functional formalism. Both the jellium model and the Newns–Anderson model are discussed.

The theoretical approaches to the electron structure of the surface of transition meals were developed by GRIMLEY [5.18,19], VAN DER AVOIRD [5.20], SCHRIEFFER and GOMER [5.21], and other authors. Both the theoretical calculations and the qualitative reasoning lead to the conclusion that, as in the

278

semiconductors, the surface levels and bands in metals generally differ from the levels and bands in the bulk. The quantum-chemical calculations of the molecular orbitals of small clusters of atoms of transition metals have been carried out in recent years. Ultimately such calculations may be extrapolated to the entire surface (Sect.5.6).

The considerable improvement of the spectral investigative techniques has allowed direct measurements of the surface levels in the transition metals and determination of the configuration of orbitals. This especially applies to the UPS technique, as well as to the method of electron-tunneling spectroscopy (Sect.2.2.4). The discovery of the surface d levels in transition metals, whose energy differs from that of the respective levels in the bulk, was one of the most brilliant developments of surface science in the recent years. These works, which will be discussed in the next section, generally confirm the concepts of outwardly projecting localized surface d orbitals.

5.3 The Real Surface of Pure Metals

The opportunities for studying the clean metallic surface, free from adsorbed gases, were opened in the sixties by the development of ultrahigh-vacuum technology. Also very advantageous was the development of Auger spectroscopy and LEED, allowing study of both the chemical composition of the surface and its crystalline structure. In recent years high-quality monocrystals of some transition metals have become available.

Although "ultrahigh vacuum" is a rather loose term, it usually refers to pressures below 10^{-7} Pa (10^{-9} torr). With modern equipment it is not hard to bring the pressure down to 10^{-8} Pa (10^{-10} torr), and some installations are capable of producing 10^{-12} Pa (10^{-14} torr). The number of collisions of gas molecules of mass M at the pressure p (Pa) and the temperature T with one square centimeter of a surface per second is given by

$$n = p/(2\pi MkT)^{1/2} \qquad\qquad (5.1)$$

At room temperature $n = 2 \cdot 10^{19}\ p/(M)^{1/2}\ cm^{-2}s^{-1}$. Metallic surfaces usually contains about 10^{15} atoms per square centimeter. Assuming that the sticking coefficient for the chemisorption equals unity we find that at 10^{-8} torr the surface becomes covered by a monolayer of gas $(M = 20-30)$ after 10^2-10^3 s; at 10^{-10} torr the same will take 24 hours. Thus the pressure of 10^{-9} torr is a sensible limit, below which one may hope to study atomically clean metallic surfaces.

Most investigations of clean metallic surfaces employed low-energy electron diffraction (LEED).

The surface by itself is a cause for the anisotropy of the properties of the metal in the subsurface layer, in particular, changing of the interatomic distances. Theoretical calculations indicate that the variations in the dynamic constants near the surface are caused by variations in the interatomic distances [5.22]. In the transition metals, the interatomic distances may sometimes be changed without drastic rearrangements; in the nontransition metals the surface often suffers irreversible reconstruction. This distinction ultimately derives from the more diffuse potential of the d shell in comparison with the sp orbitals, and thus the lower potential barriers between different surface structures. The electronic causes of the reconstruction of the (001) face of the transition metals Cr, Mo, and W are discussed in [5.23]. This face - in all BCC crystals of transition metals - is characterized by the high density of surface states near the middle of the d band. The interaction between the orbitals of adjacent atoms enhances both the bonding and the antibonding properties of the surface states. The surface d states (d_{xy} and d_{xz}) experience broadening, and their energy becomes lower. This is accompanied by a 10% contraction of the surface. If the LEED pattern shows no displacement of maxima for the upper layer, this indicates similarity of the surface structure to the structure of the bulk, i.e., the absence of rearrangement. Analysis of the relative intensity of the LEED indices often points to the variations of the amplitude of vibrations and interatomic distances in the direction normal to the surface, that is, to the anisotropy of the subsurface layer [5.24,25]. The lattice parameter parallel of the surface often remains unchanged. For the less densely packed high-index faces the compression beteen the upper layer and the subsequent layers is almost always observed. Of the low-index faces the least densely packed is the (110) face of Ni, Ag, Cu, and Ir, and for the (100) face of Mo, Fe, and Ni the compression may be as high as 10% (in the direction normal to the surface [5.26-30]).

The change in the amplitude of vibrations of the surface Mo atoms on the (100) face is reflected by the decrease in the surface Debye temperature (-120°C) with respect to the Debye temperature for the bulk (90°C for Mo). For the heavier atoms these variations are less pronounced (2-6° for (100) faces of W, Rh, Pt, Ir [5.31-34]). In the more densely packed structures - Ni(111), Ag(111), Pt(111) - the upper layer often exhibits expansion rather than contraction: by 3-5% for (111) face of Rh and by 1.5% for Pt(111). A considerable compression was observed only with Fe(111) (by 10-15%) [5.26,35-40]. The LEED pattern for the clean face W (110), which exhibits changes in the intensity but no displacement of the peaks, was also explained by a parallel shift of the upper plane [5.41].

On the other hand, LEED patterns for a large number of surfaces indicate the reconstruction of the surface layer. Such reconstruction was observed with Pt, Au, and Ni: at least some of their faces have a structure which differs from the FCC structure of the bulk.

Let us consider in more detail the structure of the metal's crystal faces, using the example of the crystal of platinum [5.16,43-50]. As indicated by LEED, the (111) face has the same

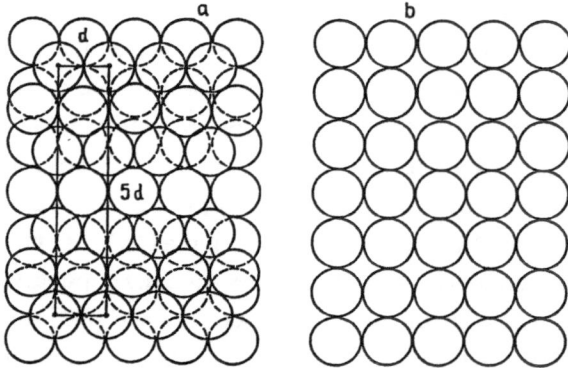

Fig.5.8. Hexagonal rearrangement of the upper layer on the (100) face of Pt, resulting in (5 x 1) structure [5.45]; (a) reconstructed layer, (b) nonreconstructed layer

structure as in the bulk. This, the most densely packed structure, persists up to the melting point. The (100) face should have been composed of square elementary cells 0.277 nm across; instead, as demonstrated in [5.46], the surface elementary cell is rectangular, one side being the same as in the bulk, and the other five times as large [denoted as a (5x1) structure]. The observed diffraction pattern was explained by the formation of hexagonal close packing, superimposed on the lower-lying layers (Fig.5.8). The distance between the surface atoms of Pt along one basic axis equals 5/6 of the interatomic distance in the bulk (the same along the other basic axis). Thus the six rows of the surface layer are precisely superimposed on the five rows of the penultimate layer. More recently this structure was shown to be more precisely defined as (5x20). This structure is very persistent. Heating to 800°C results in a slight rotation of the surface hexagonal structure by 0.7°, which is designated by Pt (100)$_{hex}$ R.0.7° and persists almost up to the melting point. The reconstruction is caused by the lowered density of free electrons on the surface and by the fact that the increased density of packing lowers the free-surface energy.

The interconversions of structure (1x1) and the hexagonal structure (5x1) on Pt(100) surface were studied recently by the tunneling electron microscopy method [5.50a].

Pt(110) has a stable rearranged structure (1x2), which in its chemical properties is similar to the (100) face of Pt [5.48]. At 830°C this structure becomes a (1 x 1) structure.

The study of the platinum neighbors Ir and Au also points to their (100) faces' having hexagonal structure, approximately described as (5x1). Actually, the hexagonal structure exhibits axial contraction [5.49,51]. The rearrangement of the surface layer was detected also for a number of other faces [5.24,52,53]. Usually the more loosely packed faces tend to assume a more dense packing with closely spaced atoms.

At first, attempts were made to explain these unusual structures by the presence of impurities. However, thorough investigations employing Auger spectroscopy capable of detecting foreign atoms in concentrations down to 10^{-3} of a monolayer failed to indicate the presence of impurities. On the contrary, in the presence of adsorbed gases (CO and hydrocarbons) the structure (5x1) quickly disappeared and was replaced by the origianl (1x1) structure of the (100) face of Pt [5.48]. The structure (1x1) pertains not to the impurities, but to the platinum crystalline face itself, although it is not stable. The same structure can be obtained by adsorbing oxygen on and desorbing it from the faces (100) of Pt and (100) of Ir. The unstable (1x1) structure quickly rearranges into the structure (5x1), shown in Fig.5.8a.

The simple low-index faces (100), (111), and (110) of a monocrystal are usually more stable. However, if the crystal is cut a small angle (5 – 15°) to one of these planes, the resulting plane corresponds to high values of Miller indices, and its surface undergoes rapid rearrangement. Moreover, LEED patterns indicate that such cuts are formed of low-index steps of nearly monatomic height. The stepped (vicinal) face of Pt, composed of (111) face 6 atoms wide and monatomic step of the (100) face, is

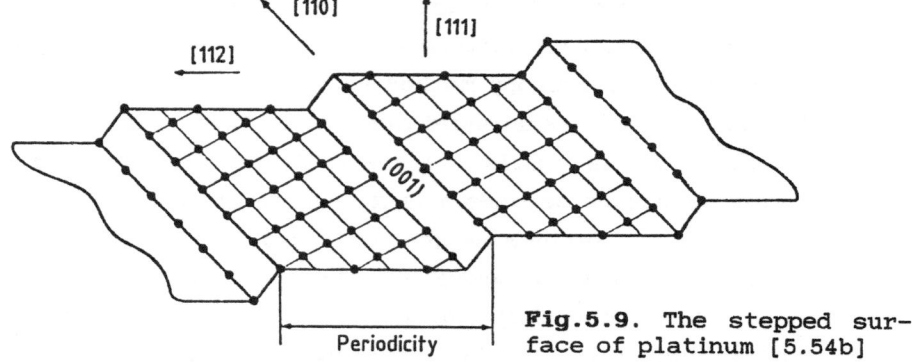

Fig.5.9. The stepped surface of platinum [5.54b]

shown in Fig.5.9. This structure - 6(111) x (100) - corresponds to the Miller indices (557). If the crystal is cut at an angle to two low-index faces at once, the resulting structure will be even more complicated: Fig.5.10 shows the surface 7(111) x (310) Pt with secondary steps (kinks) [5.54b].

Vicinal faces exhibit high thermal stability and for pure Pt and Cu in vacuum may persist almost up to the melting point. In some cases they are stable even in the presence of contaminants. Platinum faces with Miller indices (112), (113), (122), and (012) were stable in all experimental conditions, not being affected by heat treatment, adsorption of various gases, etc. BLAKELY and SOMORJAI [5.44] consider these faces, together with the low-index faces (111), (110), and (100), to be of special

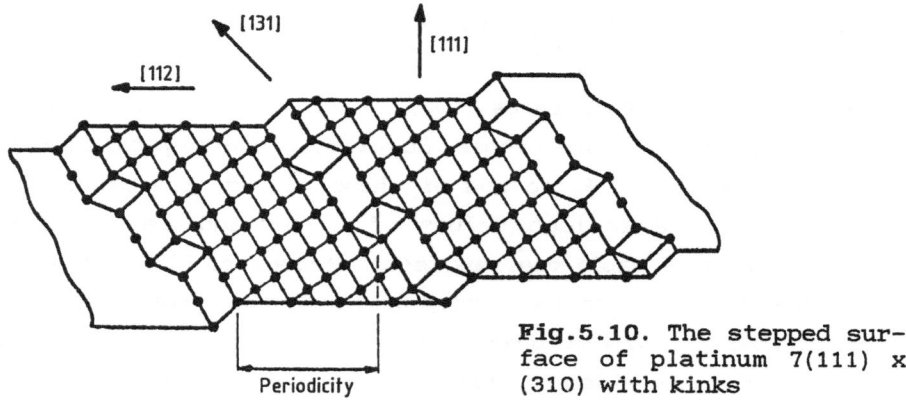

Fig.5.10. The stepped surface of platinum 7(111) x (310) with kinks

importance for catalysis. The faces with wider steps (Figs.5.9,10) are unaffected by impurities.

It must be observed, however, that the rearrangements of the surface may be triggered by foreign atoms in concentrations so low as not to be detected by the modern techniques. Using Auger spectroscopy, SAVCHENKO [5.55] has demonstrated the impossibility of obtaining a clean surface even on .999 Pt. The contaminants (Mg, Ca, Ba) after heat treatment in vacuum ($8 \cdot 10^{-8}$ Pa) tend to concentrate on the surface. Nickel and iron (0.999 pure) were found to carry sulfur and phosphorus on the surface. Alternating oxidation and reduction of iron at 1000°C failed to produce an "Auger-clean" surface.

The accumulation of impurities on the surface is a common property of metals. A number of authors have paid attention to the accumulation of carbon on the surface. This process is aided by the crystallochemical correspondence between the (0001) face of graphite and the FCC lattice of many metals. The LEED and Auger spectroscopic techniques were employed, for instance, for studying the accumulation of graphite on the (111) faces of nickel [5.56]. Three equilibrium states of carbon were detected on nickel: (a) below 790°C a multiatomic layer of graphite, epitaxially oriented by the basal plane onto the (111) face; (b) between 790 and 905°C, a monolayer, or epitaxial two-dimensional graphite phase; and (c) carbon atoms in a very low concentration. Thus, at 905°C there occurs a two-dimensional phase transition from the condensed phase to the two-dimensional gas.

A two-dimensional phase transition on the iridium surface — the melting of crystalline graphite film, giving rise to the surface carbon "gas" — was reported in [5.57].

Graphite film is also easily formed on platinum. Carbon diffuses from the bulk to the pure platinum surface at 1000°C (and at even lower temperatures), and can be detected by the characteristic peak at 276 eV in the Auger spectrum [5.16]. Contamination may change the relative stability of crystal faces.

Small quantities of sulfur cause rearrangement of the surface layer of platinum into the structure built up predominantly from (100) faces. Fully reversible structural changes at the oxidation and reduction of metallic surfaces were more than once observed by LEED.

The reconstruction of the surface may either be of a cooperative nature, or proceed at the expense of diffusion of individual atoms. The latter can be conveniently monitored by the field-emission electron (ion) microscope, or with the aid of labeled atoms. The field-emission ion microscope permitted measurement of the activation energy of self-diffusion for the individual faces of a single crystal. For tungsten, for instance, the activation energy of self-diffusion has the following values: 88 kJ/mol for the (110) face, 51 kJ/mol for the (211) face, and 84 kJ/mol for the (321) face. With the rhodium tip the values of E_D were 15 kJ/mol for the (111), 52 kJ/mol for the (311), 58 kj/mol for the (110), and 84 kJ/mol for the (100) faces; on the (111) face the self-diffusion was observed even at $-220°C$ [5.58] Attention should be called to the distinctions between E_D for W and Rh. Dense packing prevents diffusion – this is indeed the case with tungsten, but not with rhodium, since (111) is a densely packed face [5.59].

These contradictions are explained by the different mechanisms of diffusion. In those cases where E_D is the lowest, the diffusion is of a cooperative nature. The field-emission ion microscope method has demonstrated that the atoms often tend to travel in pairs. The atoms in a pair may be not too close to each other: in rhodium, for instance, the atoms in the pair are separated by 0.75 nm.

The measurements of the diffusion on the (221) face of tungsten indicate that the values of the activation energy E_D and the preexponential factor D_0 for individual self-diffusion of atoms are $E_D = 75$ kJ/mol and $D_0 = 10^{-3}$ cm^2/s, while for pairwise diffusion the same values are extremely low: $E_D = 29$ kJ/mol and $D_0 = 10^{-12}$ cm^2/s. The atoms may also associate into large clusters, which migrate over the surface as a unit. Clusters dissociate at high temperature [5.60].

All data obtained by LEED and the field-emission ion microscope pertain to single crystals. The chemical procedures of preparing metallic specimens for the investigations of catalysis and surface phenomena often produce polycrystalline surfaces. Such preparations are likely to have a large number of point defects, dislocations, domains, etc. For instance, the specimens crystallized from melt usually contain $10^5 - 10^7$ dislocations per square centimeter, and a lot of point defects. Deformation may raise the concentration of dislocations to $10^{10} - 10^{11}$ per square centimeter. The films obtained by condensation from vapor phase are, as a rule, polycrystalline. Even the monocrystalline (epitaxial) films abound with defects: silver films, for instance, were shown to contain $10^8 - 10^{12}$ dislocations per square centimeter, and a large number of other defects [5.61].

The UPS techniques made possible study of the band structure of the transition metals. The data reported in the literature usually refer to the band structure of the bulk of transition metals, mainly to the structure of inner bands. The order of occupancy of d bands generally fits in with the schemes discussed in Sect.5.1.

The improved sensitivity and, especially, the use of soft excitation sources (e.g., the helium source, Sect.2.2.2) allow detection of the surface levels in the transition metals. The d levels, differing in the energy from the bulk levels, were detected immediately below the Fermi level E_F. Using synchrotron spectroscopy and tuning the photon's energy to the appropriate transition, it was possible to unambiguously classify these levels as the surface ones. The use of ARUPS (Sect.2.2.2) has determined the symmetry of the surface orbitals.

EASTMAN and CASHION [5.62] were among the first to detect the surface states on metallic surfaces (in their case, Ni). Study of the surface of the copper (111) face by UPS and FEM

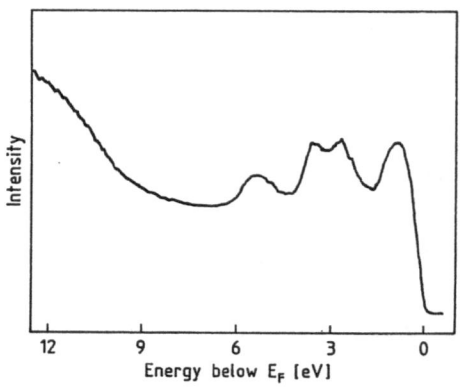

Fig.5.11. The UPS spectrum of the (100) face of W [5.68]

Intensity

12 9 6 3 0

Energy below E_F [eV]

indicates the existence of surface states about 0.4 eV below E_F [5.63,64].

A large number of investigations have dealt with tungsten [5.65-69]. Figure 5.11 shows the UPS spectrum (He source) of W(100) in the region of valence levels. Attention is drawn to a sharp peak located 0.2-0.3 eV below E_F. This peak is very sensitive to contamination and disappears almost completely after the admission of H_2 or C_2H_4. This peak was ascribed to the surface states on tungsten. A similar peak (0.4 eV below E_F) was detected with W(110). The peak at 3.4 eV below E_F pertains to the bulk d band and is little affected by adsorption. The appearance of the surface level at 0.2-0.4 eV below E_F on the (100) face of W is also confirmed by the optical diffuse reflection spectra [5.70].

Similar states, very sensitive toward contamination and adsorption, were detected by UPS on Pt(100) [5.71,72]. They also are located a few tenths of an electron volt below the Fermi level and pertain to the surface states or to the subsurface d bands with easily excited electrons.

The large width of the 5d bands of the heavy transition metals (5-8 eV) makes precise identification of UPS peaks difficult. However, the changes of the crystallographic structure of faces of Pt, Ir, and Au are found to cause considerable changes in the spectrum.

DIONNE and RHODIN [5.73] carried out a consistent investigation of surface states on different faces of Rh, Pd, Ir, and Pt using FEM spectroscopy at −195⁰C. The experimental curves R(E) were compared with the calculations of the band structure of these metals, performed by ANDERSON [5.74]. In all cases there was a definite correlation between the calculated locations of bands and the peaks on the curves R(E). For example, the peak at 0.31 eV for the (100) and (110) faces of platinum corresponds with the calculated maximum of the t_{2g} band (0.35 eV below E_F). The maximum density of surface states is almost inevitably observed near the Fermi level. These data seem to confirm the validity of the model of outwardly projecting d orbitals (Sect.5.2), although they point also to the important role of spd hybridization.

5.4 The Surface of Alloys

Interpretation of the properties of alloys containing transition metals has long been based on the concept of continuous variation of magnetic and electronic properties with the changes in the composition of the alloy. These concepts derived from PAULING's notions regarding the "d-nature of the metallic bonding", presumably determined by the proportion of d electrons taking part in the bonding [5.75]. Today these concepts are only of historical interest.

The theory of the band structure of transition metals [5.5] explains the changes in the magnetic properties of the alloys of transition metals with nontransition ones by the occupancy of the d band. Most investigations were concerned with the alloys of metals of groups VIII and IB. For instance, according to magnetic data, 9.4 states out of 10 in the 3d band of Ni are occupied by electrons, leaving 0.6 of a hole in the 3d band per one atom of Ni. Alloying of Ni with copper (a metal with an occupied d band) in the proportion 53% Cu and 47% Ni must result in the occupation of the d band by s electrons of copper. If the surface properties of the alloy exhibit correlation with its bulk properties, this composition must show a drastic change in the surface properties.

One of the first (and the most often quoted) attempts to explain the surface properties of alloys by the occupation of holes in the d band was made by DOWDEN and REYNOLDS [5.76]. The numerous discussions that followed dealt mainly with the question of the relative importance of the local and collective effects in the catalytic and adsorptive properties of alloys – which is more important: the presence of holes in the d band of the alloy, or the existence of surface transition atoms with the appropriate symmetry of d orbitals?

Experiments indicate that the phase composition of the alloy and hence its properties often depend on the preparation procedure. Even the measurements of magnetic properties, which are the bulk properties par excellence, yield different results. In particular, in Cu-Ni alloys, occupation of the d bands was attained at different percentages of copper, as derived from the magnetic data obtained by different researchers.

For the ideal solid solution A + B the heat of mixing ΔH is zero, and the entropy of mixing is calculated under the assumption of the statistical distribution of atoms of both components. The surface composition at these conditions is given by the equation

$$\frac{x_s^A}{x_s^B} = \frac{x_v^A}{x_v^B} \exp\left[\frac{(\sigma_B - \sigma_A)s}{RT}\right] , \tag{5.2}$$

where x_x^A and x_s^B are the atomic fractions of the components A and B on the surface, and x_v^A and x_v^B are the same for the bulk; σ_A and σ_B are the free-surface energy (surface tension) of the components A and B; s is the average area associated with one atom on the surface; T is the Kelvin temperature; and R is the molar gas constant.

From (5.2) it follows that the surface must become enriched
ith the metal having the lower surface tension (so-called
egregation). Unlike the bulk composition, the surface composi-
ion depends strongly on the temperature, nearing the bulk com-
osition as the temperature increases.

In the presence of the interaction between the components A
nd B the heat of mixing ΔH depends on the values of E_{AA} and
$_{BB}$ - the energies of the interaction between the nearest-
eighbor atoms A and A (B and B), determined from the heat of
ublimation of metal A or (B) - and of E_{AB} - the energy of the
nteraction between the nearest-neighbor atoms A and B, deter-
ined from the heat of formation of the compound AB. The value
f ΔH can be approximated by the expression

$$\Delta H = Nz[E_{AB} - (E_{AA} + E_{BB})/2]\, x_V^A(1 - x_V^A)\ , \qquad (5.3)$$

here N is the total number of atoms and z is the number of
earest neighbors. By virtue of (5.2,3) the composition of the
urface monolayer can be expressed as [5.77,78]

$$\frac{x_s^A}{x_s^B} = \frac{x_s^A}{x_V^B}\, \exp\left[\frac{(\sigma_B - \sigma_A)s}{RT}\right]$$

$$\cdot \exp\{(\frac{\Omega(\ell+m)}{RT}\left[(x_V^B)^2 - (x_V^A)^2\right] + \frac{\Omega\ell}{RT}\left[(x_s^A)^2 - (x_s^B)^2\right]\}(5.4)$$

here ℓ is the fraction of the closest neighbors in the same
lane (top layer), m is the fraction of the closest neighbors in
he adjacent plane; and $\Omega = Nz[E_{AB} - (E_{AA} + E_{BB})/2]$. The atom on
he (111) face has six closest neighbors; $\ell = 6/12$, m = 3/12.

Further improvements of the theory of ideal solutions were
oncerned with varying the composition of the top four atomic
ayers [5.77,78]. The layers below the fourth are assumed to
ave the same composition as in the bulk.

The results were applied to the Au-Ag alloys, which form
olid solutions in the entire range of concentrations. The cal-
ulations indicate that the surface is enriched with silver,

whose heat of sublimation is lower. From (5.4) it follows that the silver excess on the surface diminishes with increasing temperature. There is more silver on the (100) face than on the (111) face because the number of bonds which have to be ruptured to form the (100) face is greater than the number of bonds that are broken in the production of the (111) face.

As indicated by KELLEY [5.79], most thermodynamic theories of segregation disagree with experiment. In some cases the slow diffusion prevents equilibrium between the bulk and the surface of the alloy. Study of the diffusion of Al atoms toward the surface of the Ni-Al alloy indicates that the diffusion is considerably retarded when the composition of the surface reaches the values corresponding to a specifically ordered structure.

DOWDEN [5.80] has put forward a new concept of the electron structure of the surface of an alloy. In his view, the alloy AB can be presented as a set of "ensembles" consisting of several (4-9) atoms of components A and B. Each ensemble is formed at the expense of a characteristic overlapping of orbitals, similar to the overlapping in a pure metal (Sect.5.1). If the general disposition of orbitals is the same, their occupancy by electrons will depend on the number m of atoms A in the ensemble $A_m B_n$. The probability of occurrence of m atoms A in the ensemble is calculated on the basis of the statistical distribuiton. The changes in various surface properties (work function, chemisorptive and catalytic activity) can be explained by the continuous variations in the number of different ensembles on the surface.

Numerous investigations of the surface of alloys with the aid of Auger spectroscopy indicate that the surface is often dominated by one of the components. Chemical reactions between the components sometimes give rise to ordered surface structures, in which the distribution of ensembles can hardly be assumed statistical.

SACHTLER and coworkers [5.81,82] have demonstrated that the properties of copper-nickel alloys (and especially the composition of their surface) often depend on the existence of two different phases: Alloy I (80% Cu and 20% Ni) and alloy II (2% Cu and 98% Ni). At the high temperature of fusion (above 322°C) a homogeneous solid solution can be obtained in the whole range of concentrations. At 200°C the equilibrium in the bulk is established rather slowly, and the metastable solid solutions can survive for a long time. At the same time, the Ni–Cu films, deposited in vacuum and in ultrahigh vacuum, separate into two phases in the course of several hours, as indicated by the X-ray data. The segregation into the two phases depends on the preparation procedure. In vapor-deposited Cu–Ni films the surface is enriched with copper, which has a high rate of diffusion and a low surface tension as compared with Ni. The resulting film consists of the conglomerated crystals of alloy II, sandwiched between layers of crystals of alloy I.

The data on the work function for Cu–Ni alloys [5.82] can be given a fair interpretation from this standpoint. In a wide range of concentrations the work function for the Cu–Ni film remains constant and equals 4.75 – 4.80 eV, the work function being 4.60 eV for pure copper and 5.04 eV for pure nickel.

In general, similar results are obtained when the Ni–Cu alloys are prepared by other procedures: reduction of oxide mixtures and decomposition of mixed salts. Even with the single crystals of Ni–Cu alloys the surface after annealing at above 300°C is enriched with copper [5.83]. The segregation of copper on the surface of copper–nickel alloys was detected by Auger spectrosocpy [5.84,85], UPS [5.79,86,87], and low-energy ion-scattering spectroscopy [5.88].

Subsequent detailed studies of the Auger and UPS spectra of alloys for different depths of subsurface layer point to the nonmonotonic variation of the composition near the surface [5.79,84,86]. While the top monolayer in the Cu–Ni alloy is enriched with copper, the next two or three layers are depleted, and the "bulk" composition is observed only in the fifth layer and down. Similar oscillations were observed with other alloys as well.

In the fifth row the counterparts of the copper–nickel alloys are the alloys of silver with palladium. This system was found to produce both homogeneous alloys (obtained by film deposition with subsequent annealing at 400°C) and segregated alloys. The measurements of the work function were carried out on Ag–Pd films, which form solid solutions in almost the entire range of concentrations [5.89]. For almost all alloys the work function is close to that for pure silver (4.4 eV), which points to the accumulation of silver on the surface. For alloys containing over 90% Pd the value of the work function rapidly increases and nears the value for pure palladium (5.2 eV). The enrichment of the surface of Ag–Pd alloys with silver is also detected by Auger spectrosocpy [5.90–92]. The Auger spectra derive from a number of top layers, and hence the composition of the uppermost layer may be expected to differ even more from the bulk composition.

The alloys of gold with platinum [5.89–93] have properties similar to those of Ni–Cu alloys. At low temperatures the two phases are formed with subsequent segregation and enrichment of the surface by the alloy dominated by the metal with the higher mobility and the lower surface tension (in this case, gold).

According to LEED [5.93b], a superstructure is formed on the surface of the Cu–Au alloy (3:1), whose lattice constant differs from the lattice constant for the bulk.

Interesting results were obtained with Pd–Au alloys [5.84,87,89]. The surface of annealed bulky specimens, as indicated by Auger spectra, is enriched with gold, in accordance with theoretical predictions [5.77]. After exposure to intense ionic bombardment, however, the surface becomes rich in palladium. A similar phenomenon is observed after heat treatment in oxygen.

The X-ray analysis fo Pt-Ru films, obtained by vacuum deposition and annealed at 600°C, indicates the presence of two phases: the Ru-rich alloy and the Pt-rich alloy. Both phases coexist in alloys containing from 30 to 50% of Pt. The work function varies steadily from 5.11 eV (the value for pure Ru) to 5.71 eV (the value for pure Pt). These results imply that there is no enveloping of one phase by the other: there is just a slight domination of the Ru phase. The little crystals of both phases coexist on the surface.

In this connection, let us observe that when studying the polycrystalline alloy structures we have to bear in mind the not negligible effects of the transfer of mass by intergranular diffusion. The rate of diffusion along the borders of grains can be greater by many orders of magnitude than the rate of diffusion within the grian. The processes of diffusion on the interface between different metals have been given an exhaustive analysis for a large number of metallic alloys [5.94].

In the presence of impurities the compostion of the alloy surface can be quite different from the surface composition of pure specimens. In the most thoroughly investigated alloys of group VIII metals (Ni, Pd, Pt) with group IB metals (Cu, Ag, Au), contamination by S, O, and C tends to raise the surface concentration of the group VIII metal.

Currently, a great deal of attention is focused on the surface of ferromagnetic metals and alloys. The investigations in this domain have been made possible by the development of highly sensitive magneto-optical polarization techniques [5.95]. By studying the spectral and kinetic behavior of the equatorial Kerr effect, KRINCHIK with co-workers [5.96] discovered the segregation of the ferromagnetic phase in nonmagnetic catalysts ZrNi and ZrNiH$_{2.8-3.0}$ and the clustering of ferromagnetic Ni at the oxidation of the catalyst at elevated temperatures. The measurements in the UV region indicate that the layer of Ni is found in the subsurface region under the layer of Zr. This specific surface structure was shown to be responsible for the catalytic activity of the alloy in the reactions of hydrogenation and isomerization of hydrocarbons. Of special interest is the discovered effect of the external magnetic field on the segregation of Ni in the Zr–Ni alloys. The emerging ferromagnetic phase exhibited pronounced anisotropic properties. The combination of magneto-optical techniques with electron spectroscopy seems to be very promising for the study of catalysis by metals and alloys.

5.5 Supported Metals

Deposition of metals on substrates (silica gel, alumina, active carbon) in some cases allows one to obtain the metal in a state of very high dispersion. Investigations of high–dispersion metals became possible with the development of the investigative techniques of electron microscopy, X–ray spectral line broadening, magnetic techniques, etc. It has been found that by depositing salts on the high–porosity carriers (silica gel, alumogel) with subsequent reduction it is possible to get metallic crystals about the size of the pores in the carrier (1.5–3.0 nm). The size of Ni particles in the pores of silica gel was reportedly as small as 0.55 nm [5.97].

Another commonly employed technique consists in producing thin metallic films by the condensation of metallic vapor. Precipitation of Pd on crystals of α-Al_2O_3 gives particles ranging from 1.5 to 8 nm [5.98]. Ultrathin films of platinum were obtained on glass and mica, the size of particles being smaller than 1.5 nm [5.99].

Now there is the question regarding the equilibrium shape of microcrystals. VAN HARDEVELD and HARTOG [5.100] assumed that maximum stability corresponds to the lowest possible number of free valences on the surface and arrived at the conclusion that the most stable shape of small particles (600 – 700 atoms) is spherical. Indeed, the electron micrographs indicate that the small particles of a metal (Pt, Pd, and others) 1.5–2.5 nm in diameter often have a near spherical shape. Nonequilibrium particles, obtained by low-temperature reduction of metallic salts and by low-temperature vacuum condensation, may occur in a variety of shapes. The equilibrium shape with the low concentration of dislocations is assumed by the metallic particles at the high temperatures, favoring the rearrangement of the surface and rapid diffusion of surface atoms of the carrier toward high-stability centers with high coordination numbers.

For metals with the FCC lattice, the stable shape of small particles can also be represented by octahedral and cuboctahedral microcrystals [5.101,102]. According to [5.101], the stable shape of particles 2.0 – 2.5 nm in diameter is the octahedron, and the stable shape of particles bigger than 3.0 nm is the cuboctahedron. If the interaction between the metal and the carrier is strong, the particles may assume the shape of hemispheres or semicuboctahedrons, attached to the carrier by their basal planes.

POLTORAK [5.101] has proposed what he calls the "mitohedral technique", which consists in relating the measured property of the high-dispersion metal to the relative number of atoms on the surface and in the bulk of the crystal. This requires that the properties of atoms on the faces, edges, and corners of the crystal should depend only on their coordination number, that is, on the number of the nearest neighbors, and not on the dispersity. The experiment indicates that with the size of particles of 2.0 nm and bigger most physical bulk properties (except for the magnetic properties) of the metal exhibit no dependence on the dispersity. This implies that starting with this size the microcrystals develop a metallic lattice.

The relative number of atoms with low coordination numbers increases with the decrease in the particle size, as the total number of atoms in the particle decreases. Figure 5.12 shows the octahedral model of a small particle of metal of the platinum or nickel type with the FCC lattice. The edge contains 9

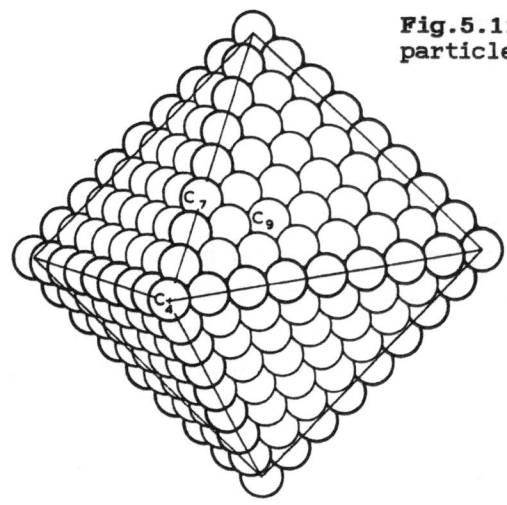

Fig.5.12. Octahedral model of a particle of 489 atoms [5.100]

atoms; the entire particle comprises 489 atoms. We can see three types of surface atoms with different coordination numbers (c.n.): sites C_9 on the octahedral (111) face, sites C_7 on the edge, and sites C_4 at the apexes (the subscripts denote the coordination number, or the number of the nearest neighbors).

The structural properties of the regular octahedral platinum crystals of varying size are listed according to POLTORAK [5.101] in Table 5.1. It is clear that the most drastic changes in the number of atoms with different coordination numbers occur in a rather narrow region (from 0.8 to 4.0 nm) in the "mitohedral region", as it is called by Poltorak.

The proportion of surface atoms with different coordination numbers (C_4, C_7, C_9) is plotted in Fig.5.13 as a function of the reduced size of the octahedral particle $d_{red} = 0.1105 \sqrt{N}$, N being the number of atoms in the particle [5.100]. The number of C_4 atoms in an octahedron is always six, and their proportion falls rapidly as the size of the particle increases. The total number of C_7 atoms increases with the increase linearly in particle size, and that of C_9 atoms quadratically. At d > 2.0 nm the surface is dominated by C_9 atoms, at 1.5 > d > 0.8 [nm]

Table 5.1. The dispersity and the surface properties of octahedral platinum crystals [5.61]

Edge length, expr. of no. atoms	expr. [nm]	Part of surface atoms	No. of atoms in crystal	Number of atoms with diff. c.n.[*] edge C_7	face C_9	bulk	Mean c.n. of surface atoms	Mean no. of bonds per atom
2	0.55	1.0	6	6	0	0	4.00	2
3	0.895	0.95	19	12	0	1	6.00	3.16
4	1.000	0.87	44	24	8	6	6.94	3.81
5	1.375	0.78	85	36	24	19	7.46	4.23
6	1.650	0.70	146	48	48	44	7.76	4.51
7	1.925	0.63	231	60	80	85	7.97	4.72
8	2.200	0.575	344	72	120	146	8.12	4.89
9	2.475	0.53	489	84	168	231	8.23	5,00
10	2.750	0.49	670	96	224	344	8.31	5.10
11	3.025	0.45	891	108	288	489	8.38	5.18
12	3.300	0.42	1156	120	360	670	8.44	5.25
13	3.375	0.39	1469	132	440	891	8.47	5.31
14	3.850	0.37	1834	144	528	1156	8.53	5.35
15	4.125	0.35	2255	156	624	1469	8.56	5.40
16	4.400	0.33	2736	168	728	1834	8.59	5.44
17	4.675	0.31	3281	180	840	2255	8.62	5.48
18	4.950	0.30	3894	192	960	2736	8.64	5.50

[*] The number of C_4 atoms (vertex of octahedron) for all crystals equals six.

by C_7, and at d < 0.5 nm by C_4. In the crystals bigger than 10–20 nm the fraction of atoms in the corners, on the edges, and on the faces becomes so small as to make the effects arising from the behavior of the coordination number on the surface unobservable (this would require the properties of the

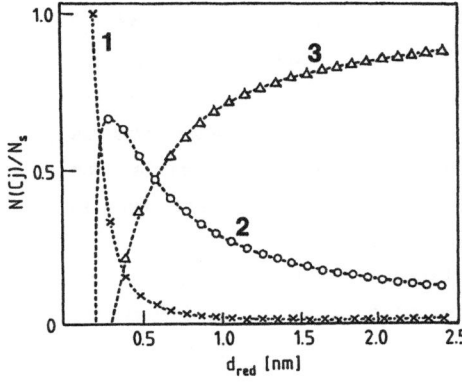

Fig.5.13. The proportions of surface atoms (1) at the tops (C_4), (2) on the edges (C_7) and (3) on the faces (C_9) of the octahedral platinum particle versus the particle size [5.100]

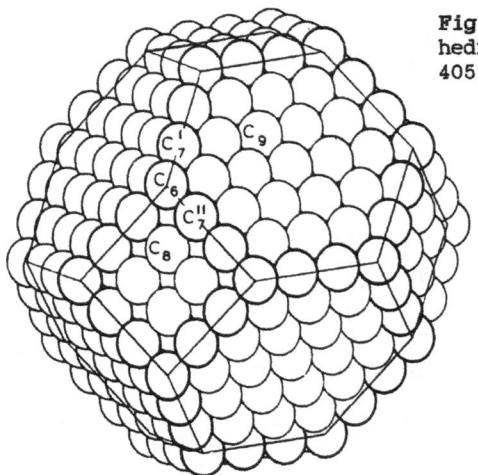

Fig.5.14. Regular cubocta-hedron model comprising 405 atoms [5.100,102]

surface atoms on the edges and faces to differ by more than three orders of magnitude).

A similar treatment was given of cuboctahedral FCC structures in [5.100,102]. A regular cuboctahedron may comprise 38, 113, 201, 405, etc. atoms. The model of a cuboctahedron made up of 405 atoms is shown in Fig.5.14. This is a high-symmetry crystal with hexagonal faces. Five types of surface atoms with different coordination numbers can be distinguished on the surface of a cuboctahedron: (1) C_9 atoms on the (111) faces, (2) C_8 atoms on the (100) faces, (3) C_7' atoms at the borders of the (111) faces, (4) C_7'' atoms at the borders of the (100) and (111) faces, and (5) C_6 atoms at the apexes. The atoms C_7' and C_7'' differ only in the positions of their nearest neighbors. In general, the proportion of atoms in the corners, on the edges, and on the faces of the cuboctahedron exhibit the same trend as with an octahedral particle; the relatively small difference consists in the persistence of the C_9 atoms on the (111) face in even the smallest cuboctahedron (N = 38, d_{red} = 3.86).

These octahedral and cuboctahedral models represent the complete crystals. Crystals can also occur in an incomplete form if the number of atoms is less than required. The growth

Fig.5.15. A model of the B_5 (113) site on the (100) face [5.102]

of a regular cuboctahedron nucleates on the (100) or (111) face and proceeds around peculiar sites designated B_5 [5.100,102-104]. A model of the stepped B_5 center (113) on the (100) face is shown in Fig.5.15. The stepped centers B_5 can be formed also on the (111) faces. Having started with B_5, the growth of the new face proceeds to completion; the new cuboctahedron is formed when its edges and corners are finally built. The sites C_7' and C_7'' are similar in structure to B_5, and BOND [5.102] designates the former as $B_5(110)$ and $B_5(113)$.

Many authors associate the B_5 sites with the peculiar catalytic and adsorptive properties. To a certain degree they are similar to the atoms on the vicinal steps of a single crystal (Figs.5.9,10). The proportion of the surface B_5 atoms is the greatest when the particle size is 2.0 – 2.5 nm, in the middle of the "mitohedral region".

SCHLOSSER [5.103] considered a spherical model of FCC microcrystals, obtained by smoothing out the edges and corners of a cuboctahedron. In this model the surface atoms constitute a smaller fraction of the total number of atoms in comparison with the true cuboctahedron. A noteworthy implication of this model is the increase (and not the decrease, as in the case of a cuboctahedron) in the number of B_5 sites with the increase in the particle size. For the particles bigger than 2.0 nm the ratio of the number of B_5 sites to the total number of surface atoms becomes constant and equals 0.5. Of course, these conclusions must be checked against reliable experimental data regarding the number of B_5 sites.

It was calculated [5.105] that the preferred shape of the small particles of metals with the FCC lattice (less than 150 atoms) is the icosahedron – a 20-sided regular polyhedron with 12 vertices. Indeed, icosahedral particles were observed experimentally in the condensation of small metallic particles from vapor. For instance, particles of gold 4 – 25 nm in size were icosahedral [5.106]. However, the difference in energy between the cuboctahedrons and icosahedrons is small, and the actual shape of particles depends on the surface structure of the support, and the presence of impurities and adsorbed gases. The electron micrographs of small particles of platinum ($\simeq 10.0$ nm) on graphite indicate that they all have the cuboctahedral shape and are oriented by the (111) face to the (0001) face of graphite [5.107]. It must be also observed that small particles cannot be imagined as a collection of incompressible spheres, like the large particles. In small particles the interatomic separation is smaller by several percent.

The study of a variety of small particles of the coating meal using the STEM (scanning transmission electron microscopy) method showed that the most abundant particle form is a truncated octahedron. There are also cubic octahedrons, hexagonal and pentagonal particles, twin forms, etc [5.108].

Highly important also is the question about the possible influence of defects on the properties of small metallic particles. As shown by POLTORAK [5.101] for the metals such as platinum, the defects constitute less than 10^{-4} of the total surface (below 500°C). This proportion increases only for small crystals (< 2.0 nm); in this case, however, we are dealing with the mitohedral region, where the fraction of surface atoms with different coordination numbers is large.

Actually, the metallic particles rarely occur in a uniform-size grade. They usually vary in size considerably, depending on the particular properties of the metal, the properties of the support, the size of pores, temperature, the presence of impurities, etc. With the same processing conditions the dispersity is higher for diluted specimens. The experimental data on the dispersity of metallic particles in supported catalysts were analyzed in [5.109]. It was found that for a large number of samples of Pt, Pd, Ni and Fe catalysts supported on SiO_3, Al_2O_3, and the like, the reciprocal dispersity, expressed as the proportion γ of atoms on the surface of the particles of the active component, exhibits a linear dependence on the surface concentration C_{surf} of the latter: $\gamma^{-1} = a + bC_{surf}$, a and b being empirical constants.

The dispersity of the supported metals was studied by XPS. The intensity of "metallic" lines in the XPS spectrum depends on the depth from which the electron comes, and thus on the microcrystal size. The strength of the lines from the carrier depends on the extent to which the surface is covered by the active material. The dispersity of platinum, determined from the ratio between the intensities of spectra of Pt and the support (Si, Al), was shown in [5.110] to correlate with the dispersity as assessed from the adsorption of H_2 and O_2.

The magnetic properties of very small particles of ferromagnets and antiferromagnets differ considerably from the properties of bulk specimens. Particles smaller than 10 nm represent individual domains and in the magnetic field behave as paramagnetic atoms with a very large dipole moment. This phenomenon got the name of superparamagnetism (collective paramagnetism, quasi ferromagnetism) and was experimentally investigated by a number of authors [5.111-115].

Some authors succeeded in demonstrating that the Curie point of a ferromagnetic metal goes down with the increase in dispersity, being, for instance, 365°C for bulk nickel and 274°C for 0.8% Ni/SiO_2 specimen (Ni particle size of 1.2 nm) [5.115].

Very small particles have no Curie point at all. For example, the reduction of mixed Ni-Zn oxalates at 230–250°C leads to the formation of supported paramagnetic Ni/ZnO catalysts composed of microcrystals smaller than 2.0 nm. Subsequent sintering results in ferromagnetic catalysts with a lowered Curie point, which is attributed to the polydispersity of such preparations. The magnetic techniques are very well suited for studying the microcrystals of Ni, in particular, these measurements determine the distribution of Ni particles according to size [5.112]. The superparamagnetic region for Ni/SiO$_2$ typically corresponds to the particles measuring from 0.8 to 4.0 nm. In the case of supported cobalt catalysts the minimum size of particles exhibiting domains of spontaneous magnetization (ferromagnetism) was found to be about 1.2 nm [5.114].

The transition of the magneto-ordered structure of the microcrystals into the superparamagnetic structure with increasing dispersity or with changes in temperature can also be studied by Mössbauer spectroscopy [5.111,116-119]. As demonstrated by Néel [5.111], the small metallic particles are never magnetically isotropic, and various energy barriers have to be surmounted at the magnetizaton. After the removal of the magnetic field the magnetization disappears in a finite relaxation time

$$\tau = \tau_0 \exp(-E_A/kT) \quad , \tag{5.5}$$

where $\tau_0 \simeq 10^{-9}$ s, k is the Boltzmann constant, T is the temperature in Kelvin, E_A is the average energy barrier, $E_A \simeq KV$, K is the constant of magnetic anisotropy, and V is the volume of the particle in cm^3.

The measured value of K for microparticles of iron is about 10^{-2} J/cm^3. The values of the relaxation time can be estimated down to 10^{-2} s by measuring the magnetic susceptibility and down to 10^{-8} s by Mössbauer spectroscopy.

Mössbauer spectroscopy was used for studying the magnetic anisotropy of the supported Fe/MgO catalysts [5.118b]. The estimated dispersity [derived from the value of E_A in (5.5)] agreed well with the values derived from electron micrographs and X-ray spectral line broadening. The size of the particles on the specimens containing 1% and 3% of iron varied from 1.5 to 4 nm. For the paramagnetic particles of this kind the values of τ and E_A were small ($\tau < 10^{-8}$s), and the Mössbauer spectrum consisted of a single peak. With the concentrations ranging from 16% to 40% of Fe the size of particles became 10 – 40 nm; such ferromagnetic particles exhibited magnetic anisotropy and were characterized by the large values of τ and E_A. The Mössbauer spectrum consisted of six lines. According to [5.118b], the calculations of the magnetic anisotropy by the Néel theory, together with other data, point to the high concentration of C_7 sites on the surface of small particles of iron (1.5 – 5nm). No electronic interaction between the particles of iron and the MgO support could be detected. A similar study of superparamagnetism and magnetic phase transitions in the small particles of iron was carried out in [5.119]. Many authors point to the possible interaction between the metal and the support.

The electron transfer from platinum to the carrier was discovered in silica-alumina supported platinum catalyst [5.120]. The number of the donor centers on the silica-alumina surface was determined by the adsorption of perylene, which produced a cation radical active in the EPR spectrum. Similarly, the concentration of acceptor centers was determined by the adsorption of tetracyanoethylene (TCNE), which forms an ion radical. In the presence of platinum the intensity of the signal from TCNE$^-$ went up, and from (perylene)$^+$ down. This points to the electron transfer from platinum to silica-alumina (charge transfer complexing between platinum and the support). To a certain degree the loss of an electron abates the metallic quality of platinum.

The interaction of metal with the carrier in the supported Pd/Al_2O_3 catalysts was detected by MARKEVICH et al. [5.121]. The treatment of this catalyst with gamma rays stabilizes a much greater number of paramagnetic defects on Al_2O_3 than does the irradiation of pure Al_2O_3.

Auger spectroscopy was used for studying the interaction between metal and the carrier in the supported Pt/Al_2O_3 catalysts [5.122]. While for the pure support $\gamma-Al_2O_3$ the Auger peaks of aluminum Al_{LMM} and oxygen O_{KLL} were separated by $\Delta E = E_{AL} - E_O = 444.4$ eV, the equivalent difference for the supported catalyst $Pt/\gamma-Al_2O_3$, reduced in H_2, was 441.3 eV. As the location of the oxygen peak in the Auger spectrum is known to be stable [5.123], we may deduce the existence of the chemical shift in the Auger spectrum of aluminum, due apparently to the change in the charge and coordination of Al. Since this effect arises with the addition of platinum, we are faced, most probably, in the interaction between platinum and aluminum, with the possible creation of pairs $Al^{\delta-}$ ... $Pt^{\delta+}$ linked either directly or via oxygen. The transfer of charge can also be accomplished over the system of fluctuating states, whose concentration in amorphous dielectrics is high.

Studies of interaction between metal and carrier often employ the XPS technique. In many cases the metal was found to acquire a positive charge (oxidation), while the cations of the support become negatively charged (reduction [5.124]). The positive charging was observed with the metallic atoms Re, Ir, Pd, and Pt, supported on Al_2O_3 and type-Y zeolites. With platinum the shift of the XPS band of 4d levels amounted to 2.4 eV (on $\gamma-Al_2O_3$), which corresponds to the oxidation of platinum to Pt^{2+}. The shifting involved not only the 4d levels, but also the 4f levels. With rhenium on $\gamma-Al_2O_3$ the shift constituted 2.6 eV (Re^{4+}). The interaction (the transfer of electrons) between the metal and the carrier, according to XPS, is the greater, the smaller the particle.

Some researchers pointed to the possible fusion of the supported metal with the metallic component of the carrier. For instance, according to [5.125], heat treatment of Pt/Al_2O_3 in H_2 at above 500°C produces the alloy Pt_3Al. Other researchers [5.126] challenge this suggestion and assume that the transfer of electrons gives rise to the reduced forms of oxides. For example, Ti_4O_7 is formed upon the heat treatment of Pt/TiO_2 in H_2, the platinum atoms in contact with the carrier assuming a positive charge.

The phenomenon of charge transfer between metal and support accompanied by reduction of the latter was defined by TAUSTER [5.127] as the SMSI (strong metal support interaction) effect. The adsorption capacity and catalytic activity of a metal suffer an abrupt change during this interaction. An abundance of recent papers has been concerned with the SMSI effect, particularly for supported metals active in the Fisher-Tropsch synthesis. One of the explanations for the changes in adsorption capacity and catalytic activity of the metal due to the SMSI effect is the metal sinking into the support on partial oxidation of the latter.

5.6 Small Metallic Clusters

A large number of work has been concerned with attempts to obtain an atomically dispersed metal on a carrier. Atomically dispersed silver, for instance, was reportedly obtained by condensation of small amounts of silver on carbon film [5.128]. Apparently, in some cases metallic atoms M^0 can be obtained by carefully reducing the metallic complexes supported on a carrier. Such atoms were detected by EPR.

In any case, the stability of isolated metallic atoms on a support is quite low. As indicated by ANDERSON [5.129], the equilibrium energy of attraction between the atoms of transition element in a pair or in a small cluster equals approximately half the heat of sublimation, and hence the equilibrium between the vapor and the isolated atoms at medium temperatures (e.g., at 450°C) is biased toward pairs and clusters. One may ask whether there are any kinetic factors favoring the prolonged existence of nonequilibrium isolated atoms on the carrier surface.

Immediately after the reduction of diluted oxide systems containing ions of a transition metal, the resulting transition atoms (e.g., Ni^0) may be located far enough from one another. However, the interaction between the metal atom and the support relies only on the van der Waals forces and is characterized by energy of no more than 35–40 kJ/mol, while the activation energy of the surface diffusion of Ni^0 on, for example, SiO_2 is about 12 kJ/mol. With the normal value of the preexponential factor the diffusion coefficient of Ni^0 atoms on SiO_2 at 450°C will be of the order of 10^{-3} cm^2/s. This ensures the high mobility of atoms and rapid clustering.

Some researchers have tried to obtain atomically dispersed metals in the zero-valence state by reducing the transition ions contained in zeolites [5.130]. Almost all of these attempts proved futile. The reduction in hydrogen of zeolites containing ions of Cd, Zn, Ni, Ag, Hg, Pt, and Pd, is accompanied by the rapid diffusion of atoms toward the outer surface of the zeolite, where they tend to form crystals 10 – 100 nm in size. These crystals are readily detected by magnetic, micrographic, and XPS techniques. The reduction may result in the breakdown of the zeolite lattice. For example, the crystallization of silver on the outer surface of AgY zeolite was explained by the scheme

$$+ \ 2\,Ag \ + \ \tfrac{1}{2}O_2 \ . \tag{5.6}$$

Evidently the bonding of the neutral atom M^0 with the lattice is weak, despite the strong electric fields inside the zeolite's cages.

The stoichiometry of the adsorption of H_2, the IR spectra of the resulting OH groups (3640 cm^{-1}), and the exchange of these groups with D_2 were studied in [5.131] on platinum-containing zeolites, obtained by a similar procedure: the exchange of Ca^{2+} for $Pt(NH_3)_4^{2+}$ in CaY zeolite with subsequent reduction. The experimental findings lead to the conclusion that under these conditions the platinum clusters, comprising no more than six atoms, stabilize in the big cages of type-Y zeolite. The clusters apparently are electron deficient due to the partial electron transfer from the cluster to the zeolite' s skeleton Y:

$$Pt_6 + Y \leftrightarrows Pt_6^+ + Y^- \ .$$

Metallic palladium in a near-atomically dispersed state was reportedly obtained in type-Y zeolite [5.132]. As indicated by EPR (and, to some extent, IR spectroscopy), after heat treatment in oxygen at 500°C of PdY zeolite almost all of the palladium occurs in the form of Pd^{2+} ions mainly at the S_1' sites; a small fraction (about 1%) is represented by Pd^{3+} ions. According to [5.132], reduction by hydrogen at room temperature results in the production of atomically dispersed palladium, stabilized somewhere near the strong Lewis centers (Al atoms). The reduced atomic palladium does not adsorb either hydrogen or oxygen, although it does interact with CO. Simultaneously, about 8% of the initial palladium converts into Pd^+ ions, which give rise to two EPR signals (g_z = 2.41, $g_{x,y}$ = 2.11 and g_z = 2.28, $g_{x,y}$ = 2.10). This is presumably caused by the partial electron transfer between Pd^0 and the zeolite's skeleton. Heat treatment in hydrogen at 200°C results in the migration of palladium toward the outer surface of the zeolite and the production of palladium crystals measuring about 2nm. Heat treatment in oxygen restores Pd^{2+} in the zeolite's cages.

The production of small metallic clusters in type-Y zeolites was investigated by means of various novel physical techniques [5.133-137]. These studies confirm the creation of ultrahigh-dispersion particles of Ag, Pt, Pd, and Rh (down to the atomically dispersed state) by the reduction of metallic complexes in the zeolite's cages. Some XPS measurements indicate the almost total absence of the positive charging of the particles of platinum in CaY zeolite [5.133,134], contrary to the results of [5.131].

The interatomic separation in the ultrahigh-dispersion metallic particles was assessed with the aid of the EXAFS technique (extended X-ray absorption fine structure spectroscopy). In the X-ray absorption spectrum the metallic atom the coefficient of absorption exhibits a drastic change ("the absorption edge") upon the transition from one level (e.g., the L level) to the next one (e.g., the K level). The fine structure of the absorption edge results from the interference between the emitted X-rays and the X-rays reflected from the adjacent atoms. By subjecting the fine structure to Fourier analysis one can determine the interatomic separation. This method can be employed with synchrotron radiation: a powerful tunable X-ray source (Sect.2.2.2).

The EXAFS technique permits measurement of the distance between the atoms in the active site of the catalysts even when this has been impossible to do using diffraction methods. In small platinum particles the interatomic separation was found to be 0.265 nm, whereas in bulk paltinum it equals 0.277 nm [5.135,136].

Another technique of obtaining small metallic clusters consists in reducing the polynuclear complexes on the surface of high-dispersion carriers. A large number of polynuclear carbonyl complexes of platinum metals have been obtained lately, e.g., $Rh_4(CO)_{12}$, $Rh_2(CO)_3(\pi-C_3H_5)_2$, $Ru_3(CO)_{10}NO_2$, $Ir_4(CO)_{12}$, and $HOs(CO)_{10}(CHCH_2)$. Many of these exhibit peculiar catalytic properties; for instance, the polynuclear complexes of rhodium are capable of catalyzing the production of ethylene glycol from CO and H_2. Such polynuclear complexes can be viewed as the intermediate class between the heterogeneous and the homogeneous catalysts.

Reduction in hydrogen of the γ-Al_2O_3-supported complex $Ir_4(CO)_{12}$ gave rise to extremely small metallic particles [5.138]. They adsorb hydrogen in the proportion of one H atom per Ir atom, and carbon monoxide in the proportion of two CO molecules per Ir atom. The supported iridium particles are supposedly extremely small (< 1 nm) and contain as little as four atoms. The complex $Rh_6(CO)_{16}$ in type-Y zeolite can be reduced in vacuum at 100°C without damage to the zeolite's structure [5.139]. The ultrahigh-dispersion flakes of Rh (one atom thick) were obtained on the surface of Al_2O_3. As indicated by electron microscopy, the flakes have a circular shape and comprise from 7 to 40 atoms [5.140]. The clusters of five rhodium atoms were directly discernible in the electron microscopy [5.141].

The clustering of a small number of atoms, bearing no semblance to metallic crystals, was observed when metallic alloys were supported on the substrates in a small concentration [5.142]. With the concentration of 1-2%, the metals may form combinations which do not occur in bulky samples. For example, by supporting Ru-Cu or Os-Cu on silica gel it was possible to obtain clusters where these metals were mixed in arbitrary proportions, although in the bulk they segregate and do not form solid solutions. The structure of bimetallic clusters was, as indicated by X-ray analysis, amorphous, while the X-ray photograph of the monometallic specimen 1% Cu/SiO_2 indicated the existence of microcrystals of copper about 20 nm in size.

In view of the outstanding importance of platinum-rhenium supported catalysts in the reforming and other conversions of hydrocarbons, the structure of binary Pt-Re clusters is currently under especially active investigation. Studies of a large number of binary complexes of group VIII metals with metals of groups IV-VII, supported on an oxide carrier, indicate that the reduction gives rise to the high-dispersion metallic particles of the group VIII element [5.143]. According to electron microscopy, the prevailing size of the metallic particles is no bigger than 1.0 nm even after reduction at 650°C. Metals of groups IV–VII do not as a rule reduce to the zero-valence state and serve as the stabilization sites for the disperse particles of the group VIII metal.

The study of small metallic particles is interesting also with a view to verifying the quantum-chemical models of clusters. Starting with a single atom of a transition metal and progressively building up the number of atoms in the cluster, it is possible to monitor the changes in the energy and location of electron levels up to the point when the further increase in the number of atoms ceases to affect the levels, i.e., when the truly metallic quality is achieved (the bulk properties). In the limit, the small clusters may serve as a model of the surface per se. The cluster comprising just a few atoms does not contain any "bulk" atoms, and hence its orbitals may be assumed to be the same as the orbitals of the surface atoms.

The approximate calculations for the clusters of 20–30 atoms of Ni, Cd, Cu, Ag, and Pd [5.144] indicate that in the small clusters there exists a large gap between the HOMO and the LUMO. The HOMO determines the electron affinity of the metal E_a, and the LUMO is characterized by the ionization potential I_n. For instance, for a cluster of three Ag atoms the HOMO is 3 eV above the LUMO. As the number of atoms in the cluster increases to about 20, the HOMO (or E_a) and the LUMO (or I_n) converge to the values of about 4.5 eV (the work function for metallic silver). Similarly, HOMO = 7.61 eV, LUMO = 2.32 eV for Ni_2, and HOMO = 6.20 eV, LUMO = 2.32 eV for Ni_6.

The quantum-chemical calculations were compared with the results of direct XPS measurements of the energy of valence levels of small clusters [5.128]. The spectra of silver, vacuum-supported on carbon film in a small concentration, correspond to isolated atoms of silver. They exhibit a strong peak eV below the Fermi level, pertaining to the 4d levels of ilver. As the concentration of silver is raised, the clusters row in size and the 4d levels are broadened into a band. With lusters of four silver atoms the 4d band splits into two ands, due apparently to the effects of crystal-field and spin-rbit interaction. With the large clusters the absorption edge hifts by 2.5 eV toward lower energies at the expense of the roadening of the 4d band and its overlapping with the 5s band f silver.

The dependence of the UPS spectra on the dispersity of pla-inum, supported on SiO_2, was studied in [5.144,145]. The spec-rum of the small crystals of platinum in the region of the alence band differed considerably from the spectrum of bulk latinum. The band at 6-9 eV below E_F, not observed with the ulk platinum, can be ascribed either to the surface states, hich multiply in number with the increase in dispersity, or to he interaction between platinum and the support. The derived and structure was correlated with the structure of the energy ands of small clusters, as calculated by the SCF-X_αSW method Sect.1.1). The electronic properties of particles bigger than 2 m seem to coincide with the electronic properties in the bulk.

The SCF-X_α-SW technique has been employed for calculating he electron structure of clusters of varying size of lithium, s well as of a number of transition metals and group IB metals 5.146].

The calculations indicate that in the eight-atom cubic clus-ers of copper Cu_8 the d band completely overlaps with the sp and, like in the bulk, although the d band is somewhat nar-ower. The transition from the HOMO (bands t_{1u} and e_u) to the UMO (t_{2g}), according to the calculations, occurs in the same nergy range (2.0-2.6 eV) as the experimentally observed band-o-band transitions, which, by the way, are responsible for the eculiar hue in the spectra of metallic copper.

In the cubic Ni_8 clusters the d band is shifted toward higher nergies, as compared with the d band of Cu_8, and exhibits con-

313

siderable broadening and splitting due to the spin polarization resulting from partial occupation of the upper e_g level. The number of holes in the d band of Ni_8 (0.25) is smaller than for the bulk Ni (0.55), in accordance with the variations in the paramagnetic susceptibility of small particles of nickel [5.147]. All these orbitals are the surface orbitals and can be visualized spatially as the lobes of d orbitals (Sect.5.2).

The electron levels in clusters occupy an intermediate position between the levels of the individual atoms and the levels of the bulky specimen. Generally, their structure is more sophisticated because of the low symmetry of clusters. The same applies to the metallic surface, simulated by clusters [5.148]. Both the experiment and the theory point to a drastic change in the properties of the cluster, when the number of atoms reaches 50-150. This is similar to the mitohedral region discussed in Sect.5.5. An abundance of new data on the structure and properties of small metal clusters is given in reviews [5.149,150].

6. Adsorption and Catalysis on the Transition Metals

It is impossible to overlook the similarities between the catalytic properties of the transition metals and their oxides and complexes. They all are capable of catalyzing the reactions of hydrogen and olefins: the reactions of H_2–D_2 exchange; hydrogenation of olefins, diolefins, and alkynes; and some reactions of polymerization and oxidation. The relative strength of bonding of various ligands in complexes corresponds with the strength of chemisorption on the same metal. For example, in the homogeneous carbonyl complexes the strength of bonding of CO with the metal is about the same as the strength of chemisorption of CO on this metal. The same applies to the chemisorption of olefins and the homogeneous complexes of olefins. The catalytic poisons for the reactions catalyzed by metals (the compounds containing the donor-type atoms O, S, N, As, P and other atoms with a free pair of electrons) are poisons also for the reactions catalyzed by complexes.

The analysis of these similarities leads to a conclusion about the common origin of the adsorptive and catalytic action of the transition metals and their compounds. Evidently the complexing of the chemisorbed molecules with the individual atoms on the surface of the transition metal via the d orbitals leads to manifestations of catalytic activity similar to the case of the oxides of transition metals (Chap.4). As regards the high conductivity of metals, this quality may sometimes be helpful for providing the required charging of the chemisorbed molecule or facilitating the smooth progress of a redox reaction. This property, however, is not the main cause of the catalytic activity of the transition metals.

6.1 The Adsorptive Complexes on the Surface of Transition Metals

The most catalytically active in the reactions of oxidation, hydrogenation, dehydrogenation are the transition metals of group VIII. The same reactions <u>in solution</u> are best catalyzed by the homogeneous complexes of the same metals (Pt, Pd, Os, Rh, Co).

Figure 6.1 shows the initial values of the heat of adsorption of hydrogen in the row of transition metals [6.1]. The heat of adsorption decreases toward the end of the period. The values of the heat of chemisorption of ethylene exhibit similar variation, although they are much larger on an absolute scale (200–400 kJ/mol). The catalytic activity of the transition metals in the hydrogenation of ethylene is plotted in Fig.6.2. It follows from the comparison of these two diagrams that the most active catalysts adsorb hydrogen and olefins in rather weak forms.

The diagrams do not include the data for Mn. There are indications, however, that the relative minimum of the catalytic

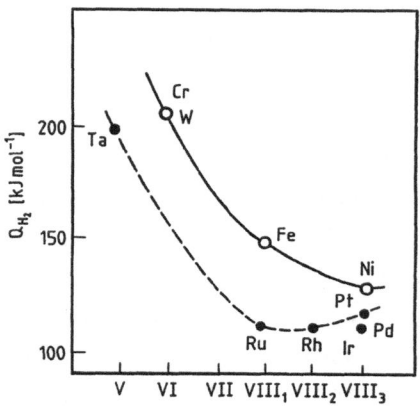

Fig.6.1. The initial values of heat of adsorption of hydrogen on transition metals [6.1]. Open circles denote the fourth-row metals; black circles denote the metals of the fifth and sixth periods

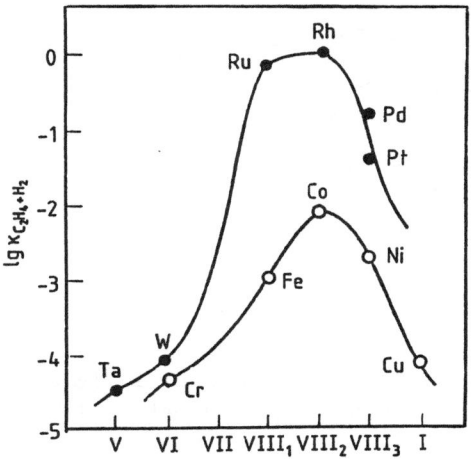

Fig.6.1. The initial values of heat of adsorption of hydrogen on transition metals [6.1]. Open circles denote the fourth-row metals; black circles denote the metals of the fifth and sixth periods

activity of the fourth-row metals falls on manganese (electron configuration d^5s^2).

Numerous investigations of the catalytic activity of alloys in the reactions of hydrogenation of olefins, H_2-D_2 exchange, and the like indicate that the occupation of the d shell by electrons (e.g., in the alloys Ni-Cu, Pd-Ag) results in an abrupt drop in catalytic activity.

These investigations launched the concept of the role of d electrons at adsorption and catalysis on metals. Of the early works in this direction the best known are those of ROGINSKIJ [6.2], BEECK [6.3], DOWDEN [6.4], and ROONEY [6.5]. ROONEY [6.5], for instance, attempted to draw a correlation between the experimental values of the heat of chemisorption and Pauling's "percentage of the d-nature of the metallic bond". Retrospective reviews of theories which explored the relationship between adsorption and catalysis on the transition metals and the electron structure of transition atoms can be found in [6.6,7].

A vivid description of molecular orbitals taking part in the chemisorption of transition metals was given by BOND [6.8]. In Sect.5.2 we have discussed the models of outwardly projecting orbitals.

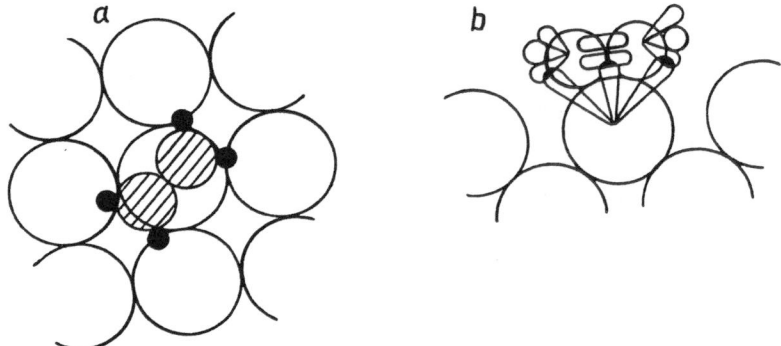

Fig.6.3. Adsorption of ethylene on the cubic face of Ni [6.8]: (a) (100) face, (b) sectional view along the C–C bond of ethylene

Toward the end of the row of transition metals the e_g orbitals become free enough to participate in σ bonding, and the occupancy of the t_{2g} orbitals is sufficient for π bonding. The configuration of orbitals determines the strength of bonding of chemisorbed particles. For instance, on the (100) face one may anticipate the relatively strong chemisorption of hydrogen, 1) at the expense of the overlapping of the 1s orbital of oxygen with the metal's e_g or d_{z^2} orbital, normal to the surface, and 2) in the octahedral interstices (four around each metallic atom), where the 1s orbitals interact with the five lobes of e_g orbitals – four of the atoms of the top layer and one of the atoms of the last–but–one layer (Fig.5.7).

Figure 6.3 shows the scheme of the adsorption of ethylene on the (100) face of nickel, which is quite similar to the schemes of π complexes (Sect.4.5). On the (110) face, C_2H_4 should be adsorbed at 45°C on the surface. The (111) face seems to be less suitable for the adsorption of olefins. One may envisage also the two-point adsorption of olefins, taking advantage of the vertical e_g orbitals on the (100) face, or the vertical t_{2g} orbitals and the 45°-tilted e_g orbitals on the (110) face.

The adsorption of a diolefin molecule requires the availability of two adsorptive (coordinative) sites. It will therefore be especially favored by the (110) faces, on which there are two outwardly projecting e_g orbitals, as well as four t_{2g} orbitals. Such adsorption can also take place on two adjacent atoms. Alkynes should be adsorbed best of all on the (110) face, where the two π bonds can interact with the surface orbitals of two adjacent atoms:

$$
\begin{array}{c}
H \\
| \\
C \\
Ni\ldots \overset{\text{III}}{} \ldots Ni \\
C \\
| \\
H
\end{array}
$$

The (100) face offers no suitable e_g orbitals, although the adsorption can take place using a single π bond. The activity of various d orbitals increases on the edges and vertices, and on the stepped and incomplete crystalline faces.

The creation of surface sites with the lowered coordination number (with respect to the bulk) was discussed in Sect.5.5. It has been assumed that the B_5 sites on the incomplete (110) face are active in the adsorption of nitrogen. According to BOND [6.9], the π complexes and the π allyl complexes on the B_5 sites easily convert into the alkyl radicals. Figure 6.4 shows the position of the hydrogen atom on the upper Ni atom of the B_5 site, relative to the ethylene molecule attached by two σ

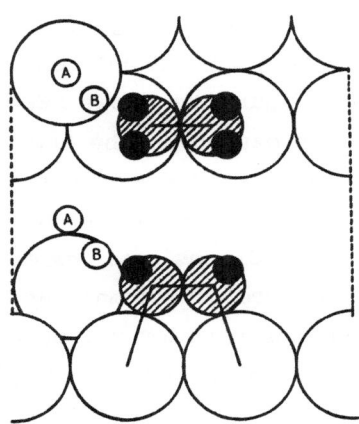

Fig.6.4. Attachment of the hydrogen atom to the ethylene molecule, adsorbed via two σ bonds on the B_5 site

bonds to the two adjacent atoms in the lower plane. In position A the hydrogen atom is bound with the e_g orbital normal to the surface and is removed by about 0.29 nm from the nearest carbon atom. However, in the intermediate position B the hydrogen atom can be bound with the t_{2g} orbital at 45° to the surface. Then it is at a distance of just 0.17 nm from the nearest carbon atom and is readily accessible for the reaction resulting in the production of ethyl radical.

These structures have much in common with the structures of complexes of adsorbed molecules on the oxide catalysts, discussed in Chap.4, as well as with the structures of the homogeneous complexes. As follows from Figs.6.3,4, the specificity of the catalytic action of metals may be due to steric factors: the increased probability of the molecule's bonding with two transition atoms. This gives rise to the bridge structures of the adsorbed CO and C_2H_4 molecules, and the uncommon structures on B_5 sites.

The considerable improvement of the physical investigative techniques has permitted verification of the structure of the surface complexes on the metals derived earlier from speculative analogies with the homogeneous complexes.

As soon as the LEED technique was put into practice 50 years ago, it was used for studying chemisorption on metals. The advances in LEED investigations have led to surprising discoveries [6.10-13]. It was found that the adsorption of gases on metallic surfaces does not proceed at random, but rather tends to form an orderly two-dimensional phase. Figure 6.5 shows LEED photographs, taken after the adsorption of oxygen on the (110) face of Ni at room temperature [6.11]. As the coverage of the surface by oxygen increases, it consecutively assumes structures (3x1), (2x1), and (3x1) - the two-dimensional oxide phases. More detailed investigations discovered occasional superstructures with the periodicity of tens of nanometers. Such structures comprise dozens of atoms in a unit, and abruptly switch from one form to another.

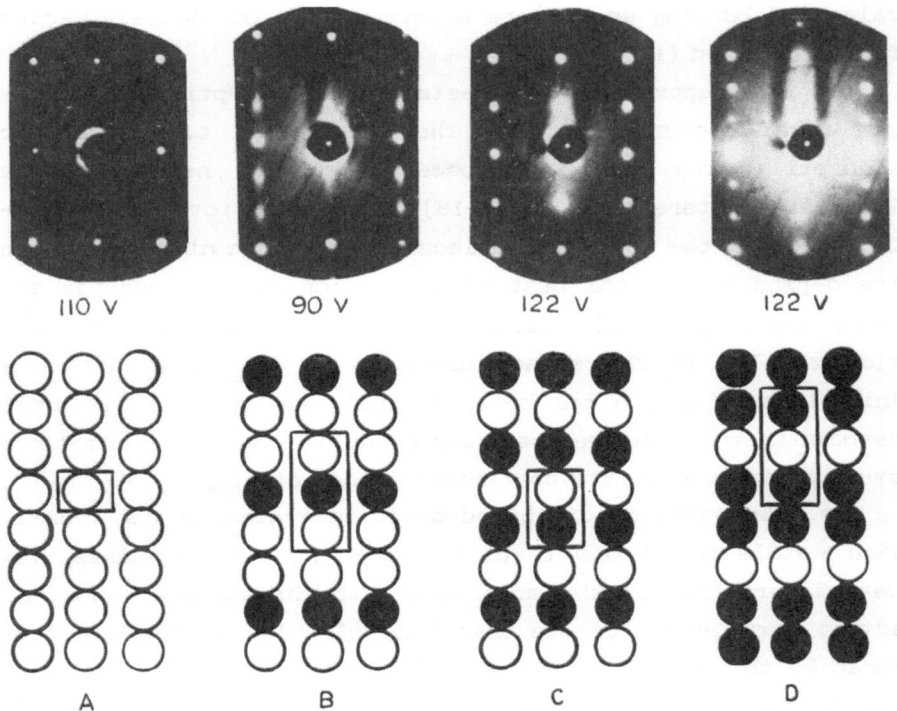

Fig.6.5. The evolution of the (110) face of Ni with the adsorption of O_2 at room temperature, according to LEED [6.11]: (a) clean surface, (b) $\theta = 1/3$, (c) $\theta = 1/2$, (d) $\theta = 2/3$. Black circles denote Ni atoms; open circles denote O atoms

These results dispel Langmuir's concept of "statistical" adsorption. The adsorption (and hence the catalysis) in such cases must be viewed as a chain of sudden transformations accompanied by the reconstruction of the surface at the expense of the energy released in the adsorptive or catalytic act. Each rearrangement alters the configuration of the surface active site, which must change the catalytic activity; on the other hand, at the instant of reconstruction the surface atoms may exhibit elevated activity and take part in the catalytic act [6.14,15].

On the high-index metallic faces the adsorbed particle is bound less firmly than on the low-index faces. The highest values of bonding energy are observed with the densely packed faces (100) and (110).

The first-approximation treatment of adsorption on metals can take into account only the interaction of the atomic adsorptive site with its closest neighbors, neglecting the long-range interactions [6.16-18]. Accounting for the interaction between the adsorbate molecules, it is possible to explain the dependence of the heat of adsorption on the coverage and the characteristic LEED patterns [6.19]. The repulsive interaction results in the steady decrease in Q_{ads} even with the initial adsorptive sites being entirely equivalent. The jumps associated with the rearrangements of the surface structures are superimposed on the continuous change of Q_{ads}.

The LEED patterns of the adsorbed particles on the transition metals allow one to distinguish two extreme cases: (a) weak interaction of the molecules with the surface, when the actual arrangement of the adsorbed molecules is little influenced by the metallic lattice; and (b) strong interaction, when the arrangement of the adsorbed particles is determined by the periodicity of the metallic lattice. In the latter case the creation of superstructures is associated with the interaction between the adsorbate molecules via the electron gas. The electronic interaction between the adjacent atoms of the transition metal gives rise to the repulsion between the particles adsorbed on these atoms. The energy of the repulsive interaction is usually low and constitutes several kilojoules per mol. This is nevertheless sufficient for the creation of superstructures and produces jumps on the curve depicting the dependence of the heat of adsorption on the coverage.

For instance, the heat of adsorption on the Pd (100) face drops abruptly as the coverage reaches half of a monolayer [6.18]. Under equilibrium conditions the structure (2 x 2) is formed, in which none of the sites adjacent to the adsorptive site is occupied. Further on, the adsorption proceeds with a much lower heat of adsorption, because of the repulsion of the adjacent particles. The spectrum of CO, thermally desorbed from the (100) face of tungsten, exhibits three maxima α, β_1, and β_2 [6.20]. The examination of the XPS spectrum of the adsorbed CO indicates that the α states differs from the states β_1 and β_2, although no difference whatever could be detected between the states β_1 and β_2 in spite of the difference in E_{des} ($\simeq Q_{ads}$). The areas beneath the peaks β_1 and β_2 are equal; from the chemical point of view these two peaks correspond to one and the same state.

The difference between these peaks is due to the varying structural environment. The (1x1) structure is present on the surface before the start of desorption: a half of the desorbing CO molecules (state β_1) leave the surface via one site. There remains structure (2x2) which is later desorbed as the β_2 peak.

Naturally enough, on other metallic faces the lateral repulsion will result in other peculiarities in the dependence of the adsorption heat on coverage. Although at the full coverage of the surface all the adsorbed particles are indistinguishable in their properties, they exhibit a certain (discrete) nonuniformity in terms of the heat of adsorption and the activation energy of desorption (and hence in terms of catalytic activity). Observe that the real surfaces, owing to the presence of impurities and the existence of a large number of crystallographically dissimilar atoms, often exhibit a broad nonuniform distribution with respect to the heat of adsorption and the activation energy of desorption [6.21].

The creation of the two-dimensional phases depends on the coverage and the temperature, and can be described by the constitutional diagrams, similar to the three-dimensional phase diagrams. At low temperatures and low coverages the two-dimensional structure is not formed, and the molecules are adsorbed at random, in accordance with Langmuir's law. A rise in the temperature causes a transition of the order-disorder type, which can be viewed as two-dimensional vaporization. For example, such a transition for the (2 x 2) structure of adsorbed H on the (110) face of W at the coverage $\theta = 0.50$ occurred at about 230°C [6.18].

In a series of experiments USTINOV [6.22] has proved that the two-dimensional phase formation is a common feature of the chemisorption of simple gases on metals. Only at low temperatures and low coverages does the adsorbate form an "amorphous" layer with a random arrangement of molecules. At the temperatures which ensure sufficient mobility of the adsorbate, the chemisorption passes through two successive steps: (1) sticking of the gas molecules to the surface, and (2) migration toward the patch of the ordered adsorbate. The probability of the initial nucleation of the two-dimensional phase is determined by the supersaturation of the adsorbed layer; this probability is highly increased by the existence of defects, dislocations, steps and kinds on the metallic surface.

New experimental techniques determine directly the orientation of bonds in the chemisorptive complex. Electron-stimulated desorption ESD [6.23], which consists in the bombardment of the surface by low-energy electrons (10 - 1000 eV), stimulates the desorption of neutral molecules, as well as molecular fragments (radicals), ions, and labile particles. Most easily detected – directly by the mass spectrometer – are the ionized particles. However the relative yield of different products of desorption is determined by the cross sections of the interaction of electrons with the chemisorbed particles, and does not reflect directly their relative concentration on the surface.

An interesting version of this technique is the ESD with angular distribution of ions (ESDIAD) [6.24-28]. Bombardment of the (100) face of tungsten with a collimated electron beam gave rise to the sharply focused beams of O^+ ions, producing visible spots on the screen [6.28]; the direction of the ionic beam may characterize the angle of the chemical bond between the adsorbed molecule and the surface (Fig.6.6). Similarly, by measuring the angle of departure of the H^+ ion from the (100) face of Ru after the adsorption of water, it was possible to

324

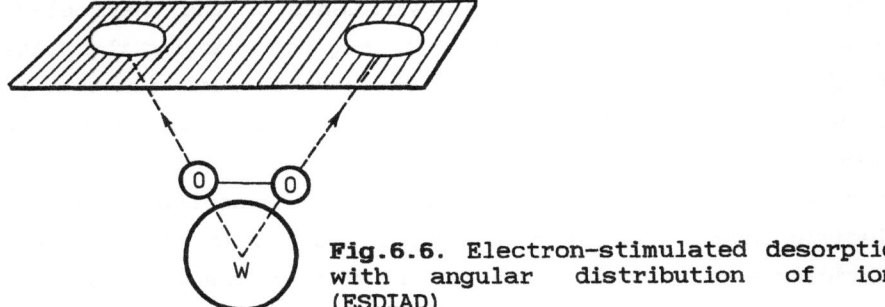

Fig.6.6. Electron-stimulated desorption with angular distribution of ions (ESDIAD)

demonstrate that water is adsorbed in the nondissociated form at the expense of the free pair of electrons of the oxygen atom (Fig.6.7). The H-O-H angle in this complex is $116\pm10°$ (for comparison, the same angle in the free water molecule is $104.5°$, and $110.6°$ in the free H_2O^+ ion). It is noteworthy that a similarly large deformation of water molecules (the increase in the interproton separation) was detected by NMR for the adsorption of water on zeolites containing the ions of transition elements [Ref.6.14,Chap.4]. According to ESDIAD, the adsorption of cyclohexane on the (001) face of Ru at low temperature follows the three-point mechanism [6.27].

The symmetry of adsorption sites and the geometry of the individual chemisorptive bonds of particles with the surface is studied also by the angle-resolved UPS technique (ARUPS) [6.29-31]. This technique demonstrates that CO on the (100) faces of Ru and Ni is adsorbed normally to the surface [6.30,32].

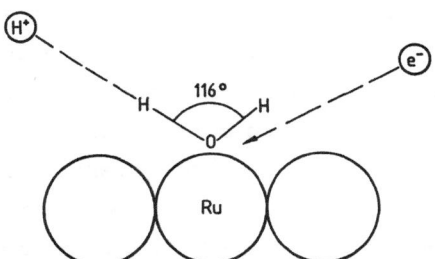

Fig.6.7. Adsorption of H_2O on the (100) face of Ru, according to ESDIAD

In the fifties and sixties measurements of the thermoelectronic work function were popularly used for studying the state of adsorbed molecules. The thermoelectronic work function ϕ_t suffers a considerable change at adsorption because of the change in the double electric layer. The difference in ϕ_t prior to adsorption and after it characterizes the additional double electric layer, created by the dipole moments of the adsorbed particles - the so-called dipole component of the work function ϕ_D. For submonolayer coverage

$$\phi_D = 4\pi\mu_n n_0 \; , \tag{6.1}$$

where μ_n is the projection of the dipole moment onto the normal to the surface, and n_0 is the concentration of adsorbed molecules.

Generally, μ_n does not coincide with the intrinsic dipole moment of the molecule. Many nonpolar molecules (H_2, O_2, etc.) give rise to considerable variations in ϕ_t. By μ_n in (5.1) we mean the effective normal component of the dipole moment of the surface chemical compound, produced at chemisorption [6.33-35].

The changes in the work function of individual faces of the tip of the field-emission electron microscope due to adsorption $\Delta\phi_t$ are monitored by the varying intensity of luminescence of the screen, or directly by measuring the electron current. Together with the overall change in the intensity of luminescence, the admission of a gas often produces very bright dots on the screen, attributed to the individual adsorbed molecules [6.36].

The particular structures arising at the chemisorption of a number of simple gases on metals and verified by physical investigations will be discussed below. In general, the concepts of complexing and the role of the d orbitals on the metallic surface for adsorption and catalysis have been confirmed.

The cryochemical synthesis of complexes of transition metals, which has greatly advanced in recent years, was very helpful for interpreting the spectra of surface compounds on metals [6.37]. By condensing the vapor of metal M together with the vapor of ligand L in the matrix of inert gas at near-liquid-helium temperatures, it is possible to obtain various complexes M_xL_y, which cannot be synthesized by conventional procedures. In this way the superunsaturated complexes containing several atoms of metal M were obtained. Such complexes may serve as models in the study of chemisorption and catalysis on metals.

For example, the condensation of Ni with CO gives rise to clustered complexes Ni_nCO, presumably having the following structures:

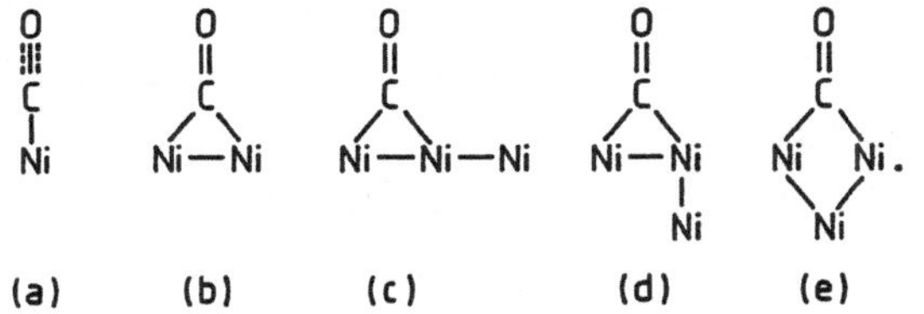

$$
\begin{array}{ccccc}
O & O & O & O & O \\
\text{\small ⫶} & \| & \| & \| & \| \\
C & C & C & C & C \\
| & \diagup\diagdown & \diagup\diagdown & \diagup\diagdown & \diagup\diagdown \\
Ni & Ni\!-\!Ni & Ni\!-\!Ni\!-\!Ni & Ni\!-\!Ni & Ni\quad Ni. \\
 & & & | & | \\
 & & & Ni & Ni \\
(a) & (b) & (c) & (d) & (e)
\end{array}
\qquad (6.2)
$$

The frequency ν_{CO} in the IR spectrum is 1996 cm^{-1} for complex (6.2a), 1973 cm^{-1} for (6.2b), 1969 cm^{-1} for (6.2c), 1963 cm^{-1} for (6.2d), and 1938 cm^{-1} for (6.2e). Plotting ν_{CO} against $1/n$, we can extrapolate this curve and get the value of ν_{CO} for "pure" metal. This value (1905 cm^{-1}) is close to the experimental value of ν_{CO} on bulk nickel at small coverages [6.38].

The IR bands of the bridge cluster $Rh_2(CO)_8$ are characteristic of the linear (2060, 2043, 2080 cm^{-1}) and the bridge (1852, 1830 cm^{-1}) complexes of CO. These values are very close to the frequencies ν_{CO} for the adsorption of CO on pure rhodium film.

The following complexes of oxygen with metals were obtained: Rh_2O_2 (ν_{O-O} = 1266 cm^{-1}), $Rh_3(O_2)_2$ (1126 cm^{-1}), and $Rh_2(O_2)_4$ (1076 cm^{-1}). All these complexes exhibit ν_{O-O} within the range of normally expected frequencies of coordinated $O_2^{\delta-}$, where $1 \leq \delta \leq 2$. The complexes with different numbers of O_2 molecules may be viewed as models of "low" and "high" coverage of a metallic surface by oxygen:

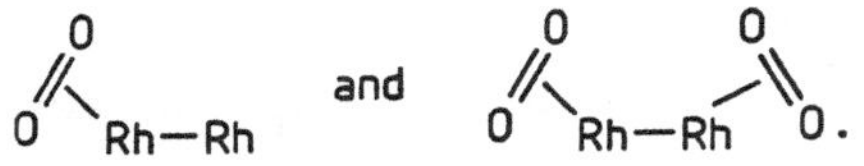

$$
\begin{array}{ccc}
O & & O \qquad O \\
\diagdown\!\| & & \diagdown\!\| \quad \|\diagup \\
O\quad Rh\!-\!Rh & \text{and} & O\quad Rh\!-\!Rh\quad O.
\end{array}
$$

Study of the spectra of complexes of transition atoms with olefins indicates that the simple π complexes $Ni \cdot C_2H_4$, $Cr \cdot C_2H_4$, and $Co \cdot C_2H_4$ have a relatively low stability and convert into binuclear complexes. For instance, $Cu_2(C_2H_4)$ and $Cu_2(C_2H_4)_2$ are much more stable than $Cu \cdot C_2H_4$. The optical spectra of these complexes show a much greater dependence on the stoichiometry than do the vibrational spectra. For example, the UV absorption band of $Cu(C_2H_4)$ lies at 375 nm, of $Cu_2(C_2H_4)$ at 240 nm, and of $Cu_2(C_2H_4)_2$ at 215 nm. The respective variations in the frequency of C–C vibration in the 1500 cm^{-1} range constitute about 20 cm^{-1}. The frequencies of C–C and C–H vibrations of the olefin in the complex are close to the corresponding frequencies for the adsorbed olefin on metal.

These data confirm the validity of the models of localized electrons for describing the chemical bonding of adsorbed particles with a metallic surface, and the usefulness of the cluster models for the description of adsorption and catalysis.

6.2 The Energy Levels of Adsorbed Particles on Metallic Surfaces

Despite the large number of studies dealing with the interaction of the adsorbed particles with the metal, there still remains an abundance of conflicting opinions. Detailed surveys of quantum-chemical work dealing with adsorptive bonding on metals can be found in [6.6,19,39,40]. Here we shall briefly outline some of the most important questions.

The energy diagram of the complex consisting of the metal and the adsorbed atom is shown in Fig.6.8 [6.39,41,42]. If the atom is at a large distance from the metallic surface (the right-hand side of the diagram), the atom and the metal represent separate quantum-mechanical systems, and the atom's levels are discrete; I is the ionization potential of the atom, and A is its electron affinity. If I < φ, φ being the metal's

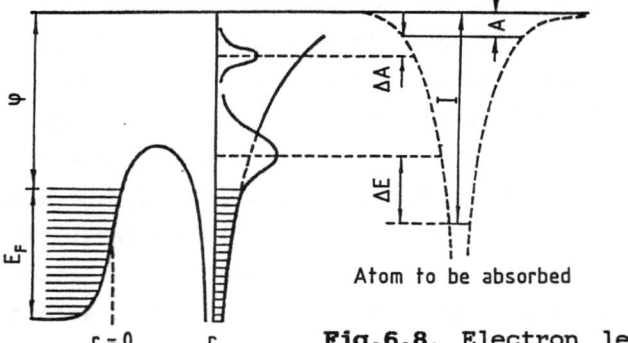

Fig.6.8. Electron levels of the metal and the adsorbed atom

work function, the adsorbed atom, after transferring a valence electron to the metal, becomes a positively charged ion; if A > ϕ, then a negatively charged ion is produced. If I > ϕ, and A < ϕ, then the adsorbed atom is stable in the neutral state. For a hydrogen atom I = 13.6 eV, A = 0.7 eV. Since for most metals ϕ = 4–6 eV, hydrogen should be expected to be adsorbed as a neutral atom. For alkali–metal atoms I varies from 5.39 eV for Li to 3.89 eV for Cs; hence for most metals I < ϕ and one can expect the adsorbed alkali–metal atoms to become positively charged ions [6.41]. For adsorption on metals A < ϕ almost always, so the production of negatively charged ions is hardly possible.

On adsorption of an atom the electrons may tunnel between the adsorbed particles and the metal, leading to the broadening of levels, as indicated in the left-hand side of the diagram (Fig.6.8). The level of the valence electron (the maximum of its wave function) is raised by ΔE, and the level of the additional (excited) electron is lowered by ΔA. As a result, in the first case the main density of the wave function may lie above the Fermi level E_F, and the particle will only be partly ionized (charge $\delta+$). In the second case, the "tail" of the wave function may fall below E_F, and the particle will carry the charge $\delta-$. We see that the overall density of states of the adsorbed atom is the sum of states of these two broadened levels: one

329

with a maximum at $(-I+\Delta E)$ and the other with a maximum at $(-A-\Delta A)$. The example given in Fig.6.8 corresponds to the predominantly ionic bonding: the maximum of the wave function is above E_F. A maximum below E_F would correspond to covalent bonding.

DOWDEN [6.43] in one of his pioneering works on chemical bonding on metallic surfaces used the concepts of the band theory to obtain simple equations that relate the concentration of particles ionized at adsorption to the location of adsorbate's levels with respect to the Fermi level of metal. According to these equations, the creation of positive ions on the surface is favored - apart form high ϕ and low I - by the high positive values of the density of states $d\ln N(E)/dE$ at the Fermi level, where $N(E)$ is the distribution function of density of states over energy E for the metal. The production of negative ions is facilitated, apart form at low ϕ and high I, by the high negative values of $d\ln N(E)/dE$ at the Fermi level. The covalent bonding is favored, apart from high ϕ, by the high positive values of $d\ln N(E)/dE$ at the Fermi level together with the availability of free atomic orbitals. These conclusions were confirmed by experimental results [6.4] which indicate a loss in the activity of Cu-Ni alloys in the reaction of hydrogenation of benzene at the proportion 60% Cu + 40% Ni, i.e., when the d band becomes occupied and the value of $d\ln N(E)/dE$ becomes small.

Treatment of the electron levels of the adsorbed particles on metals is often based on Goodenough's model of transition metals, discussed in detail in Sect.5.1, and on the concept of spatial distribution of d orbitals over the metallic surface (Sect.5.2).

Figure 6.9 shows the scheme of electron levels on the (100) face of Ni or Co - according to DOWDEN [6.44], who used the concepts of Goodenough - and the interaction of the hydrogen atom with these levels. The respective disposition of the hydrogen atoms is illustrated in Fig.6.10; this geometry corresponds to the particular symmetry of d orbitals. The hydrogen atoms can

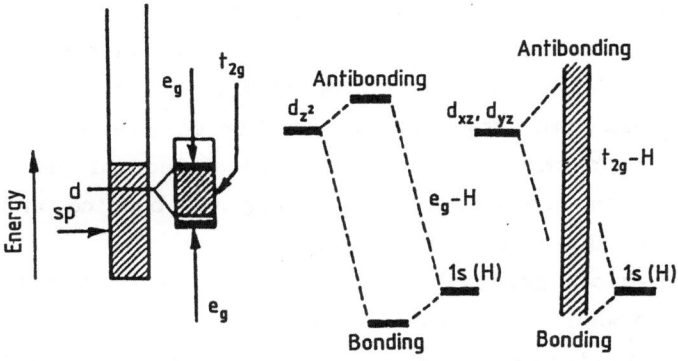

Fig.6.9. Electron levels on the (100) face of Ni or Co, and the interaction of the hydrogen atom with these levels [6.44]

be adsorbed in "linear" positions, interacting with the localized d_{z^2} orbital, and in the bridge positions, interacting with the d_{xz}, d_{yz} orbitals which form the t_{2g} band (Fig.6.9). When the hydrogen atom (I = 13.6 eV) nears the metallic surface along the z-axis, the localized bonding orbital will be below 13.6 eV, and the antibonding orbital will be high, probably above the E_F (the work function for Ni and Co is about 4–5 eV). The production of the two-electron bonding H–Ni requires the transfer of 0.86 of an electron from the 1s orbital of hydrogen

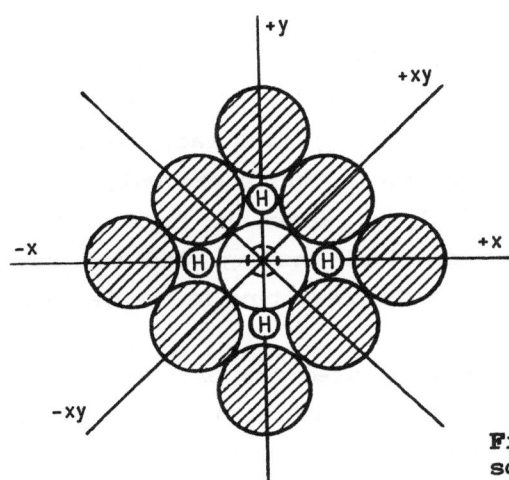

Fig.6.10. The geometry of adsorbed hydrogen atoms on the (100) face of Ni or Co

331

to the Fermi level to fill the holes in the 3d orbitals of Ni. For adsorption in the bridge positions (Fig.6.10), the 1s orbital of oxygen overlaps with the four lobes of d_{xz}, d_{yz} orbitals, the result being the same: the transfer of 0.86 of an electron to the Fermi level. In this case, however, the molecular orbitals combine to form the (possibly overlapping) bonding and antibonding bands (right-hand part of Fig.6.9).

The quantum-chemical analysis of the interaction of the 1s orbital of the hydrogen atom with the localized d_{z^2} and $d_{x^2-y^2}$ orbitals of Ni was performed in [6.45] using the LCAO-MO method according to Mulliken-Wolfsberg-Helmholtz version (MWH). The treatment was based on the Goodenough-Bond model described earlier. The local electron levels of adsorbed hydrogen atoms were obtained neglecting the interaction with the collectivized electrons of the metal. According to [6.46], however, the adsorption is accompanied by the complete rearrangement of the bonds of the adsorbent atom, contrary to the presumptions of Bond.

The concepts regarding the interaction of hydrogen with metals can be checked against measurements of the magnetization of Ni and Co on the adsorption of hydrogen [6.6,47]. The decrease in the saturation magnetization after the adsorption of one atom of hydrogen on Ni equals 0.56 Bohr magneton, and 0.54 Bohr magneton for the adsorption on Co, which is much lower than that required for the occupation of the d orbitals of Ni and Co (0.86 and 1.78, respectively). This fact was explained by the creation of surface levels, which are only partly submerged under the Fermi level, as shown in Fig.6.8. Some electrons participate in the chemical bonding, and the rest fill the holes in the 3d and 4s bands, resulting in the decrease in magnetization.

The LCAO-MO-MWH method was also used for calculating the energy levels for the interaction of oxygen with the surface of nickel [6.48], assuming that the oxygen molecule is engaged in a local interaction with two atoms of nickel. The configuration of the outer orbitals is $3d^8 4s^2 4p^0$ for an isolated Ni atom, and $3d^{9.2} 4s^{0.5} 4p^{0.3}$ for the atom in the bulk. Thus, to the surface atoms were ascribed the intermediate configurations $3d^{8.5} 4s^{1.5} 4p^0$ and $3d^{8.5} 4s^{0.8} 4p^{0.7}$. The calculations indicate that the existence of molecular and atomic adsorbed forms of oxygen on nickel depends on the occupancy of the 3d levels by electrons. As the occupancy of the 3d levels increases (which corresponds to involvement of the Ni atom in the crystalline bonding), the 3d level of Ni comes close to the π^* level of oxygen, the barrier is lowered, and the transition into the atomic form is facilitated. The decrease in occupancy of the 3d orbitals causes the decrease in the occupancy of the level of O_2^-, which lies above the 3d levels of Ni, and the molecular form of adsorption of O_2 becomes totally impossible.

The Coulomb interactions between the electrons in the adsorbed particles play a most important role in the formation of the energy structure at adsorption. They result in a large shift of the energy levels, as the occupancy of levels is changed. ANDERSON [6.49] used the Hartree-Fock self-consistent field approximation to account for the Coulomb interactions in the description of magnetic effects in diluted alloys. A number of authors [6.50-54] have subsequently used the formal Anderson model for describing the chemisorption on metals.

The Anderson model is based on the one-electron approximation. The metal is described by the wave function $\Phi(r)$ of the Bloch states in the bulk, and the exponential wave function of the orbital directed into vacuum. The adsorbate is described by the wave function ϕ, participating in the chemisorptive bond. The total wave function of the system ψ is a linear combination of Φ and ϕ. In the Anderson model it is assumed that the self-consistent potential depends only on the occupancy of the orbital ϕ of the adsorbate, being independent of the occupancy of states Φ — in other words, the model does not take into account the changes in the metal caused by adsorption. For the hydrogen atom, the Coulomb interaction puts the level of H^- ion

I-A = 12.4 eV above the level of H atom (Fig.6.8). The occupancy of the metal's levels is of less importance because of the effects of screening, and partly also because of the large size of metallic atoms in many systems.

GRIMLEY [6.51] used the Anderson model for analyzing the chemisorption of H atoms on the (100) face of tungsten. The calculations indicate that the interaction involves only those d orbitals of W which are the closest to the hydrogen atom: e.g., the $d_{x^2-y^2}$ orbitals of W interact with the 1s orbital of hydrogen in the linear complex WHW, or in the planar (square-pyramidal) W_4H. What is formed is a kind of "surface molecule" consisting of the adsorbed particle and a small number of surface atoms and involving the localized d levels of metal. The essential difference from the conventional molecule is that on the surface we are dealing with an open system with weak interaction between the localized d electrons and the delocalized s electrons of the metal. It is important that the occupancy of surface levels is determined not by the number of electrons in the bulk, but rather by the energy and the symmetry of these levels and their interaction with the lattice.

The Anderson method (the self-consistent field approximation SCF) was further improved by GRIMLEY [6.19,51] and by SCHRI-EFFER and GOMER [6.41]. It yields correct results in the case of strong bonding, but overestimates the importance of the Coulomb interaction between the adsorbate's electrons in the case of weak bonding. If the gap $\Delta\epsilon$ between the bonding and the antibonding orbitals (or the "width of the adsorptive level") is small, and $(I-A)/\Delta\epsilon > 1$, then the correlative effects of the electron-electron interaction should be taken into account; they can be important for the adsorption of, e.g., the H_2 molecule. The effects of interelectronic correlation have been taken into account in a number of studies concerned with the quantum-chemical treatment of adsorption on metals [6.41,42,54,55].

The quantum-chemical treatment (based on perturbation theory) of the ionic bonding in the adsorption of alkali ions on metals was given by GADZUK and coworkers [6.56-58]. The calculations do not indicate the appearance of the local surface levels.

The chemisorption of O, Li, and H on metals was analyzed on the basis of the "jellium" model of metal [6.40]. This model employs the concept of a homogeneous electron gas in the solid, neutralized by the field of uniformly distributed positive charges.

BONCH-BRUEVICH and GLASKO [6.59] put special emphasis on the fact that the problem of the interaction of adsorbing particles with the metal is a many-electron problem, and should be approached with adequate means, e.g., using Green's functions. Theory indicates two ways of solving this problem. In the first case the influence of the adsorbed atoms is reduced to just causing small variations in the electron density; the electrons do not localize near the adsorbed atom, and the chemisorption takes the form of weakly bound ions. In the second case the wave function has its maximum near the adsorbed ion; in other words, local levels are created that fall below the Fermi level. It is noteworthy that even at high coverages the levels differ little from the level of a single adsorbed atom owing to the strong screening effect. The boundary layer effects in a pure metal are totally absent. As the Fermi level in a metal is fixed, it would appear according to [6.59] that metals, in contrast to semiconductors, are very much like the classical Langmuir adsorbents. Naturally, this model disregards the inevitable atomic rearrangements due to the adsorption.

The levels of adsorbed atoms on metals can also be analyzed with the aid of $SCF-X_\alpha-SW$ method discussed earlier. The application of this method to the levels of hydrogen atom adsorbed in the center of tetrahedral cluster Pd_4 indicates the creation of hydrogen levels near the bottom of the d band of palladium [6.60].

A detailed analysis of the main assumptions of cluster methods for calculating the chemisorptive complexes on the surface of transition and non-transition metals, as well as a synopsis of computed spatial and energetic distributions of the electron density for a number of adsorptive systems, can be found in the reviews [6.40,61]. The experimental basis of the cluster models has already been outlined in Sects.5.6,6.1. The theoretical calculations of clusters were based both on the most simple Hückel-type methods, and on more sophisticated method X_α-SW. As pointed out by MESSMER [6.40], the best of these is the SCF-X_α-SW technique, which satisfies the requirement of self-consistency and includes the Coulomb interactions between the electrons, between the nuclei, and between the electrons and the nuclei, as well as the local or nonlocal exchange potential.

We see that the development of quantum-chemical models of adsorption on transition metals goes in more than one direction. So far, no universally accepted models have evolved to describe the experimental findings as nicely as does the method of molecular orbitals in the case of isolated complexes or in the case of adsorption on isolated atoms of transition metals (Chap.4). Most models are based on the assumption of adsorptive bonding using the localized d electrons, and the weak influence of the metal's collectivized electrons. The levels calculated by the SCF-X_αSW technique fit in well with the experimentally derived levels of adsorbed atoms and molecules.

KNOR [6.62] has proposed a model of chemisorption which takes into account both the local electrons of the metal (d electrons) and the free electrons (s,p electrons, Sect.5.2). After the bonding of the molecule with the d electrons of the metal and the reconstruction of the initial molecule, the nearly free electrons relax into the new state around the incipient complex. The effects of screening weaken the bonds not only within the initial molecule (the "activation" of bonds), but also the bonds with the metal atoms; in other words, the role of d electrons in establishing a strong chemisorptive bond is subdued. This conclusion agrees with the results of some theoretical work on the adsorption of gas atoms on metallic clusters [6.61].

From the viewpoint of catalysis, according to [6.62] the optimum interaction between the molecules and the surface of the transition metal can be achieved at a certain proportion of localized and free electrons. For instance, as indicated by UPS, the metals Fe, Ni, and Co have a narrow d band and hence the more localized d orbitals, the d band of Pt, Pd, and Ir is wider (5-8 eV), and the density of the delocalized (metallic) states is higher. The latter usually are better catalysts. The catalytic activity of the metals with a narrow d band can be improved by raising the density of free metallic electrons – for example, by doping Fe with alkali metals.

The direct detection of the surface adsorptive levels (as well as the nonadsorptive, Sect.5.2) on metals became possible with the development of UPS technique. First in this field was the work of EASTMAN and CASHION [6.63], who took the UPS spectra of the (100) face of Ni, both pure and with the adsorbed gases O_2 and CO. This pioneering work was followed by a number of publications concerned with the UPS spectra of adsorbed molecules [6.29,30,64-70].

A typical UPS spectrum of the adsorbed molecules (CO on Pd) is reproduced in Fig.6.11 [6.71]. The study of adsorption is usually based on the differential spectra; the spectra of adsorbed moelcules are compared with the respective spectra of molecules in the gaseous phase. Usually the locations of maxima differ by several tenths of an electron volt or even by several electron volts. The difference between the peaks (bond energies E in the gaseous and the adsorbed phases) is

$$\Delta E = \phi_{sp} + \Delta E_R + \Delta E_B , \tag{6.3}$$

where ϕ_{sp} is the work function of the spectrometer, ΔE_R is the change in the energy of the relaxation (including mainly the

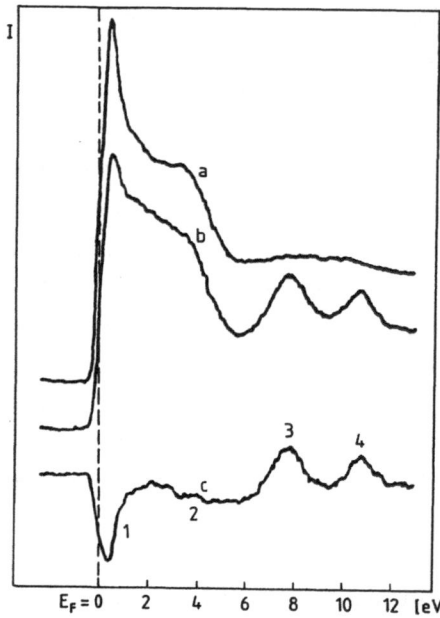

Fig.6.11. The UPS spectrum of CO adsorbed on the (110) face of Pd: (a) clean (110) face, (b) after the adsorption of CO, (c) the differential spectrum ($h\nu$ = 40.8 eV)

relaxation of the surrounding atoms after the emission of the electron), and ΔE_B is the change in the energy level of the chemical bond.

The minima 1 and 2 characterize the decrease in emission as a result of adsorption. They may be caused by the decrease in the escape depth of the photoelectrons due to adsorption, by the changed conditions of scattering of the outgoing electrons on the surface of the solid, or by the redistribution of electrons in the subsurface layers when the chemical bond is formed (elimination of intrinsic surface states). The third cause is the most important for adsorption on the transition metals. The deep minimum 1 directly below the Fermi level corresponds to the levels of d electrons and is almost always observed with the transition metals, which demonstrates the importance d orbitals in chemisorptive bonding.

The occupied valence levels in chemisorption on metals were studied also by XPS, FEM, and INS techniques (Sect.2.2). The vacant levels, which are also often involved in chemisorptive

338

bonding, were investigated by internal reflection spectroscopy in the near UV range and by the characteristic electron–energy–loss spectroscopy. Some results of the investigations of the surface levels in the chemisorption of simple gases are presented below.

The spectroscopic investigations of the surface levels have only just begun, but they have brought very important results: for one thing, they have confirmed the hypothesis about the role of d levels in adsorption. Elucidation of the detailed mechanism of chemisorption and catalysis will depend on the future advances of experimental techniques.

6.3 Adsorption of Hydrogen on Pure Metals

The majority of investigations of the adsorption of have been performed on the metals of group VIII. These metals exhibit the greatest catalytic activity in the reactions of hydrogenation and dehydrogenation (Fig.6.2). With them, the heat of chemisorption of H_2 is the lowest. The data presented in Fig.6.1 (Q_{H_2} = 100–120 kJ/mol) pertain to the polycrystalline metallic specimens; the value of Q_{H_2} for the low–index faces of these metals is still smaller (50–100 kJ/mol).

The use of the LEED technique for studying the mechanism of adsorption of hydrogen on metals is obstructed by the fact that the hydrogen layer by itself fails to produce a diffraction pattern. In some cases the adsorption of hydrogen does not even change the intensity of the diffraction maxima, to say nothing about producing new ones [6.72]. When the adsorption does change the diffraction pattern, these changes find a natural explanation in the reconstruction of the surface triggered by the adsorption of H_2.

Let us consider the specific features of the adsorption of H_2 on different faces of a platinum monocrystal – a metal with unique catalytic and adsorptive qualities.

After the adsorption of H_2 at low temperature on the face Pt(100), the curve of thermal desorption exhibits five peaks between -20 and +160°C [6.73]. In Sect.5.3 we have pointed to a peculiar feature of the (100) face of platinum. The simple cubic face (100) of Pt with (1x1) structure is a nonequilibrium structure and can only be obtained by taking special precautions, whereas at equilibrium conditions the same face has a hexagonal (5 x 1) structure (Fig.5.8). The equilibrium (5x1) and the nonequilibrium (1x1) structures were found to possess highly different chemisorptive properties. At room temperature the equilibrium (5x1) structure does not chemisorb either H_2 or O_2, while the same gases are chemisorbed almost instantly on the (1x1) structure: the sticking coefficient is by at least three orders of magnitude greater than for the (5x1) structure. The adsorption on the (5x1) (100) face at elevated temperatures is accompanied by the reconstruction of the surface: the structure (1x1) is formed after the adsorption [6.74]. These differences can be explained on the basis of the concept of outwardly projecting d orbitals, which apparently are present on the (1x1) face (100) and absent on the (5x1) structure (Fig.5.7). In the hexagons of the (5x1) structure the d orbitals overlap to form rings similar to the benzene ring, and only the reconstruction makes chemisorption possible:

$$
\begin{array}{ccc}
\begin{array}{c}
Pt\!-\!Pt \\
Pt \quad Pt \quad Pt \\
Pt\!-\!Pt
\end{array}
+ H_2 & \longrightarrow & Pt
\begin{array}{c}
Pt\!-\!Pt \\
\quad H \\
Pt \\
\quad H \\
Pt\!-\!Pt
\end{array}
Pt
\end{array}
\qquad (6.4)
$$

The hypothesis of outwardly projecting d orbitals is confirmed also by the UPS spectra. The chemisorption of H_2 on the (100) face of platinum gives rise to the UPS peak 5.5 eV below E_F, which corresponds to the Pt-H bond, and lowers the strength of the band directly below E_F, which is explained by the involvement of the surface d orbitals in the chemical bond.

340

The curve of thermal desorption from Pt(110) exhibits two peaks for hydrogen, and the LEED spectrum indicates the existence of a weak structure at near-monolayer coverage. Hydrogen is apparently adsorbed in the troughs between the platinum atoms, which form "ridges" on the surface [6.11].

The adsorption of H_2 on the face Pt(111) was studied by the methods of thermal desorption, UPS, LEED, FEM, and work function measurements [6.13,39,73-83]. These investigations interpreted the classical results of MIGNOLET [6.34], who was one of the first to measure the variations in the work function $\Delta\phi$ with the adsorption of hydrogen on metals. According to him, the adsorption of H_2 on polycrystalline platinum results first in an increase in ϕ (negative charging), and then in a decrease in ϕ (positive charging). The negative charging was attributed to the atomic adsorption Pt-$H^{\delta-}$, and the positive charging to the molecular adsorption Pt-$H_2^{\delta+}$. LEED points to the creation of a (1x1) structure [6.82].

FEM investigations performed on the Pt(111) face [6.39,77,78] generally confirm the results of Mignolet. Apparently, the surface of the polycrystalline specimen was dominated by the (111) face. Different authors report from one to three peaks of thermal desorption between -50 and 150°C; the kinetics of thermal desorption is described by the second-order curve. This indicates that all the hydrogen on the (111) face of platinum (with both negative and positive charging) is adsorbed in the atomic form. Upon adsorption of H_2 on the face Pt(111) the EELS spectrum (Sect.2.2) exhibits peaks at 550 and 1230 cm^{-1}. According to [6.83] these peaks are associated with the adsorption of hydrogen on the site of the three-fold symmetry.

This question was further elucidated by studying stepped surfaces [6.76-79,84,85]. Thermal desorption from the smooth (111) face of Pt exhibited only two peaks, and almost all the adsorbed hydrogen was charged positively. However, the thermal desorption from the stepped platinum surface 9(111) x (110) exhibits a third high-temperature peak, corresponding to nega-

341

tively charged hydrogen. This hydrogen is apparently adsorbed near the edges of the stepped surface (Fig.5.9). Further adsorption of H_2 occurs in a weakly positive atomic form:

$$\begin{array}{c} Pt\ Pt\ Pt \\ Pt\ Pt\ Pt\ Pt\ Pt \end{array} \xrightarrow{H_2} \begin{array}{c} H^{\delta-} \\ H^{\delta-}\ Pt\ Pt\ Pt \\ Pt\ Pt\ Pt\ Pt\ Pt \end{array} \xrightarrow{H_2} \begin{array}{c} H^{\delta-}\ H^{\delta+}\ H^{\delta+} \\ H^{\delta+}H^{\delta-}\ Pt\ Pt\ Pt \\ Pt\ Pt\ Pt\ Pt\ Pt. \end{array} \quad (6.5)$$

It is exactly these stepped faces that are catalytically active in the reactions of hydrogenation and H_2–D_2 exchange [6.12]. The active sites are apparently represented by the centers of strong chemisorption of H_2, that is, the platinum atoms on the edges of steps. According to HREELS data obtained at the adsorption of H_2 at –196°C on the stepped platinum surface 6(111) x (110), the adsorbed hydrogen occurs in two states: one form (threefold symmetry) on the planar surface, and the other bridge in the inner corners of steps [6.85]. The heat of adsorption of the second form is 13 kJ/mol lower than that for the first form [6.84].

Observe that the variations in the work function $\Delta\phi$ on the adsorption of hydrogen, measured by the method of contact potential difference, exhibit similar regularities for most transition metals [6.34]. The adsorption of hydrogen, at least at coverages below one-half of a monolayer, acts to increase the value of the work function ϕ, with $\Delta\phi$ amounting to 0.48 eV for W, 0.45 eV for Fe, 0.35 eV for Ni, and 0.14 eV for Pt. This points to the creation of a layer of atomic surface-bound hydrogen, with the partial transfer of an electron to the hydrogen atom: $M^{\delta+}$ ... $H^{\delta-}$. This is accompanied by a drop in electroconductivity.

Most models of the adsorption of hydrogen on transition metals assume that on the (100) face d_{z^2} orbitals normal to the surface participate in the bonding; however, the bonding may also involve the octahedral interstices at the expense of over-lapping of four lobes of $d_{x^2-y^2}$ orbitals with the hydrogen atom (Fig.6.10). The adsorbed states of hydrogen with $\Delta\phi_1 = 0.40$ eV

and $\Delta\phi_2 = 0.08$ eV were detected on Ni. The former possibly pertains to the outer layer of H⁻, while the latter corresponds to
the layer slightly sunk into the surface. The increased electrical resistance after the chemisorption can be explained by the
participation of s electrons of the metal in the hybridized dsp
orbitals. Observe that a similar effect — an increase in ϕ —
will be brought about by the adsorption of hydrogen in protonic
form, if the protons are dissolved in the subsurface layer of
the metal.

Among the d orbitals projecting from the (111) face one can
discern the e_g orbitals at an angle of 35°16' to the surface,
and the t_{2g} orbitals at an angle of 54°44' (Fig.5.7). For geometric reasons the molecular adsorption of H_2 favors the t_{2g}
orbitals, while the atomic adsorption tends to employ the e_g
orbitals. At the e_g site the hydrogen atom overlaps with all
three e_g orbitals at a distance on the order of the Bohr radius;
at the t_{2g} site the overlapping is possible at much greater
distances.

Measurements of thermal desorption detected five forms of
adsorbed hydrogen on pure nickel film [6.86]. The most thoroughly investigated has been the adsorption of hydrogen on the
(110) face of Ni, where, according to LEED [6.11,13,87], a (1x2)
superstructure is formed. Two models have been proposed for
explaining this structure. According to one of them, a complete
reconstruction of the surface results in the creation of parallel ridges in the [10] direction. The mobile adsorbed atoms of
hydrogen are on the slopes of these ridges, which represent the
rudimentary (111) faces. Their exact location is not determined. The heat of adsorption of H_2 on Ni(110) (120 kJ/mol) is
quite sufficient for such reconstruction. According to the alternative model, the Ni atoms of the upper layer shift by just
0.02 nm toward the chemisorbed hydrogen atom.

The data on the adsorption of hydrogen obtained by the FEM
technique [6.88] seem to confirm the validity of the first
model — the profound reconstruction of the (110) face. The

adsorption of hydrogen entirely changes the emission pattern of electrons for the (110) face, but has no effect on the emission from the (100) and (111) faces.

The adsorption of H_2 on the (110) face of Ni exhibits yet another interesting and enigmatic feature [6.18]. The adsorption isotherms of H_2 on Ni(110) have an S-shaped form: a rapid increase in the equilibrium adsorption starts after having reached a certain coverage. This points to the mutual attraction of adsorbed particles, as opposed to the repulsion observed with all the other systems. At low coverages the heat of adsorption of H_2 on (110) Ni tends to increase rather than decrease. This effect also points to the reconstruction of the surface due to the adsorption of hydrogen.

A vast amount of data obtained by LEED for the adsorption of hydrogen on various metals [6.13,82] indicate that the (110) face is more prone to reconstructions than other faces. Together with the (1x1) structure, it is capable of assuming structures (1x2), (1x3), etc. The predominant structure of the (111) face is (1x1); the (100) face exhibits structures (1x1) and (1x2).

The adsorption of H_2 on the (100) face of W has been investigated by LEED techniques (including its polarized version PLEED), ARUPS, and HRLEELS [6.64,89-91]. These investigations indicate that the pure (100) face of tungsten is unstable, and below 80°C has the (2x2) structure. The adsorption of H_2 stabilizes the (2x2) structure at higher temperatures up to the coverage $\theta = 0.5$ (state β_2, corresponding to $2.5 \cdot 10^{14}$ molecules per square centimeter). With increased coverage the surface assumes the (1x1) structure of the bulk; this state (β_1) is saturated at the concentration of $1.7 \cdot 10^{14}$ cm^{-2}. The HREELS spectra (the peaks at 80, 130, and 160 meV) indicate the creation of bridge W...H...W structures for all θ. This is confirmed by ARUPS: the peak at 2 eV below E_F is ascribed to the t_{2g} orbitals of W, involved in the bonding of the H atom. According to [6.64], the HREELS peak at 155-160 meV may also correspond to the onefold symmetry, i.e., to the adsorption of one atom of hydrogen over one atom of tungsten.

The adsorption of H_2 on W(110) gives rise to two UPS peaks (2.0 and 4.0 eV below E_F). The former corresponds to the adsorption in the bridge position (β_2), the latter to the three-fold site (β_1) [6.92].

Very interesting are investigations of the surface levels by electron-tunneling spectroscopy (Sect.2.2.4). The technique of FEM spectroscopy was used for studying the adsorption of hydrogen and nitrogen on the faces of monocrystalline tips made of Rh, Ir, Pt, and Pd [6.93].

The projecting t_{2g} and e_g orbitals were detected in the range from 0 to 2 eV below E_F. The emission maximum, observed with the pure (111), (100), and (110) faces, corresponds to the occupation of the surface levels, or, more precisely, to the excess occupancy of the surface levels with respect to the bulk levels. The e_g orbitals are higher; the upper t_{2g} orbitals have an antibonding nature and embrace a greater portion of the surface dipole layer. On the other hand, on the high-index faces (210) and (211), these orbitals are partly sunk into the surface and interlocked with one another, lowering the emission.

The adsorption of hydrogen on the (100) face resulted in a sharp drop in the emission in the range from 0 to 2 eV below E_F. The bonding then relies mainly on the e_g (i.e., d_{z^2}) orbital, and the electron density is displaced into the space between the metal atoms and hydrogen atoms. The t_{2g} orbitals also take part in the bonding by virtue of the overlapping of the four lobes with the s orbital of hydrogen. The t_{2g} orbital is also slightly compressed, and the emission reduced. The adsorption of hydrogen on the (110) face leads to a density of surface states which is practically the same as in the bulk.

Some very surprising results were obtained in the UPS investigations of adsorption of H_2 on the fourth-row transition metals. We have already pointed out the considerable reduction in the emission from the surface 4d and 5d levels (0 to 2 eV below E_F) with the adsortion of hydrogen on platinum metal. The same is observed for the adsorption of hydrogen on tungsten. At the same time, the adsorption of H_2 on the 3d metals Ni and Fe results not in a decrease, but rather in an increase in emission [6.81]. The maximum observed at 5.5 eV is similar to the maxima observed after the adsorption on platinum. Attempts have been made to explain this observation from the quantum-chemical viewpoint [6.50,94,95]. The calculations indicate that the H-M bonding involves 4s electrons rather than 3d electrons.

However, the 3d electrons can hardly be assumed not to participate in the chemisorption of hydrogen. Their participation is indicated, for instance, by the highly different adsorptive and catalytic properties of the metals of groups VIII and IB. The former adsorb molecular hydrogen in a dissociative manner; the latter (Cu, Ag, Au) adsorb only atomic hydrogen. The catalytic and chemisorptive properties of 3d metals differ not too much from the properties of 4d and 5d metals. It can therefore be assumed that 3d orbitals take part in the early stage of interaction with H_2 and "catalyze" chemisorption, which further proceeds at the expense of 4s electrons.

Apart from being adsorbed, hydrogen may also be dissolved in the bulk of the metal. Especially large quantities of hydrogen can be absorbed by palladium – up to the creation of palladium hydride phase.

6.4 Adsorption of Oxygen on Pure Metals

The keen interest in the adsorption of oxygen on pure metals arises not only from its being catalytically important, but also from the fact that adsorption of oxygen represents the intial stage of the oxidation of metals.

The adsorption of oxygen assumes a much greater variety of forms than does the adsorption of hydrogen: O^-, O^{2-}, O_2^-, O_2^{2-}, as well as neutral atomic and molecular forms. This variety is reflected in the wide range of values of the heat of chemisorption: from 80 kJ/mol (and lower) for silver to 800 kJ/mol for tungsten. The adsorption of oxygen can be both activated and nonactivated, the former often having very high activation energy (up to 250–300 kJ/mol) associated with the diffusion of oxygen into the bulk.

Before the introduction of ultrahigh-vacuum investigative techniques (LEED, Auger spectroscopy) information regarding the status of adsorbed oxygen on metals was gained chiefly from measurements of contact potential difference [6.34]. The same experiments – with the preliminary cleaning of the surface in ultrahigh vacuum – allowed revision of the earlier conclusions and correlation of the variations of the work function $\Delta\phi$ with the structural changes in the surface layer, detected by LEED.

6.4.1 Two-Dimensional Layers of Adsorbed Oxygen on Transition Metals

The initial stages of the adsorption of oxygen on the clean surface of transition metals at about room temperature are usually accompanied by a considerable increase in the value of the work function. For example, on a nickel film the increase in coverage up to one-half of a monolayer is accompanied by a linear increase in the work function, wit $\Delta\phi$ ultimately reaching +1.0–1.2 eV [6.96,97]. On platinum, ϕ increases linearly with increase in the coverage, reaching +1.0 eV at $\theta = 0.4$ of a monolayer [6.98,99]. The activation energy of adsorption is about 80 kJ/mol, and some authors believe it to be much lower. With iron, the initial value of $\Delta\phi$ is +0.2 eV with near-zero activation energy [6.100]. For molybdenum at room temperature $\Delta\phi$ = +1.0 eV, and slight heating raises $\Delta\phi$ to 1.74 eV [6.34,101].

The use of the FEM and XPS techniques with single crystals whose surface has been cleaned in ultrahigh vacuum permitted measurement of $\Delta\phi$ due to adsorption of O_2 on individual faces.

On the faces (100) and (110) of Ni at $\theta = 0.35$ the work function increases by 0.3 to 0.5 eV [6.102,103]. For copper $\Delta\phi = 0.125$ eV for the (111) face, 0.39 eV for (100), and 0.075 eV for (110) [6.104]. The sticking coefficient of O_2 for low-index faces is smaller than for the polycrystalline samples.

The large increase in ϕ at $\theta < 0.5$, the low activation energies, and the high sticking coefficients were originally explained by the molecular form of adsorption. For example, the study of thermal desorption from platinum by the "flash" method revealed the existence of two forms of adsorption of oxygen, one of which ($Q = 25\text{-}60$ kJ/mol) was identified with molecular adsorption [6.98,105]. The molecular form of adsorption of O_2 was deduced also from FEM and FIM data, and the results of electrochemical measurements [6.106-108].

Despite the abundance of hypotheses, direct proof of the existence of stable molecular oxygen on the surface of noble metals is still lacking. Analysis of XPS spectra indicates that the bonding energies (oxygen bands 2p, 2s, 1p) are typical of the atomic, partly ionized forms of oxygen [6.103,109]. Considerable amounts of molecular oxygen are detected on the surface of transition metals of the platinum type only below -100°C [6.110-112].

Molecular (weakly adsorbed) oxygen is a short-lived precursor of atomic adsorbed oxygen:

$$O_2(gas) \leftrightarrows O_2(ads.) \leftrightarrows 2O(ads.) .$$

This follows, in particular, from the measured dependence of the sticking coefficient s on the coverage θ. For the adsorption of oxygen and other biatomic molecules on metals, the sticking coefficient s often shows no dependence on θ in a wide range of coverages (Fig.6.12, Curve 1). Should the adsorption proceed according to Langmuir's law, s would be a linearly decreasing function of θ (Curve 2). The independence of s from θ is explained by the existence of a mobile precursor state, located

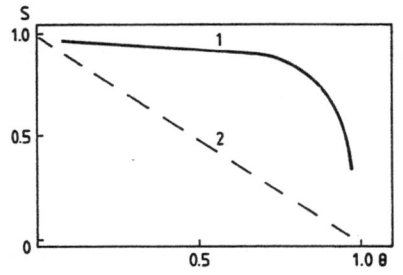

Fig.6.12. The sticking coefficient s of simple molecules versus the coverage θ:
(1) experimental curve, (2) calculated Langmuir curve

above the chemisorbed layer and taking some time before going into the immobilized state.

The precursor is not a physically adsorbed molecule. Experiment indicate that the state of strong chemisorption can be preceded by dozens and hundreds of jumps performed by the molecule on the surface. The precursor's lifetime is much longer than the lifetime of physically adsorbed molecules, as calculated by Frenkel's formula $\tau = \tau_0 \exp(Q/RT)$. Apparently, the precursor molecules are in the vibration-excited or electron-excited state [6.113,114]. Then the long lifetime of the precursor may be due to its need to give away its energy in small portions to the lattice phonons (≈ 0.01 eV) in order to go into the state of strong chemisorption (2-3 eV).

The regularities of the sticking coefficient of O_2 as a function of coverage for different transition metals are discussed in [6.115]. The deeper the potential well (the greater the Q_{O_2}), the wider is the interval of independence of s from θ. The most rapid decrease in s is observed with the metals with low Q (Ag, Cu). According to the principle of microscopic reversibility, the presursor state must be observed not only at adsorption, but also at desorption. Indeed, at the adsorption-desorption of O_2 on W(110), the desorption was found to have a very low value of E_{des} (7 kJ/mol), and $E_{des} < Q_{ads}$ [6.112]. This can be associated with the two-stage process of desorption

$$O_2(gas) \text{ or } 2O(ads.) \xrightarrow{k_1} O_2(precursor) \xrightarrow{k_2} O_2(gas) ,$$

if $k_2 \gg k_1$ – that is, if the probability of desorption from the precursor state is higher than the probability of losing the vibrational energy and regaining the ground state.

The sticking coefficient of oxygen is especially high on the polycrystalline specimens and stepped surfaces [6.12,116–118]. The adsorptive properties of the smooth (111) face of Pt and the stepped surface 12(111) x (111) of platinum were compared in [6.116]. For the stepped surface at room temperature s = 0.4. As the coverage increases, the value of s drops down to 0.01, which corresponds to the value of s for the smooth face (111) at θ = 0.5. If, however, we relate the initial value of s not to the entire surface area, but rather only to the atoms on the edges of steps (Fig.5.9), it will be even greater than unity. Apparently all O_2 molecules, having struck the flat (111) face, are further transferred to the edges and suffer dissociative chemisorption.

According to LEED and Auger spectroscopy [6.24,110,116,119–121], the adsorption of O_2 on the stepped surface of Pt, Au, and W first gives rise to the two-dimensional structures on the edges. For example, on the edges of the platinum surface 12(111) x (111) the oxygen atom is bound with every other atom of the edge. The creation of the ordered structure on the stepped surface of tungsten was proved by ESDIAD of O^+ ions. If the stepped surface has kinks, the oxygen occupies these first [6.122]. Kinks and edges are active in the catalyzed oxidation reactions.

In most cases – even at low temperatures and low coverages – the LEED patterns are perfectly clear, which can only be explained by adsorption in the atomic form, or, to be more precise, in the ionic form $O^{\delta-}$. By contrast with the adsorption of hydrogen, the adsorption of oxygen, according to LEED, gives rise to a great variety of structures, reflecting the reconstruction of the surface and the subsurface layer. These structures do not reduce to a simple succession α, β_1, and β_2 (Sect.6.1). The most common structure on the cubic faces is the

(2x2) structure (a) at low coverages and the c(2x2) structure (b) at medium coverages:

$$
\begin{array}{ll}
\begin{array}{lllll}
\text{M} & \text{M} & \text{M} & \text{M} & \text{M} \\
\text{M} & \text{O} & \text{M} & \text{O} & \text{M} \\
\text{M} & \text{M} & \text{M} & \text{M} & \text{M} \\
\text{M} & \text{O} & \text{M} & \text{O} & \text{M}
\end{array} \quad \text{(a)}
&
\begin{array}{lllll}
\text{M} & \text{O} & \text{M} & \text{O} & \text{M} \\
\text{O} & \text{M} & \text{O} & \text{M} & \text{O} \\
\text{M} & \text{O} & \text{M} & \text{O} & \text{M} \\
\text{O} & \text{M} & \text{O} & \text{M} & \text{O}
\end{array} \quad \text{(b)} \; .
\end{array} \qquad (6.6)
$$

In accordance with the stoichiometry, the maximum intensity of the LEED pattern is observed at $\theta = 0.25$ for structure (6.6a) and at $\theta = 0.5$ for structure (6.6b). In the (2x2) structure the oxygen atoms are arranged directly over the metal atoms. This explains the high increase in ϕ at the beginning of adsorption of O_2. In structure (6.6b), denoted by c(2x2) the oxygen atom is located centrally above the square formed by four metal atoms. This also results in the increase in ϕ, which, however, is smaller than with he (2 x 2) structure.

The transition from the structure (2x2) to the structure c(2x2), i.e., from structure (6.6a) to structure (6.6b), represents the transition of the oxygen atom from the location above the surface to the location "sunk" into the lattice, where the atoms of metal and oxygen lie in the same plane and form bridges M...O...M. Kinetic studies of the adsorption of O_2 on different faces of the Ni single crystal point to the creation of ordered structures, or phases. Within each structure the sticking coefficient s exhibits a monotonic change, but prior to the transition from one structure to another [at $\theta = 0.25$ and 0.5 for the (100) face] there is a sudden drop in s (Fig.6.13). According to [6.123], each new phase is formed via the precursor molecular state O_2.

The structures (2x2) and c(2x2) were detected for the adsorption of O_2 on the cubic faces of Ni [6.123–129], Cu [6.110], Co [6.130], Fe [6.131–133], Pd [6.134], and Mo [6.135]. At high coverages the (1x1) structure is formed with one oxygen atom per atom of metal.

The adsorption of O_2 on the (100) faces of Pt and Ir gives rise to structural rearrangement similar to the rearrangement

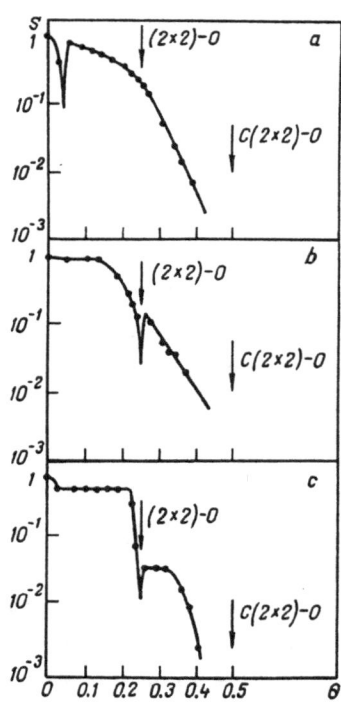

Fig.6.13. The sticking coefficient of oxygen s versus the coverage O for the (100) face of Ni [6.123]: (a) at 22°C, (b) at 100°C, (c) at 150°C

that takes place at the adsorption of H_2 on the same face (Sect.6.3). The structure (5 x 1) or (5 x 20), for Ir typical of the pure surface, goes through a series of other structures: (2x2), c(2x2), (1x1), (2x2), (2x1), c(2x2), (4x1), etc. [6.13,136–138]. The structures (4x1), (2x2), (2x1), and (5x5) were discovered on the (100) face of tungsten [6.13]. The data for the (100) face of W were compared with the results of investigations of the surface structure of W by secondary ion mass spectroscopy (SIMS) [6.139] and ESDIAD [6.140]. These results point to the reconstruction of the surface at coverages above 0.25.

On the (111) faces of Ni, Pt, Pd, Ir, and Rh the adsorption of O_2 at low coverages and low temperatures proceeds at random. At room temperature the structure (2x2) is formed [6.13,110,116,141–143], and is accompanied by a slight expansion of the surface layer. Study of the adsorption of O_2 on the

(111) face of W by LEED and ESDIAD indicates that oxygen is at first adsorbed in the bridge position W...O...W, and then on top of W atoms [6.144]

The (110) face is characterized by a great variety of elongated structures. Examples of such structures on (110) Ni were already given in Sect.6.1. On the (110) faces of Pt [6.13,145,146], Ni[6.13,147-149], Cu[6.150,151], Nb[6.152] and W [6.153-155] the adsorption of oxygen was found to give rise to a (2 x 1) structure at $\theta < 0.5$. This structure arises from the adsorption of oxygen atoms in the troughs between the ridges of metal atoms in the crystallographic [10] direction. Oxygen, slightly protruding above the surface, forms a bridge between the two metal atoms. For the (2 x 1) structure on the (110) face of Ni

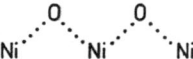

the bond length is 0.192 nm [0.197 nm in the (2 x 2) structure on the cubic face]. The lengths of M-O bonds, as determined experimentally for the chemisorption of oxygen on metals, are very close to the sum of covalent radii of M and O (within an accuracy of 10%) [6.156]. This points to the predominantly covalent nature of chemisorptive bond.

Further adsorption of O_2 on the (110) face may give rise to more elongated structures with large values of the lattice parameter in one direction. For instance, the adsorption of O_2 on Ni(110) at 300°C in a very narrow range of coverages (0.555 < $\theta < 0.625$) gives rise to four superstructures – (9x1), (7x1), (5x1), and (8x1) – having a periodicity of several nanometers, which is much greater than the bulk lattice parameter.

Similar phenomena were observed for the adsorption of O_2 on the (110) faces of Cu, Pt, Pd, and Fe [6.120, 157-161]. The simultaneous employment of different investigative techniques (adsorption measurements, LEED, XPS, Auger spectroscopy, FEM, work-function measurements) has produced irrefutable evidence in favor of the deep reconstruction of the surface upon the adsorption of oxygen. According to MAY [6.11], by adopting the concept of surface reconstruction, it is easier to explain the constancy of the heat of adsorption of oxygen in the wide range of coverages. Only reconstructions can ensure the identical the crystallographic surrounding for each added atom of oxygen.

An interesting proof of the surface reconstruction upon adsorption of oxygen on tungsten was reported in [6.101]. The (2x1) structure, formed after the adsorption of O_2 and subsequent slight heating, persists to 1300°C. The electron-stimulated desorption of oxygen results in a considerable reduction in the work function, but does not affect the (2x1) structure, as indicated by LEED. This time, however, even a slight heating destroys this structure and restores the structure typical of the pure tungsten surface.

The growth of the two-dimensional phase often depends on the migration of particles to an initial nucleus. The creation of elongated nucleus domains has been detected on the (110) face of platinum; their average size is 0.9 nm at θ =0.1 and 3.5 nm at θ = 0.3. In the catalyzed reaction between the adsorbed oxygen and H_2 or CO, the reactive oxygen atoms are those located on the domain boundary.

The genesis and reactivity of two-dimensional phases of adsorbed CO and O_2 on Pd(111) was carefully studied by CONRAD et al. [6.160]. It was demonstrated that in some cases the kinetics of oxidation of CO on Pd can be explained by assuming that the reaction takes place on the boundary between the adsorbed phases of O or CO. The less strongly bound component migrates to the "patch" of the more strongly bound component.

At sufficiently high coverages the conversion of one structure into another may take the form of a second-order phase transition. For example, on the (110) face of W at θ = 0.25 there is only the (2x1) structure, and only the (2x2) structure at θ = 0.75. These structures are separated by a phase transition [6.154]. The intervals between the ordered structures consist of less-ordered structures, which, in particular, produce oblong streaks in the LEED pattern instead of the usual circular reflections. These intermediate structures may be viewed as an irregular mixture of the two ordered structures, e.g., the simultaneous existence of (2x1) and (2x2) patches. Consequently, the two-dimensional phase transition may be a time-consuming process.

There are also some interesting results concerning the relative stability of different metallic faces in an oxygen atmosphere. Investigations of adsorption of O_2 on different faces of single crystals of platinum at temperatures up to 1100°C indicate the stability of the low-index faces (100), (110), and (111). Apart from these, only a few of the other faces are stable [(211), (311), (331), (221), (210)]; all the rest suffer profound rearrangement.

6.4.2 Initial Steps of the Oxidation of Transition Metals

The initial rapid increase in the work function is often followed by a drop in ϕ. Such drops, which approach and sometimes equal the initial rise, were observed at the adsorption of O_2 on iron [6.100,161], tungsten [6.162], ruthenium [6.157], copper [6.158,159], and nickel [6.102,103]. Slight heating makes this effect even more pronounced: the drop may actually exceed the initial rise. For the adsorption of oxygen on Mo, Co, Nb, and Mn the value of ϕ began to decrease almost at once, without any noticeable initial rise [6.100,147,162-165].

The most popular explanation of this effect is due to BURSHTEIN and co-workers [6.166,167]. According to this theory, oxygen penetrates ("creeps") beneath the uppermost layer of metal atoms; this results in the production of an electric double-layer, the positive side outermost, and the work function decreases.

LANYON and TRAPNELL [6.168] proposed a similar explanation, which, however, accounts for the reconstruction of the surface layer:

$$
\begin{array}{c}
\begin{array}{l}
\text{0 0 0 0 0 0} \\
\text{M M M M M M} \\
\text{M M M M·M M}
\end{array}
\xrightarrow[\text{quick}]{\substack{+O_2 \\ (+\Delta\varphi)}}
\begin{array}{l}
\text{0 0 0 0 0 0} \\
\text{M M M M M M} \\
\text{M M M M M M}
\end{array}
\xrightarrow[\text{slow}]{\substack{+O_2 \\ (-\Delta\varphi)}}
\end{array}
$$

$$
\xrightarrow{}
\begin{array}{l}
\text{0 0 M M 0 0} \\
\text{M M 0 0 M M} \\
\text{M M M M M M}
\end{array}
\xrightarrow{(+\Delta\varphi)}
\begin{array}{l}
\text{0 0} \\
\text{0 0 M M 0 0} \\
\text{M M 0 0 M M} \\
\text{M M M M M M}
\end{array}
\tag{6.7}
$$

According to HALL and MEE [6.100], the adsorption on iron is more likely to follow the Lanyon-Trapnell mechanism with the reconstruction of the surface. The adsorption on Co and Mn, which exhibit a decrease in ϕ directly after the beginning of adsorption, is more likely to follow the mechanism of simple "creeping" of oxygen atoms under the top layer. The different reaction of these materials to adsorption is explained in the peculiarities of their crystalline structure. In the surface layer of manganese there are "holes" 0.45 nm wide and 0.17 nm deep which can accept atomic oxygen (0.12 nm in diameter). Similar holes exist on the surface of cobalt (0.41 nm across and 0.03 nm deep), and atomic oxygen (but not O^{2-}, whose diameter is larger) can get there with a certain amount of activation energy. The lattice of iron is packed more tightly: the "holes" are only 0.16 nm in diameter and 0.02 nm in depth, and the penetration of oxygen is impossible without the reconstruction of the lattice.

The drop in ϕ due to intrusion of oxygen or reconstruction of the surface layer is usually followed by a considerable increase due to the build-up of the new layer of atomic oxygen on the surface (cf. the last stage in the Lanyon-Trapnell scheme). This process is in essence the beginning of the creation of oxide film on the surface and goes via a chain of atomic rearrangements detected by LEED. An example is given in Fig.6.14, which shows the adsorption diagram of O_2 on a tungsten film at

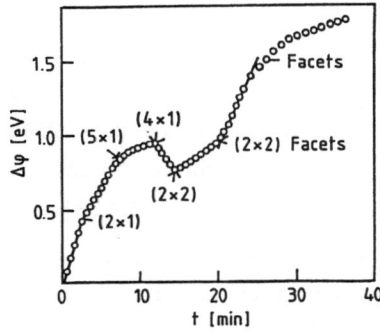

Fig.6.14. Changes in the work function $\Delta\phi$ in the course of adsorption of oxygen at 760°C on a tungsten film [6.101]. Arrows indicate the corresponding surface structures, as derived from LEED data. Facets are microportions of different crystalline faces

760°C [6.101]. Each of the clear-cut steps of the curve corresponds to the creation of a new structure, detected by LEED.

The adsorption of O_2 on Ru also resulted in the rise, drop, and rise again in the value of ϕ [6.162]. The LEED pattern steadily indicates one and the same (2x1) structure, although the intensity of the spots varies in line with the changes in the work function.

When the metal surface is oxidized and becomes covered with the oxide film, the change in the work function may be as large as 1-2 eV. The oxide film may be several atoms thick, and sometimes – e.g., in the adsorption of oxygen on Pd at above 250°C – the oxide film is hundreds of atoms thick [6.169]. On easily oxidized metals (Ni, Co, Fe, Mo, W) the adsorbed oxygen at first usually forms a monolayer (sometimes two or three). This is followed by the reconstruction of the surface and nucleation of the oxide phase. Further growth of the oxide phase exhibits no dependence on the crytallographic directions. The process of oxidation of Ni has been studied by SIMS [6.170–173].

The yield of ions O^-, O_2^-, Ni_2^+, Ni_3^+, NiO^- and others was correlated with the particular structures of the surface layer. The initial rate of build-up of the oxide layer is the greatest on the (110) face of Ni. When the film of NiO is complete (2-3 monolayers), microcrystals of NiO start growing independently of the orientation of the underlying face. The layer of Ni_2O_3 is produced at high pressure of O_2.

Study of the oxidation of the (100) face of tungsten by the SIMS and LEED techniques reveals that as the first monolayer is completed, a (4 x 1) structure is formed, which represents the layer of WO_2. Accordingly, WO_2^+ ions are detected in the SIMS spectrum. After this, the three-dimensional WO_3 phase begins to grow, and WO_3^- ions are observed in the SIMS spectrum [6.138].

The creation of an oxide layer is confirmed by the appearance of XPS, UPS, and Auger bands, pertaining to the ionic forms of oxygen (O^-, O_2^-) and metal (M^{n+}). In the initial stages the spectral lines of metal correspond not to the oxidized, but rather to the metallic (zero-valence) state.

6.4.3 Surface Levels of Adsorbed Oxygen

The UPS technique was used for studying the surface levels of oxygen adsorbed on Ni [6.63,174–177], Co [6.178], Fe [6.64,133], Cu [6.178], Mo [6.98], W [136], Ir [6.179,175,180], Pt [6.133,180,181], and Pd [6.175,182]. The results of investigations of different systems verify the mechanism of adsorption of O_2 and oxidation of transition metals described above. The initial stages of adsorption produce a very slight lowering in the intensity of the peak of the d levels in the range of 0–2 eV below E_F, and give rise to a strong peak at 5.6–5.7 eV, pertaining to the 2p levels of adsorbed oxygen. Additional peaks arise at the stages of reconstruction and oxidation pertaining to the oxidized metallic atoms.

For example, the bands occuring in the adsorption of O_2 on Ni resemble the UPS spectrum of NiO: a sharp peak at 2 eV below E_F, a wide maximum at 5–6 eV and a small peak at 10 eV. The analysis of the octahedral cluster Ni_6^{10-} by the SCF-X_α-SW method indicates that the first peak is associated with the photoemission from the highest occupied levels of Ni: $e_g\uparrow$ and $t_{2g}\downarrow$. The wide band pertains to the 2p band of oxygen, and the peak at 10 eV to the emission from the bonding $2p\sigma$-4s orbitals from the bottom of oxygen's 2p band [6.63]. For platinum, the peak of oxygen 2p level is less well defined than with Fe, Co, and Ni. This is possibly associated with the fact that oxygen on Pt, Pd, and Ir is bound not only with the d band of metal, but also with the s and p bands [6.181,182]. Study of the adsorption of O_2 on Mo(110) by the SPI technique (Sect.2.2.4) reveals the existence of a strong peak at 6 eV below E_F, pertaining to ionization of the 2p level of adsorbed oxygen [6.183]. In all the cases of oxidative adsorption of O_2 the differential UPS spectra indicate a very strong suppression of emission from d levels located near E_F. Evidently, oxygen is the strongest acceptor of all the adsorbates studied.

6.4.4 Oxygen Adsorption on Silver

Silver occupies a unique position with respect to adsorption of oxygen: it seems to be the only metal capable of adsorbing oxygen in the molecular form at high temperature. This unique property of silver is responsible for its special catalytic property: only silver is capable of catalyzing the reaction of oxidation of ethylene to ethylene oxide.

At room temperature the adsorption of O_2 on Ag has a low value of the heat of adsorption (60–120 kJ/mol) and is accompanied by a considerable increase in the work function. As indicated by FEM spectroscopy [6.184], the faces of the Ag single crystal are occupied by oxygen in the following order: first (111), (100), (211), (533); then (110); and finally (210), (310), (320). The main evidence of the molecular form of adsorption comes from the IR spectra [6.185,186]. The band at 870 cm^{-1}, observed after the adsorption of O_2 on Ag, is ascribed to the stretching vibration ν_{O-O}, which is further confirmed by using isotopic molecules $^{18}O_2$ and $^{18}O^{16}O$.

According to [6.185], the initial rapid adsorption is dissociative and can be described as the interaction of the O_2 molecule with four adjacent Ag atoms:

$$4Ag + O_2 \rightarrow 4Ag^+ + 2O^{2-} \quad .$$

The atomic forms of adsorption of O_2 on the Ag surface, pre-cleaned in ultrahigh vacuum, have been detected by LEED, XPS, and UPS. The structures (2x1), 3x1), and (4x1) were discovered on Ag (110) [6.187,188]. The UPS spectra taken for adsorption of O_2 on Ag (110) point to the lowering of emission at 5–7 eV below E_F (which corresponds in energy to the 4d band of Ag), and exhibit three peaks between this band and the Fermi level. This spectrum is explained by the interaction of the 2p orbitals of oxygen with the 5s and 4d orbitals of silver in the cluster Ag_4O, calculated by SCF-X_α-SW method [6.189].

When, however, the major portion of the surface is already covered with oxygen ions (coverage close to unity), four uncovered silver atoms are not likely to be found close together. Accordingly, the nondissociative adsorption becomes more probable, e.g.,

$$O_2 + Ag \rightarrow Ag^+ + O_2^- \ .$$

As a matter of fact, the adsorption of O_2 on silver supported on porous glass was found to give rise to an EPR signal with $g_z = 2.034$, $g_y = 2.012$, and $g_x = 2.004$ [6.190]. This signal corresponds in parameters to the ion radical O_2^- adsorbed on oxides (Sect.3.4). It was therefore assumed that the O_2^- ion is stabilized not on the neutral silver atom Ag^0, but rather on the oxidized silver $Ag^{\delta+}$. At $\theta = 0.44$ the form O_2^- constitutes a small fraction of the adsorbed oxygen (about 0.02). In this connection let us observe that at coverages close to a monolayer, the oxygen on the surface of silver occurs in the form of Ag_2O_3 oxide, which can be viewed as $Ag_2O + O_2$ [6.191].

6.5 Adsorption of Carbon Monoxide on Metals

6.5.1 The Structure of Surface Carbonyls

The mechanism of adsorption of carbon monoxide by transition metals was deduced mainly from the data supplied by IR spectroscopy [6.192-211]. Similar to the adsorption on oxides, the adsorption of CO on metals allows observation of the bands of C-O vibrations in the range of 1800-2200 cm^{-1}.

According to EISCHENS and PLISKIN [6.193], the absorption bands above 2000 cm^{-1} pertain to the linear carbonyls M—C≡O, and the bands below 2000 cm^{-1} to the bridge carbonyl structures $\overset{\displaystyle M}{\underset{\displaystyle M}{\diagdown}}$C—O. The stronger bands associated with the adsorption of CO are 1960 cm^{-1} with Fe, 1920 cm^{-1} with Pd, and 1990 cm^{-1} with Ru, the weaker bands are observed near 2100 cm^{-1}. Only the bands at about 2100 cm^{-1} are, as a rule, observed with the metals of group IB: 2120 cm^{-1} with Cu, and 2170 cm^{-1} with Ag. The IR spectra – obtained by the technique of multiple internal reflection for the adsorption of CO on gold films, precleaned in ultrahigh vacuum – exhibit the band at 2120 cm^{-1} with the beginning of adsorption and at 2115 cm^{-1} with θ = 0.6 [6.196]. The adsorption of CO on platinum gives rise to one high-frequency band at 2075 cm^{-1}. Experimental data concerning the frequency of M-C vibrations in the surface carbonyl compounds are not abundant. The band at 435 cm^{-1}, observed in the IR spectrum of CO adsorbed on Ni, was attributed to the stretching vibration of the M-C bond [6.194]. Figure 6.15 illustrates the adsorption of CO on the nickel (100) and (110) faces [6.1]. On Ni(100) the chemisorption of CO is strong and produces bridge structures (Fig.6.15a); the ratio of CO molecules to

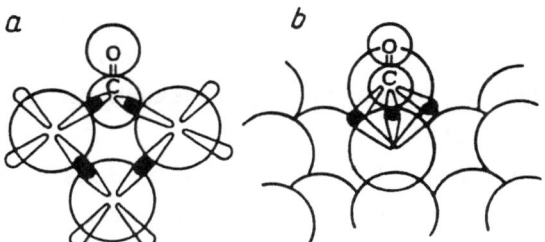

Fig.6.15. Adsorption of carbon monoxide on Ni: (a) on the (100) face (normal to the plane); (b) on the (110) face (at 45º); CO molecules lie in the plane of the diagram

surface Ni atoms is one to two. On the (110) face of Ni the adsorption is weaker (linear, Fig.6.15b) with [CO]/[Ni] = 1.

The interpretation given by EISCHENS and PLISKIN [6.193] was challenged by BLYHOLDER [6.194] and other researchers, who maintain that the frequencies both above and below 2000 cm^{-1} pertain to the linear carbonyl structures. In the former ($\nu>2000$ cm^{-1}) the back donation between the metal and CO ligand is suppressed, and the C-O bond is relatively strong; in the latter ($\nu<2000$ cm^{-1}) the M-C bond is stronger and the C-O bond weaker. This point has been discussed in detail in Sect.4.6.1.

It seems that these two viewpoints in many cases are not mutually exclusive. In the linear carbonyl structures M—C≡O the back donation involving t_{2g} orbitals is weaker, and the donor-acceptor interaction (involving e_g orbitals) stronger than in the bridge structures. In both linear and bridge structures the frequency of C-O vibrations goes down as the number of metallic atoms taking part in the bonding increases.

The IR spectroscopic data correlate well with the results of other techniques, in particular, with the measurements of surface potentials [6.200,212-215]. In most cases the adsorption of CO is accompanied by an increase in the work function: the maximum value of $\Delta\phi$ is +1.33 eV for Fe, +1.35 eV for Ni, +0.82 eV for W, and +0.67 eV for Ta. For the metals of group IB, $\Delta\phi$ is negative (a decrease in the work function); $\Delta\phi$ amounts to -0.28, -0.31, and -0.92 eV for the adsorption of CO on Cu, Ag and Au respectively. For adsorption of CO on tungsten the increase in ϕ reaches 0.82 eV at θ =0.5, and then ϕ goes down a little [6.214]. For adsorption on copper the work function generally decreases, although it keeps fairly constant in the range $0.1 < \theta < 0.5$ [6.200,207,215].

The adsorption of CO on platinum leads to the variations in ϕ that are quite small and do not exceed several hundredths of an electron volt. With other platinum metals $\Delta\phi$ (negative) increases with separation in the periodic table from platinum; for adsorption on iridium $\Delta\phi$ = -1.35 eV. The magnitude of ϕ depends

362

on the structure of the surface. The decrease in ϕ is greatest for the "open" surfaces (110) and (210); it is smaller for (111) and (100) faces in FCC structures [6.207].

There exists a very conspicuous correlation between the variations in the work function and the location of ν_{C-O} in the IR spectrum. The shifting of the C-O band toward high frequencies is accompanied by negative charging of the surface, and shifting toward low frequencies (< 2000 cm^{-1}) is accompanied by positive charging. This effect fits in well with the model of molecular orbitals of surface carbonyl structures discussed in Sect.4.6.1. The increased contribution of metal-carbonyl back donation results in the displacement of the electron density from metal to CO, the charging of the carbonyl [CO]$^{\delta-}$, a decrease in ν_{C-O}, and strengthening of M-C bonding. Conversely, the increased contribution of the donor-acceptor bonding between metal and carbonyl results in the displacement of the electron density from CO to metal, positive charging [CO]$^{\delta+}$, increase in ν_{C-O}, and strengthening of the C-O bond, since the 5σ orbital of CO, which donates electrons, is weakly antibonding.

Many researchers have observed the shift of ν_{CO} toward higher frequencies as the coverage of the surface by CO increases. For example, on Pd(100) the frequency ν_{CO} varies from 1895 cm^{-1} at low θ to 1949 cm^{-1} at $\theta = 0.5$; on the (111) face of Pd, ν_{CO} varies from 1823 cm^{-1} to 1936 cm^{-1} at $\theta = 0.5$. Further increase in the coverage drives ν_{CO} above 2000 cm^{-1}. At low coverages the threefold and twofold structures of CO are possible [6.205]. The shift toward high frequencies at high coverage is explained in [6.204] by the compression of the adsorbed layer, due presumably to the dipole–dipole interaction.

The bridge structures are usually the first to appear in adsorption and the last to vanish in desorption. According to [6.197], in the adsorption of CO and Pd, first to appear is the band at 1900 cm^{-1}, ascribed to the isolated bridges (6.8a) (see below). As the coverage rises to $\theta = 0.30$, this band splits in two. One of the resulting bands (1970 cm^{-1} with a wing at 1900 cm^{-1}) is ascribed to the bridge structures (6.8b); the other (2070 cm^{-1}) presumably pertains to the linear structures. As the coverage nears a monolayer, the two bands reach saturation and a third band appears at 2095 cm^{-1}; brief pumping out restores the band at 2070 cm^{-1} and shifts the band at 1970 cm^{-1} to 1980 cm^{-1}. According to [6.197], the structure of type (6.8c) is then formed. At desorption the bands disappear in the reverse order.

$$
\begin{array}{cc}
\underset{\text{Pd...Pd...Pd...Pd}}{\overset{\overset{\displaystyle O}{\|}}{\underset{C}{\diagup\;\diagdown}}} \;(a) &
\underset{\text{Pd...Pd...Pd...Pd}}{\overset{\overset{\displaystyle O}{\|}\qquad\overset{\displaystyle O}{\|}}{\underset{C\qquad C}{\diagup\diagdown\diagup\diagdown}}} \;(b)
\end{array}
$$

$$
\underset{\text{Pd...Pd . . . Pd...Pd}}{\overset{\overset{\displaystyle O}{\|}\;\overset{\displaystyle O}{|}\;\overset{\displaystyle O}{|}\;\overset{\displaystyle O}{\|}}{\underset{C\;C\;C\;C}{\diagup\diagdown\diagup\diagdown}}} \;(c) \qquad\qquad (6.8)
$$

For the adsorption of CO on Ni(111), ν_{CO} = 1817 cm^{-1} at θ = 0.05; according to [6.208] this frequency corresponds to the adsorption on threefold sites. The frequency 1910 cm^{-1} at θ = 0.30 corresponds to the adsorption on the twofold site, or to the simple bridge; at θ = 0.57 the adsorption takes the linear form with ν_{CO} = 2045 cm^{-1}.

The new method of vibrational spectroscopy HREELS (Sect.2.2.9) was employed in investigations of adsorption of CO on the faces of single crystals [6.205,216–220]. An important achievement was that not only the frequencies of C–O vibrations could be measured, but also the frequencies of vibrations of the M–C bond. With Pt(111) at low coverages the peaks were located at 58 MeV (474 cm^{-1}) and 258 MeV (2075 cm^{-1}; at θ > 0.2 there appear extra bands at 45 MeV (367 cm^{-1}) and 232 MeV (1870 cm^{-1}). The attribution of the band at 2075 cm^{-1} to the linear carbonyl and the band at 1870 cm^{-1} to the bridge carbonyl is confirmed by the analysis of the stretching vibrations of the Pt–C bond (474 and 367 cm^{-1}). For the low-coordination sites, where the C atom is bound with only one Pt atom, the frequency ν_{C-O} must be higher. As θ increases, the surface of platinum becomes covered first by linear and then by bridge carbonyls [6.216,217].

For the (100) and (110) faces of Ni the measured frequencies of Ni-C vibrations were in the range of 450-460 cm^{-1}. Accordingly, the frequencies in the range of 1880-1935 cm^{-1} were ascribed to the bridge, and in the range 1990-2015 cm^{-1} to the linear forms of adsorbed CO. Only the bridge forms (1880-1910cm^{-1}) were detected on Ni(111), while the linear forms (2080 cm^{-1}) dominated on the surface of carbidized nickel [6.220).

The investigations using SIMS seem to confirm the conclusions of EISCHENS and PLISKIN [6.193] regarding the linear and the bridge structures in adsorption of CO on metals. After the adsorption of CO on Cu the SIMS spectra indicate the presence of only the ions Cu^+ and $CuCO^+$, agreeing with the IR spectroscopic data pointing to the creation of the linear form of adsorbed CO [6.221]. For adsorption on nickel, the SIMS spectrum indicates the existence of bridge Ni_2CO^+ and threefold Ni_3CO^+ ions, together with the linear ions $NiCO^+$. The relative yield of each particular ion correlates with the coverage of the surface [6.221-225].

Yet another new technique ESDIAD (Sect.6.1) was used for studying the orientation of adsorbed CO molecules. After the adsorption of CO on the (001) faces of Ru and W at low temperatures, the beams of electron-desorbed CO^+ and O^+ ions are directed more or less normal to the surface, indicating to the upright orientation of the adsorbed CO molecules with respect to the surface [6.226]. At high coverages the linear form of CO is normal to the surface.

6.5.2 Structure of Adsorbed Layers of CO on Transition Metals

Extensive LEED investigations have led to reliable interpretation of a number of surface structures of CO on transition metals [6.13,11,141,200,204,207,215,227-248]. It must be observed, however, that many results obtained by LEED relate, strictly speaking, only to the structure of the adsorbed CO layer, and fail to characterize correctly the underlying metallic layer. It has been possible, for instance, to compare the LEED data for adsorption of CO on the (111) face of Ni [6.229,235,244,247], with data obtained by other investigative techniques. As the coverage increases to $\theta = 1/3$, the work function exhibits a linear increase ($\Delta\phi$ =1.1 eV), the heat of adsorption goes down from 125 kJ/mol to 110 kJ/mol, while the LEED pattern displays the growing intensity of spots pertaining to the peculiar structure denoted by $(\sqrt{3} \times \sqrt{3})R30^\circ$. This structure arises due to the mutual attraction of adsorbed CO molecules, which results in the creation of clusters – e.g., rhombic structures of four CO molecules:

$$
\begin{array}{ccc}
 & \overset{\displaystyle O}{\underset{\displaystyle \parallel}{C}} & \\
\overset{\displaystyle O}{\underset{\displaystyle \parallel}{C}} & & \overset{\displaystyle O}{\underset{\displaystyle \parallel}{C}} \\
 & \overset{\displaystyle O}{\underset{\displaystyle \parallel}{C}} &
\end{array} .
$$

Their orientation with respect to the substrate is not specified although each CO molecule is assumed to form a bridge between two Ni atoms. Further increase in coverage (to θ=0.53) results in some extra increase in the work function ($\Delta\phi$ =1.31 eV), the heat of adsorption (due to the repulsive interaction) goes down to 85 kJ/mol, and the sticking coefficient is reduced from unity to almost zero. As the packing becomes denser, the structure $(\sqrt{3} \times \sqrt{3})R30^\circ$, as indicated by LEED, transfers into the structure composed of random-oriented c(4x2) domains. At low temperature (-70°C) and $\theta > 0.57$ there appears yet another structure (hexagonal close packing), in which the CO molecules occur in both bridge and linear positions. The separation between CO molecules in this structure (0.33 nm) is even smaller than the diameter of CO molecule in the gas.

The structure $(\sqrt{3} \times \sqrt{3})R30^\circ$ is very common with octahedral faces, and was observed for the adsorption of CO on (111) faces

of Pt [6.13,245], Ir[6.13], Pd[6.13], Cu[6.13], Rh[6.248], a
some other metals. Also observed was the structure c(4x2).

According to LEED, the adsorption of CO on Ni(10C
[6.215,242] at $\theta < 0.5$ gives rise to a simple structure c(2 $\times$
2), which can be explained by the creation of pairwise CO
bridges between two atoms of Ni. At $\theta = 0.61$ this structure is
replaced another regular structure, which suffers contraction
as the coverage increases to $\theta = 0.69$. These structures were
correlated to the phases α, α_0, and β, which become manifest in
thermal desorption. These data were used for plotting the
"phase diagram", showing the coverage of the surface by CO
versus temperature. An interesting feature of this diagram is
the existence of a triple "eutectic" point, where the phases
α[c(2x2)], β (hexagonal), and α_0 (disordered) exist in equilibrium.

In the adsorption of CO on Ni(110) there first arises the
structure c(2x2), which further transforms into (4x2) and
finally into (2x1) [6.231]. The values of heat of adsorption
differ little (within 10 kJ/mol) for different faces of Ni. The
variations in the work function also are about the same for
different faces: $\Delta\phi_{max}$ varies from +1.2 to +1.4 for different
faces. This indicates that the nature of bonding between CO and
Ni is the same for all faces. At low temperatures the variations in the work function, as well as changes in IR and LEED
spectra, are reversible; above 200°C decomposition, $2CO \rightarrow CO_2 +$
C, takes place.

According to LEED data [6.207,234], the initial structure
formed in the adsorption of CO on Cu(100) is c(2 x 2). In accordance with the IR spectra, this structure was attributed to the
linear adsorption of a CO molecule on every other metallic
atom:

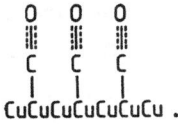

On Cu(110), the adsorption of CO takes the bridge form; the structures (2x1) at low coverages and c(1.3x2) at high coverages are observed [6.200].

As indicated by the IR spectra of homogeneous carbonyl complexes, the bridge carbonyls are more likely to occur in the complexes of metals of the first transition period [6.192]. With the second and third period transition metals, because of the large atomic radii, the linear complexes become more probable. Nevertheless, according to the LEED data [6.204,227,233,249], the adsorption of CO on palladium gives rise to structures very similar to those formed on palladium's electronic counterpart, nickel. At low coverages a disordered structure – a two-dimensional gas or a two-dimensional liquid (an aggregate of clusters) – may exist on any crystal face. On Pd(111), as on Ni, the structure $(\sqrt{3}x\sqrt{3})R30^{\circ}$ may be formed at θ < 1/3. On Pd(110) as the coverage increases, the structure c(2x2) transfors into (4x2) and then into (2 x 1), similar to the adsorption on Ni(110). The initial values of the heat of adsorption are slightly lower than for Ni [103, 113, and 109 kJ/mol for the faces (100), (110), and (111), respectively], and as the coverage increases, the heat of adsorption is reduced by 30 – 35 kJ/mol.

Several forms of adsorption of CO were discovered on platinum [6.13,228,236,238,243,250,251]. At very low coverages the heat of adsorption of CO at room temperature is as high as 160 kJ/mol, which corresponds to the dissociative adsorption of CO presumably on the edges or steps of the crystal. At 0.1 < θ < 0.5 the heat of adsorption is 110-130 kJ/mol, and as θ increases further, the heat of adsorption drops rapidly to 40 kJ/mol; at the same time the sticking coefficient falls from 0.1 to 0.01. The structures (1x1), (2x2), and c(4x2) are formed on Pt(111); c(4x2), c(2x2), and (1x1) on the (100) face; and (2x1) and c(2x1) on the (110) face. Before the structure (2x1) has formed on the (110) face, the adsorption of CO is random, but the face itself reconstructs into a collection of (100) "facets". This indicates that the energy of interaction of CO with the (100) face is larger than with the (110) face.

The adsorption of CO on Pt(100), similar to the adsorption of H_2 and O_2, converts the structure (5x1) or (5x20) into the structure (1x1). The kinetics of desorption of CO from this face cannot be described by simple first- or second-order equations because of the sophisticated dynamic balance between the structures (5x1) and (1x1), involving the reconstruction of the surface. Thermal desorption from this face exhibits three peaks with E_{des} = 117, 132, and 138 kJ/mol [6.252,253].

A considerable amount of dissociation, as indicated by LEED, takes place in the adsorption of CO on Mo, Ta, and W [6.13,232,238-241]. For example, three states of adsorption of CO were discovered [6.232] on W(100): the α state with heat of adsorption (Q) of 100 kJ/mol, β_1 with Q = 220 kJ/mol, and β_2 with Q = 430 kJ/mol. The first of these seems to pertain to the simple bridge form of adsorption ($\Delta\phi$ = +0.29 eV). Heating to 700°C reverses $\Delta\phi$ to −0.17 eV, and the disordered structure converts into c(2x2). Apparently, the β states relate to the dissociative adsorption of CO, β_2 with a deep reconstruction of the surface, and β_1 without. The LEED data are supported by the analysis of HREELS spectra of adsorbed molecules. With W(100), the low-energy peaks at 68 and 78 meV, observed at the early stages of adsorption of CO, are ascribed to the vibrations of adsorbed atoms C and O respectively, bound to the surface of W [6.216,219].

6.5.3 Surface Levels of Adsorbed CO on Transition Metals

The changes in the surface levels of metals upon adsorption of CO, and the nature of the interaction of CO molecules with the surface d orbitals, were studied by means of UPS. Since 1974, an enormous number of reports of the use of UPS for studying the adsorption of CO on metals have been published. Today this system is by far the best investigated of all the systems of this kind.

369

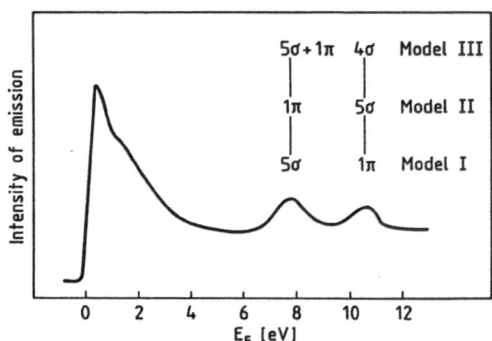

Fig.6.16. The UPS peaks for the adsorption of CO on a transition metal

The UPS technique was used for studying the adsorption of CO on nickel [6.64,174,227,229,254-262], iron [6.87,259,263-265], copper [6.235,256,257,265-268], cobalt [6.269], tungsten [6.96, 270], ruthenium [6.32], palladium [6.71], iridium [6.92,271-274], platinum [6.74,260,274-279], and other metals. Figure 6.11 shows how the UPS spectrum of adsorbed particles can be derived from the spectra of adsorbate + metal and pure metal. A typical UPS differential spectrum for the adsorption of CO on transition metals is reproduced in Fig.6.16. The spectra of adsorption of CO on different transition metals resemble one another, except for the difference in the intensities of individual bands and the difference in the relative intensity of bands arising when the excitation sources vary in energy. The essential feature of these spectra is the presence of two peaks: at 8 ± 0.6 eV (α peak) and at 11 ± 0.5 eV (β peak) below E_F (Fig.6.16). The slight variations of the peaks are associated with the variations in the work functions of the metals in question. Being almost independent of the particular kind of metal, these peaks are most likely to arise from the emission of electrons directly from the levels of adsorbed CO.

Three models of adsorption of CO, accounting for the peculiarities of the UPS spectrum, have been proposed. According to the Eastman-Cashion model [6.63], the α peak pertains to the chemisorptive level, corresponding to the molecular 5σ orbital of gaseous CO (Sect.4.6.1, Fig.4.26) and the β peak pertains to the molecular 1π orbitals of CO. In the Batra-Robaux model the order of orbitals is the reverse [6.256]: the α peak pertains to 1π, and the β peak to the 5σ orbitals. Finally, according to the third model [6.71], based on the quantum-chemical calculations of DOYEN and ERTL [6.227], the α peak corresponds to both the orbitals ($5\sigma+1\pi$), and the β peak corresponds to the 4σ orbital of CO.

The last model has gained general recognition now. Indeed, in the high-resolution UPS spectra the α peak actually consists of two peaks. For example, in the spectrum of CO adsorbed on Ir(100), the α peak had local maxima at 8.5 and 7.2 eV [6.271]; the first of these was identified with 1π and the second with the 5σ orbital. More detailed knowledge about the order of occupation of orbitals and their symmetry was obtained with the aid of tunable-energy synchrotron spectroscopy and ARUPS. These investigations reveal [6.29,64,69,257,270,272,280] that the CO molecule is adsorbed perpendicular to the surface, and the bonding relies mainly on the 5σ orbital of CO. In most cases the orbitals, according to their energies, stand in the following order (from low to high energies with respect to E_F): 5σ, 1π, and 4σ. However, there are indications that the order of the 5σ and 1π orbitals can be reversed. For example, according to [6.262], at the adsorption of CO on Ni(100) the 5σ orbital is 8.4 eV below E_F, and the 1π orbital is 7.8 eV below E_F. The 3σ orbital of CO is in the range of 25–30 eV below E_F (Fig.4.26).

The transitions of valence electrons become manifest in the EELS spectra. This technique is more sensitive to adsorptive centers than UPS; it can measure the energy of the electron transitions from the 5σ, 1π, and 4σ levels of adsorbed CO to the 2π level of CO, not observed in the UPS spectra. With Cu [6.281], Pd [6.282], and Ni [6.283] this level is located 5–6 eV above the Fermi level.

As already indicated, the adsorption of the CO molecule is accompanied – similar to the case of homogeneous carbonyl complexes – by the creation of a donor-acceptor bond at the expense of the occupied 5σ orbital of CO (HOMO) and vacant d_{z^2} orbital of metal, and a back donation bond due to the interaction of the vacant 2π orbital of CO (LUMO) with the occupied d_{xz} and d_{yz} orbitals of metal (Fig.4.26). Consequently, of the three orbitals (5σ, 1π, 4σ) detected by UPS, the 5σ orbital must suffer the greatest change on adsorption, while the location and intensity of bands of the 4σ and 1π orbitals should show little variation.

Experiment indicates [6.277] that the width of the α peak decreases as the coverage of the surface by CO increases; this may be attributed to the slackening of stabilization of the 5σ orbital of CO. With the increase in temperature, the α peak shifts by 1 eV toward low energies; the width and location of the β peak remain unchanged.

Interesting experiments were done on the simultaneous adsorption of CO and pyridine on the (100) face of Pt [6.277]. Pyridine is a very good donor of electrons, and its adsorption results in the filling of vacant metallic levels which can then take part in the bonding with the 5σ orbital (i.e., a lone pair of electrons) of the CO molecule. As a result, a smaller amount of charge is transferred from CO to the metal at the coordination, the intensity of the α peak (reflecting the occupancy of the 5σ orbital) increases, and the α peak is broadened and shifts toward the Fermi level, while the β peak exhibits no change. Similar results were obtained for the co-adsorption of CO and propylene, pointing to the high donor capacity of propylene.

ALLYN [6.257] indicates that the gap between the 4σ and 1π orbitals (according to UPS measurements) remains the same (about 3 eV) whether the CO molecule is adsorbed on Ni(100), is found in the $Ni(CO)_4$ complex, or occurs in the gas. At the same time, the gap between the 5σ orbital and the 1π orbital is 2.9 eV for gaseous CO, +0.5 eV for $Ni(CO)_4$, and −0.5 eV for CO adsorbed on Ni(100). ALLYN et al. [6.257] draw the conclusion that the 4σ and 1π orbitals take no part in adsorption, which depends only on the 5σ orbital.

However, BRODEN and RHODIN with co-workers [6.269], having compared the UPS spectra of CO adsorbed on a large number of metals, discovered a systematic variation in the energy difference Δ(1π −4σ) between the non-bonding 1π and 4σ orbitals. With the metals located to the left of Co, Ru, and Re (i.e., those on which the adsorption of CO at high temperatures takes the dissociative path), Δ is larger than 3.15 eV and can be as high as 3.6 eV. With the metals in the right-hand side of the periodic system this difference is 2.6–3.1 eV, being 2.75 eV for a free CO molecule. The 4σ orbital is not likely to take part in the bonding; most probably, therefore, the variations in Δ are caused by the participation of the 1π orbital in the adsorptive bonding between CO and metal. The value of Δ may vary by 0.2 eV for different faces of one and the same metal. The authors of [6.269] draw a direct correlation between the value of Δ and the reactivity of the CO molecule on the surface; in particular, its tendency to dissociate: the higher the value of Δ, the lower the potential barrier for the dissociation of CO.

In a later work RHODIN with coworkers [6.273] used the experimental dependence of the UPS spectra of CO adsorbed on (111) faces of metals on the coverage for proving the participation of 1π orbitals in the chemisorptive bonding. At θ < 0.5 the 1π peak is split in two; this was explained by the attachment of the CO molecule at an angle of about 35⁰ to the surface. This orientation makes the participation of 1π orbitals in chemisorptive bonding possible.

A third peak (Fig.6.16), directly below E_F, pertains to the surface d states of the metal itself. Its intensity is greatly reduced by the adsorption of CO, as indicated by UPS and XPS measurements [6.271,277]. Apparently, the localized occupied surface orbitals, located near E_F - at least on the (100) face of metals studied - have the symmetry of the t_{2g} orbitals. They overlap extensively with the underlying antibonding π orbitals, facilitating the back donation (charge transfer from metal to ligand). Simultaneously the work function grows.

According to quantum-chemical calculations of DOYEN and ERTL [6.227], the chemisorptive bonding between CO and transition metals is established chiefly at the expense of the dative bond, i.e., back donation of electrons from the metal to the $2\pi^*$ levels of CO. This interaction involves the upper occupied d levels of metal. It was found that the wave functions, corresponding to the higher occupied d levels, are even more localized than those pertaining to the d electrons of free atoms. The energy of adsorption is determined by the location of the highest occupied chemisorptive level relative to the highest occupied d states of the metal. Calculations indicate the existence of three chemisorptive levels: two occupied and one unoccupied. For Ni(110) the upper occupied d level coincides with E_F, $\epsilon_\alpha = 0$. Assuming that the heat of adsorption of CO is equal to 125 kJ/mol, and the work function $\phi = 4.85$ eV, the authors of [6.227] obtained the following energy values of chemisorptive levels: $\epsilon_1 = -8.7$ eV, $\epsilon_2 = -0.63$ eV, and $\epsilon_3 = -13.6$ eV with respect to E_F. These values agree well enough with the values derived from UPS measurements. The calculations of the occupancy of adsorptive levels indicate that the adsorptive complex CO on Ni(110) must carry a charge of about $0.02e$ (e being the charge of electron). This agrees with the experimental findings concerning the negative charging of the Ni surface at the adsorption of CO, and the lowering of the ν_{CO} frequency in the IR spectrum on adsorption.

According to the calculations, the bridge complexes are more strongly attached to the metal surface than the linear complexes. In the linear complexes, the CO molecule interacts with three d orbitals (d_{z^2}, d_{xz}, d_{yz}), and altogether with seven (including the s and p) orbitals. The orbitals d_{xy} and $d_{x^2-y^2}$ do not take part in the bonding. The most careful high-resolution UPS measurements reveal that the adsorption of CO affects not only the intensity, but also the location of the peak of the d

374

levels, which sometimes also splits in two. These variations can be attributed to the participation of different d orbitals in the adsorption of CO.

6.6 Adsorption and Catalysis on Alloys

In Sect.5.5 we analyzed the possible correlations between the composition of metallic alloys and their surface and bulk structure. Melting of metals A and B may result in various structures: (a) solid solutions A + B; (b) separate monocrystals of A and B; (c) microcrystals B enveloping phase A; (d) an AB alloy, whose surface layer is rich in A; (e) two-dimensional surface phases, whose composition differs from the composition of the bulk of AB alloy; or (f) several phases, each having the same composition of the surface and bulk. For each of these structures the dependence of adsorptive and catalytic activity on the composition will be different: for instance, some kind of linear dependence should be expected for the mixture of sep-arate microcrystals of A and B; with the specific surface or bulk phases the curves of activity versus composition may exhibit bends and breaks, since different phases may have dif-ferent activity (Fig.6.17).

In adsorption and catalysis the equilibrium between the surface of the alloy and its bulk may be different from the

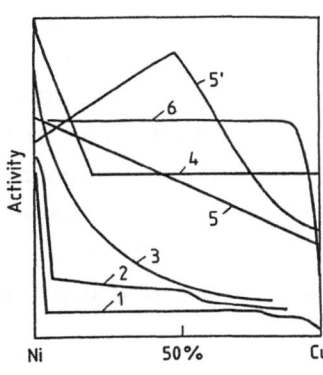

Fig.6.17. The adsorption and catalytic activity of Cu-Ni alloy versus the composition of the alloy [6.286, 295–302]: (1) adsorption of H_2, (2) hydro-genation of benzene, (3) hydrogeno-lysis of ethane, (4) H_2-D_2 exchange, (5,5') hydrogenation of ethylene, (6) hydrogenation of butene-1

equilibrium between the clean surface in vacuum and the bulk. For example, the surface of a Ag-Pd (Cu-Ni) alloy in an atmosphere of CO becomes enriched with Pd(Ni), the metal which strongly adsorbs carbon monoxide. Conversely, in an oxygen atmosphere, the surface of these alloys becomes enriched with Ag(Cu). The surface of a Au-Pt alloy in the atmosphere of hydrogen becomes rich in platinum, and the surface of a Cu-Ni alloy becomes enriched with nickel [6.284-288]. What happens can be expressed by the following simple rule: the surface becomes enriched with that metal which establishes the stronger bonding with the given adsorbate [6.287-289].

Sometimes, equilibrium between the bulk and the surface is not attained; the composition of the surface will then be governed not by the thermodynamic, but rather by the kinetic factors, e.g., by the relative rate of diffusion of atoms A and B toward the surface. For example, in the 50% Cu-Ni alloy the equilibrium between the bulk and the surface at 500°C is established (at the expense of diffusion from the bulk) in less than 10 seconds, while at 300°C the same process takes about 24 hours [6.290].

Very important from the viewpoint of the theory of adsorption and catalysis is the question of the relative role of the "collective" and the "local" properties of metals and alloys. The key role of the collective properties was first emphasized by COUPER and ELEY [6.291], who studied the para-ortho conversion of H_2 and H_2-D_2 exchange on Pd-Au alloys. The sharp rise in the activation energy and the drop in catalytic activity, observed when the proportion of gold in the alloy reaches 40%, are often interpreted as an indication of participation of the unoccupied d band in catalysis. In pure palladium the occupancy of the d band is about 9.6 electrons per atom, i.e., the number of holes in the d band of Pd is about 0.4 per atom. It has been assumed that the alloying of Pd with 40% Au (metal of group IB) must result in filling the d band.

In order to verify this assumption, a large amount of experimenting was done on the alloys of group VIII metals with metals of group IB. The relationship between the activity and the composition will be different depending on which is more important: the collective or the local interactions. If the activity depends mainly on the vacancies in the d band, then the full occupation of the d band ("collective properties") should result in an abrupt drop in the activity. With Cu-Ni alloys this turning point corresponds to 53% Cu + 47% Ni, since the s electrons of copper go to the d band. On the other hand, if the chemisorption and catalysis depend on the individual properties of surface atoms, then the activity of Cu-Ni alloys should go down steadily as the surface content of nickel decreases, in accordance with the composition of the ideal solid solution - in other words, the activity must be a linear function of the percentage of one of the components.

The growing interest in the adsorptive and catalytic activity of alloys arises from their great practical importance. Especially important are the alloys of paltinum with other metals, inactive per se. For example, the binary Pt-Re alloys are excellent selective catalysts of the reforming of hydrocarbons.

As pointed out in the reviews [6.286,289-292], the accumulated bulk of experimental data on the catalytic and adsorptive properties of alloys of group VIII and group IB metals seems to support the notion of activity being linked with the local rather than the collective properties. The atoms of each component tend to retain their individuality on the surface. This is also confirmed by investigations of the valence bands of alloys (Sect.5.4).

Assuming as a first approximation that the activity is determined by the properties of individual atoms (or DOWDEN's "ensembles" [6.293]), the activity may be related to the particular distribution of atoms or ensembles on the surface. In general, the regularities derived in this way are similar to the regularities displayed by the dilute oxide systems (Sect.4.7). If the adsorption follows the one-point mechanism [e.g., structure (6.9a) for hydrocarbons or structure (6.9b) for CO], the dilution of an active metal (Pt, Ni) by an inactive one (in the absence of other effects) will result in linear lowering of adsorptive activity with the increase in the concentration of inactive metal. The fall of activity will be steeper with the two-point adsorption [structures (6.9c and d)] and even more so with the three-point adsorption [structures (6.9e and f)]:

$$
\begin{array}{ll}
\mathrm{H_3C}\underset{\underset{\mathrm{Pt}}{|}}{\overset{\mathrm{CH_2}}{\diagup}}\mathrm{CH}\overset{\mathrm{CH_2}}{\diagup}\mathrm{CH_3}\overset{\mathrm{CH_3}}{\diagup} & \text{(a)} \qquad
\underset{\mathrm{Ni}}{\overset{\mathrm{O}}{\underset{|}{\overset{\|\!\|}{\mathrm{C}}}}} \quad \text{(b)}
\\[2em]
\mathrm{H_3C}\underset{\underset{\mathrm{Pt}}{|}}{\overset{\mathrm{CH_2}}{\diagup}}\mathrm{CH}\underset{\underset{\mathrm{Pt}}{|}}{\overset{\mathrm{CH_2}}{\diagup}}\mathrm{CH_3}\overset{\mathrm{CH_3}}{\diagup} & \text{(c)} \qquad
\underset{\mathrm{Ni}\quad\mathrm{Ni}}{\overset{\mathrm{O}}{\overset{\|}{\mathrm{C}}}} \quad \text{(d)} \qquad\qquad (6.9)
\\[2em]
\mathrm{H_3C}\overset{\mathrm{CH_2}}{\underset{\mathrm{Pt}}{\diagup}}\underset{\underset{\mathrm{Pt}}{|}}{\mathrm{CH}}\overset{\mathrm{CH_2}}{\underset{\mathrm{Pt}}{\diagdown}}\mathrm{CH_3} & \text{(e)} \qquad
\underset{\mathrm{Ni}\ \underset{\mathrm{Ni}}{|}\ \mathrm{Ni}}{\overset{\mathrm{O}}{\overset{\|}{\mathrm{C}}}} \quad \text{(f)} \qquad .
\end{array}
$$

All three patterns of activity have been observed experimentally, depending on whether the adsorption followed the one-point, two-point, or three-point mechanism.

BOUDART [6.294] suggested distinguishing between "structure-insensitive" reactions, in which the activity of a catalyst shows no dependence on the mode of preparation, and "structure-sensitive" reactions, in which the activity of a catalyst depends on the details of the preparation procedure.

The structure-sensitive and insensitive reactions were found to be affected differently by the dilution of the active group VIII metal by an inactive group IB metal. The rate of structure-insensitive reactions per atom of active metal is not changed by the dilution. Such are the reactions of dehydrogenation of paraffins and hydrogenation of olefins, hydrogenation of C=O and C≡N bonds, and isomerization and metathesis of olefins, which follow the one-point mechanism. By contrast, the structure sensitive reactions of hydrogenolysis of hydrocarbons (etherification of alcohols involving the rupture of C-O bonds, methane synthesis from CO + H_2, Fischer–Tropsch process, hydrogenation of NO_2, hydrogenation of benzene) follow the multipoint mechanism [structures (6.9c,d)], and the catalytic activity per active atom drops quickly as the active component is diluted by the inactive one [6.286,292].

Let us now describe a few typical examples of chemisorption and catalysis. The catalytic activity of the Cu–Ni alloy versus composition is plotted in Fig.6.17 on the basis of data obtained by different authors [6.286,295-302]. In most cases the alloys were represented by Cu–Ni films, pretreated in different environments (hydrogen, vacuum) and at different temperatures. Nickel is always more active than copper, so the catalytic and chemisorptive activity of an alloy should be more naturally related to the proportion of nickel. From the diagram it becomes clear that the observed dependences of activity on concentration were as different as can be. In some cases a sharp rise in activity was observed at low concentrations of nickel, in others the activity was the greatest with almost pure Ni, and in still others [6.300] the activity passed through a maximum. Commonly detected was the existence of a plateau (i.e., constant activity over a wide range of concentrations), which can be explained by the formation of surface phases of more or less permanent composition. The structure of these phases was studied by UPS and Auger spectroscopy [6.290,291]. The general trend consists in the reduction in activity in going from Cu to Ni.

The thermal desorption spectra [6.291] indicate that hydrogen is adsorbed in one and the same form whether on Ni or on Cu–Ni alloys. The rate of adsorption of hydrogen is proportional to the fourth power of the concentration of Ni atoms on the surface, which can be explained by the fact that the fixing of the H_2 molecule on the surface requires a four-atom ensemble of Ni. Similarly, the shape of Curve 2 (hydrogenation of benzene) was explained by the adsorption of benzene on three-atom ensembles [6.82].

Curve 3 (hydrogenolysis of ethane) was explained [6.298] by the chemisorption of ethane on a "doublet" of nickel atoms. Curve 5 (hydrogenation of C_2H_4) is typical of one-point catalysis. A similar dependence was observed for the adsorption on Cu-Ni alloys. It was attributed [6.302] to the corrosive reconstruction of the surface in the presence of CO, equalizing the composition of the surface and the bulk of the Cu-Ni alloy. In the series of consecutively taken spectra of thermal desorption of CO from a Cu-Ni alloy, the peak corresponding to the adsorption of CO on Ni is found to increase with each spectrum, while the CO/Cu peak decreases, as if the adsorption of CO on Cu-Ni alloy pulled the atoms of Ni out to the surface [6.286]. The SIMS spectra taken after the adsorption of CO on a Cu-Ni alloy, detect the ions Ni_2CO^+, $NiCO^+$, and $CuCO^+$, characteristic of the adsorption individually on Cu or Ni, but never $NiCuCO^+$ [6.221].

Thoroughly investigated was the adsorption of CO on the alloys of Pd with Cu, Ag, and Au. The variation in the work function $\Delta\phi$ for the adsorption of CO on Pd was 0.6 eV, and zero for adsorption on Au. The curve exhibited a break at the point corresponding to the filling of the d band of Pd in a Pd-Au alloy (about 60% Pd). However, after subjecting the alloy to heat treatment in CO, the $\Delta\phi$ curve went close to $\Delta\phi$ of the Pd + CO system in a very wide range of concentrations, which can be explained by the enrichment of the surface by palladium in the presence of CO [6.303-305].

For the adsorption of CO on Pd-Ag and Ni-Cu alloys the IR spectra displayed two groups of bands: above 2000 cm^{-1} (the so-called linear complexes) and below 2000 cm^{-1} (the so-called bridge complexes). The low-frequency bands (1915-1980 cm^{-1}) prevail in adsorption on Pd, and the high-frequency bands (2020 cm^{-1}) in adsorption on Ni. The admission of CO fails to produce any IR bands with Ag, and gives rise to only a weak band at 2130 cm^{-1} with Cu. As Pd is being diluted by Cu or Ag, (or Ni by Co), the relative intensity of the high-frequency bands goes up, and of low-frequency bands down. This indicates that the dilution results in an increase in the number of isolated Pd atoms surrounded by Ag, or that of isolated Ni atoms surrounded by copper. These data testify in favor of the local interaction. If the interaction of metal with CO had depended on the supply of electrons from the bulk, a shift in frequency ν_{C-O} should have been observed. However, the shifts of the IR bands of C-O at the adsorption on alloys are quite small (0 - 25 cm^{-1}), about the same as with pure metals. At the same time the intensity of absorption exhibits a very strong dependence on the composition of the alloy [6.306-309].

The activation energy of desorption E_{des} of CO from the linear complexes on Ni-Cu alloys in a wide range of concentrations (including pure Ni) was 140 kJ/mol; E_{des} = 120 kJ/mol for Ag-Pd alloys and 170 kJ/mol for pure Pd. The constancy of E_{des} points to the maintained identity of active sites [6.306].

To some extent, the effects of occupancy of the d band in Cu-Ni alloys become manifest in the poisoning of their surface by the bulk hydrides. The creation of the latter results in filling of the d band of Ni (or Pd in Pd-Ag alloys) [6.310].

It was demonstrated that in the adsorption of H_2 and CO on Pt-Au alloys [6.311-313] the activity of gold is low; the activity of the alloy remains constant in a wide range of concentrations and shows a sharp increase with 2% Au + 98% Pt. It has been assumed [6.311] that in this case also the adsorptive activity depends chiefly on individual atoms or on ensembles of several atoms. On the surface there exists a two-dimensional phase 17% Pt + 83% Au. The active sites are the same as on pure platinum, although their distribution is different. In the alloy the platinum ensembles are separated by patches of pure gold, which prevent the migration of adsorbed substances. At the adsorption of CO on Pt-Au alloys at 100°C [6.313] the

activity of alloys of different proportions was found to increase with time, approaching the activity of pure Pt. This is
explained by the enrichment of the surface by platinum in the
presence of CO. The drawback of almost all investigations of
this kind is lack of data on the structure and composition of
the alloy surfaces at the steady-state catalytic conditions.

6.7 Adsorption and Catalysis on Supported Metals

The study of the catalytic and adsorptive properties of supported metals is of great practical importance. Among the most
commonly employed industrial catalysts are the supported platinum catalysts (Pt/Al$_2$O$_3$) of reforming, platinum on refractory
supports used in automobile exhaust converts and the supported
nickel catalysts of hydrogenation.

The supported metals are interesting also from a theoretical
viewpoint because of the possibility to obtain very small
metallic particles on the carrier. Noteworthy are the quantum-
chemical calculations of the bonding between the adsorbed
molecule and the small metallic clusters or domains. A matter
of major importance is to determine the number of atoms in the
domain, starting with which the energy of orbitals and the
bonding energy between the metal and the adsorbate assume
values typical of the bulk samples of the same metal. The calculations done for small clusters were mentioned in Sect.5.6.
Calculations indicate [6.314] that the donor properties of
small clusters and metallic particles on the surface are
higher, and the acceptor properties lower, than those of the
bulk metal. The energy of bonding of adsorbate with metal increases with the increase in the cluster size.

Many researchers [6.9,294,315-319] have come to the conclusion that catalytic activity will be most affected by dispersity in the range form 1 to 4 nm. In this range (termed the
"mitohedral region" by POLTORAK [6.316]) the number of surface

atoms in a particle becomes close to the number of "bulk" atoms; in the same range the proportion of surface atoms with lowered coordination numbers increases (Sect.5.5).

Along with the electron microscopy and X-ray data the average size of metallic microcrystals on a carrier is often assessed by chemisorptive techniques. These estimates are based on the assumption that each atom on the surface of metal M, disregarding the dispersity, adsorbs one atom of adsorbate; for instance, in the adsorption of O_2 and H_2 the numerical ratio H/M = 1 and O/M = 1; for the adsorption of carbon monoxide CO/M = 1 [6.320]. Similar techniques are based on the isotopic exchange of deuterium with pre-adsorbed hydrogen [6.321].

POLTORAK [6.316] calls attention to the fact that the chemisorption of oxygen on high-dispersion metals is usually characterized by highly non-Langmuirian isotherms, which do not show signs of saturation at H/M = 1. The study of adsorption of hydrogen on Pt/Al_2O_3 reveals that as the concentration of platinum is reduced, the ratio H/Pt increases and with low concentrations (about 0.1% by mass) may become as high as 1.5. With platinum-containing zeolites the ratio H/Pt can be as high as 2 [6.315]. These data led to the assumption that with very low concentrations of Pt adsorptive activity is displayed by very small aggregates and even by isolated atoms of Pt. The intercomparison of adsorption isotherms for various supported platinum catalysts indicated that at saturation each surface atom of platinum takes two atoms of hydrogen [6.316]. For the Rh/Al_2O_3 catalyst, according to adsorptive data , H/Rh = 1.5.

The considerable decrease of the stoichiometric H/M ratio with adsorption on supported metals is due to the effect SMSI (Sect.5.5). This is particularly true for metals on a readily reducible support, such as TiO_2, ThO_2. For instance, a ratio H/M < 0.04 was found for H_2 adsorption on Pt/TiO_2 [6.322]. As shown by the EXAFS method, shortened M-Ti or M-Th bonds will form thereby. It was found for Rh/TiO_2 that a Rh-Ti bond 0.252 - 0.256 nm long, i.e. shorter than the Rh-Ti bond (0.268) nm in the RhTi intermetallide was formed under action of H_2 [6.323].

The adsorption of oxygen shows much greater dependence on dispersity than does the adsorption of hydrogen. For the adsorption of oxygen at 52°C [6.324] on low-dispersion platinum supported on Al_2O_3, the ratio O/Pt is 0.86, while the adsorption of O_2 on high-dispersion platinum has O/Pt = 3.7. At the same time, the ultrahigh-dispersion platinum, namely, the clusters of six platinum atoms in the pores of type-Y zeolite, was less capable of adsorbing oxygen than were the bigger microcrystals. This is explained by the electron-acceptor properties of small clusters of platinum [6.325,326]. This conclusion was confirmed by other experiments. It was demonstrated that while for Pt/Al_2O_3 the ratio O/Pt is greater than unity, the same ratio is less than unity for platinum in 5A zeolite [6.327]. The study of adsorption of CO indicates that in a wide range of dispersities the ratio CO/Pt equals unity [6.328].

With ultrahigh-dispersion flakes of Rh, supported on Al_2O_3 (each comprising about 20 atoms), the ratio CO/Rh is 1.7 [6.329]. Sometimes the reported values of CO/M are quite high; with Ru particles less than 1.6 nm in size, supported on Al_2O_3, the ratio CO/Ru can be as high as 4 or 5 [6.330].

The study of interatomic separation in the particles of supported metals by the EXAFS technique (Sect.5.6) indicates that the process of adsorption is associated with the relaxation of interatomic distance in the small metallic particles (< 0.1 nm): the interatomic separation becomes close to that typical of the bulk metal. For instance, in the adsorption of H_2 on platinum particles (smaller than 1.2 nm) contained in type-Y zeolite, the Pt-Pt distance increases from the initial value of 0.265 nm to 0.274 nm [6.331]. In the adsorption of O_2 on similar clusters the bonding Pt-Pt disappears altogether, and is replaced by the Pt-O-Pt structure with the platinum atoms separated by 0.332 nm [6.332].

A detailed investigation of IR spectra for the adsorption of N_2 and CO on supported Ni/SiO_2 catalysts was carried out in

[6.315]. On some extinction isotherms the absorption band of nitrogen at 2200 cm^{-1} reaches its maximum when the coverage is far less than a monolayer. Maximum extinction on the absolute scale corresponds to the microcrystals of Ni 1 – 3 nm in size. Accordingly, the intensity of bands of adsorbed oxygen has been proposed as a measure of the number of B_5 sites on Ni (Fig.5.15). This number varies from $6 \cdot 10^{12}$ cm^{-2} with crystals about 20 nm in size to $8 \cdot 10^{13}$ cm^{-2} with crystals measuring about 2 nm.

The IR spectrum of CO adsorbed on Ni/SiO_2 exhibits a number of bands between 1810 and 2080 cm^{-1}. The authors of [6.315] challenge the interpretation proposed by EISCHENS and PLISKIN [6.193] (Sect.4.6.1), according to which the IR bands above 2000 cm^{-1} pertain to the bridge complexes. They maintain that there is no direct relationship between the intensity of any band and the concentration of B_5 sites, as determined from the IR spectra. They give more support to the hypothesis of BLYHOLDER [6.194], according to which all the bands ν_{C-O} pertain to the linear-bound CO. Different frequencies characterize the various coordinations of surface atoms on which the molecules of CO are stabilized. For instance, the bands at 2030 and 2075 cm^{-1} pertain to CO adsorbed on C_9 atoms on the (111) face of Ni; the band at 1950 cm^{-1} belongs to CO on C_8 atoms [face (100)]; the band at 1920 cm^{-1} pertains to CO on C_7 atoms, located along the crystal edges and on the steps of faces (110) and (113); and the band at 1810 cm^{-1} pertains to CO adsorbed on C_6 atoms, located at the vertices of the crystal (Sect.5.5, Fig.5.13). Furthermore, the band at 2057 cm^{-1} pertains to the gaseous carbonyl compound $Ni(CO)_4$ presumably adsorbed on the surface, and the bands at 2050 – 2080 cm^{-1} pertain to the forms

$$Ni\diagup^{CO}_{\diagdown CO} \quad \text{and} \quad Ni\diagup^{CO}-CO_{\diagdown CO}$$

created on atoms C_6 and C_7. On small crystals (< 2nm) atoms C_8 and C_9 are practically absent. However, these data do not seem to invalidate the interpretation given by Eischens: the experiment indicates that the increase in the crystal size is accompanied by the increase in the integral intensity of bands below 2000 cm^{-1}, which can be ascribed to the increasing proportion of bridge forms of adsorption of CO on planar faces. Quite likely there is some connection between the number of B_5 sites and the number of C_6 and C_7 centers with a lowered coordination number (Sect.5.5). According to [6.333], the spectra of the linear complexes of CO adsorbed on atoms C_6, C_7, etc. must have more or less similar forms.

For adsorption on Pt/SiO$_2$ a rise in intensity of C-O bands above 2000 cm^{-1} and a drop in the intensity of bands below 2000 cm^{-1} accompanied a reduction of size of platinum microcrystals. For adsorption on Ir/SiO$_2$ only the high-frequency bands were observed, independent of the dispersity of iridium.

With Ru/Al$_2$O$_3$ catalysts, the adsorption of CO on large particles gives rise to only one band at 2048 cm^{-1}, and adsorption on smaller-size particles gives rise to the bands at 2086 and 2148 cm^{-1}, which pertain to CO adsorbed at edges and vertices: $-Ru(CO)_2$, $Ru(CO)_3$, and even $-Ru(CO)_4$ [6.334]. For the adsorption of CO or NO on Pt/Al$_3$O$_3$ the frequencies ν_{C-O} and ν_{N-O} were found to increase, as the size of particles became smaller. This observation was explained by assuming that the back donation from d orbitals to π orbitals of adsorbed molecule depends on the collective properties of the metal [6.335].

In the preceding section we have described BOUDART's classifications of structure-insensitive and structure-sensitive reactions [6.294,319,336]. The reactions of the former type do not exhibit any dependence on the dispersity or the structural features of the given metallic catalyst. Such reactions may comply with the principle of constant specific activity (i.e., constancy of reaction rate per square centimeter of surface area of crystals of one and the same composition), proposed by

BORESKOV [6.337]. The rate of reactions of the second type depends on the dispersity and surface structure of the cata- lyst. These processes seem to comply with Taylor's concept of active sites and with the consideration regarding the inhomo- geneity of active surfaces.

The difference between the structure-sensitive and the in- sensitive reactions can best be explored by studying the depen- dence of the catalytic activity on the dispersity of supported metal. The most remarkable result is that the activity turns out to be independent of the size of catalyst particles in a great many reactions. For example, in going from particles of Pt on a carrier 20 nm in size to particles 1 nm in size the specific catalytic activity remains more or less constant in the reactions of dehydrogenation of cyclohexane and isopropyl alcohol; hydrogenation of benzene, cyclohexene, cyclopropane, and allyl alcohol; hydrogenolysis of pentane; hydrogen exchange in pentane; and isomerization on n-hexane and methylcyclopen- tane [6.316,294,319,338,339]. The activity of platinum in these reactions does not depend on the nature of the carrier, and when reduced to unit area of platinum surface turns out to be the same, whether on the faces of a single crystal, or on sup- ported platinum catalysts. Observe that this list of reactions includes typical "sextet" reactions involving aromatic and cyclic molecules, assumed by the multiplet theory to comply with the six-point mechanism of adsorption on the plane. Appar- ently, the concept of π complexing with single atoms of plati- num is more plausible: the "sextets" of atoms can hardly be available on small crystals of platinum.

The increased rate of hydrogenolysis of n-hexene, 2- and 3-methylpentane, and methylcyclopentane on very small parti- cles (1.5 nm) of Pt/SiO$_2$ catalyst is viewed in [6.340] as an argument in favor of π complexing with corner atoms.

The rate of hydrogenation of benzene on Ni/SiO$_2$ catalysts was found to fall off a little with an increase in dispersity [6.315,341]. At the same time, the rate of hydrogenation of benzene and styrene on Ni particles contained in type-Y zeolite (per atom of Ni) was by two orders of magnitude slower compared with the same reactions on Ni/SiO_2, although the overall rates of hydrogenation of olefins were about the same [6.342]. As indicated in the preceding section, the study of catalysis by alloys points to the ensembles of three Ni atoms as active sites of hydrogenation of benzene. The existence of activity maxima in H-D exchange in benzene, corresponding to a particle size of 2 - 4 nm, was attributed by some researchers to the reaction's taking place on B_5 sites. An example of a structure-sensitive reaction is given by the isomerization of neopentane to isopentane. The size of reduction of platinum particles in supported catalysts brought platinum's specific activity down by two orders of magnitude [6.339]. This reaction was also differently catalyzed by two catalysts of the same preparation and dispersity, only heat pretreated at different temperatures [6.294]. These results are support for the multisite mechanism of adsorption. It was assumed, for instance, that neopentane should be adsorbed on the site comprising three atoms of Pt on the (111) face.

The structure-sensitive reactions also include many oxidation reactions (including oxidation of methanol, ethanol, acetaldehyde), decomposition of hydrogen peroxide [6.316], oxidation of propylene [6.343], and synthesis of ammonia [6.344]. As the size of particles of SiO_2-supported platinum was reduced from 3.0 to 1.1 nm, the platinum's specific catalytic activity in oxidation of alcohols dropped by two orders of magnitude. A similar effect was observed for the decomposition of H_2O_2 on platinum.

The oxidation rate of propylene per unit area of metallic surface on Pt/Al_2O_3 goes down as the dispersity of the catalyst is increased. This was explained in [6.343] by the fact that propylene is more readily oxidized on the planar faces of platinum (terraces) than on the edges or kinks of stepped surface (Figs.5.9,10). The multisite adsorption of O_2 and C_3H_6 results

388

in stronger bonding on terraces, which ultimately ensures the higher reaction rate.

By studying the mechanism of oxidation of methanol <u>in solution</u> on platinum with electrochemical techniques VASIL'EV and BALETSKIJ [6.345] succeeded in demonstrating that the reacting particles in the intermediate state are each bound to six surface atoms, as in the following scheme:

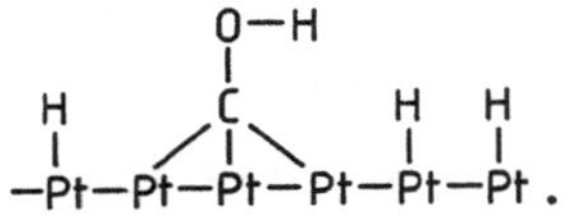

```
           O—H
            |
    H       C       H   H
    |      /|\      |   |
  —Pt—Pt—Pt—Pt—Pt—Pt .
```

In going from massive platinum to supported platinum the reaction rate decreases. Apparently the high-dispersion platinum is less likely to furnish a flat portion of surface containing an active site of several platinum atoms.

The oxidation of CO is an exception to this regularity. The rate of this reaction was found to be the same on faces (100), (110), and (111) of palladium, allowing one to classify it as a structure-insensitive reaction [6.319,346]. This may be due to the corrosive chemisorption of O_2, which levels all sites. The same applies to the oxidation of H_2 on Pt/SiO_2 in an excess of oxygen. At the same time, the oxidation of H_2 in excess hydrogen is a structure-sensitive reaction [6.347]

The size reduction of metallic particles in Rh/SiO_2 catalyst results in a drastic decrease in the rate of hydrogenolysis of ethane to methane [6.294].

The data for various catalyzed reaction rates as a function of the dispersity of metal correlate well with the results obtained in studying the reactions on stepped faces of single crystals. It was demonstrated in [6.348] that the structure-insensitive reaction of dehydrogenation of cyclohexene into benzene takes place on the (111) face of Pt and does not depend on the concentration of steps. The structure-sensitive reaction of hydrogenolysis of cyclohexane to n-hexane takes place on the edges of steps and maybe even on kinks.

Multisite adsorption is one of the possible reasons for the structure-sensitivity of catalytic reactions. Another reason may lie in the decline of "collective properties" of very small metallic particles, which suggests that the oxidation reactions rely on the transport of electrons via the catalyst's lattice. Platinum particles 1.0 - 1.5 nm in diameter with adsorbed oxygen must be viewed as highly doped semiconductor particles, containing 10 - 20 excess metal atoms, rather than being treated as a metallic particle covered with oxygen [6.316].

BOUDART [6.319] cites yet another reason for the structure sensitivity of oxidation reactions. These reactions release a lot of heat, spent on the reconstruction of the surface layer, which enhances the catalytic activity of bulky catalysts. If the metallic particles are small, they may be short of the number of atoms required for building the active sites of the appropriate structure.

The dependence of activity on the dispersity of supported metal may also have a quite different origin in cases when the metal is engaged in chemical interaction with the support, or when there exists diffusion of adsorbed molecules from the metal to the support.

The diffusion of hydrogen from the metal to the support has been actively investigated [6.347,349-351] and was termed the "spillover". Spillover of hydrogen was first discovered by NEIKAM and VANNICE [6.349], who studied the adsorption of hydrogen on the Pt/WO_3 catalyst. This adsorption was found to lead to the reduction of tungsten oxide at the expense of the initial adsorption of hydrogen on platinum and further diffusion to WO_3. At the same time, the reductive adsorption of hydrogen was not observed on pure WO_3. The reduction of Ce^{4+} to Ce^{3+} on platinum-containing CeY zeolite was observed in the adsorption of perylene, which serves as a bridge for transporting the active hydrogen from platinum to the zeolite. A similar bridge is formed by perylene in the system $Pt/NaY + H_2$, which displayed the production of metallic sodium, possibly according to the reaction

$$\left[\underset{O}{\overset{O}{\text{O}}}\text{Al}\underset{O}{\overset{O}{\diagup}} \quad \underset{O}{\overset{O}{\text{O}}}\text{Si}\underset{O}{\overset{O}{\diagup}} \right]^{Na^+} + H \xrightarrow{Pt} \underset{O}{\overset{O}{\text{O}}}\text{Al}\underset{O}{\overset{O}{\diagup}} \quad \overset{\overset{Na}{\overset{|}{H}}}{\underset{O}{\overset{O}{\text{O}}}\text{Si}\underset{O}{\overset{O}{\diagup}}} .$$

The same was observed with the supported catalysts Pt/SiO_2, Pt/RuO_2, and Pt/ZrO_2. In some cases the spillover was facilitated by the presence of water, which can also serve as a bridge.

Prolonged exposures to hydrogen were found to lead to the additional chemisorption of hydrogen on platinum supported on active carbon [6.350]. Here the spillover of hydrogen from platinum to the carrier also took place.

The dilution of Pt/Al_2O_3 by pure alumina raises the specific rate of hydrogenation of benzene at the expense of the diffusion of hydrogen through Al_2O_3 to distances up to 0.5 mm. At $-52^{\circ}C$ the surface of Al_2O_3 can be covered by oxygen up to $\theta = 3 \cdot 10^{-2}$ [6.349]. A similar effect was observed in [6.351] for the adsorption of pentene-1 on Pt/SiO_2 and Pt/Al_2O_3. However, SANCIER [6.352] did not find any acceleration of the reaction of hydrogenation of ethylene when Pt/SiO_2 was mixed with Al_2O_3.

The effect of spillover may be due to the creation of an active particle (precursor) on the metallic phase at the instant of adsorption (Sect.6.4, Fig.6.12). The precursor can migrate not only over the metallic surface, but also over the surface of the carrier, until it goes into the chemisorbed state or reacts with another molecule.

The spillover process has been studied in situ by the FTIR method [6.353]. D_2 was adsorbed on Pt/SiO_2 and it then diffused from Pt to the support as D atoms that exchanged with the OH groups of SiO_2 to form OD groups. Though the deuterium transport from Pt to the support is endothermic (about 40 kJ/mol), spillover occurs due to the high concentration gradient between Pt-D and D on the support.

7. Conclusion

In this book we have considered the properties of the surface of transition metals and their oxides, and the main regularities of adsorption and catalysis taking place on these surfaces. The transition elements - whether in homogeneous complexes, isolated ions contained in the solid matrix, or in the metallic form - were shown to be much superior in their chemisorptive and catalytic properties to the nontransition elements. As early as in 1960 ROGINSKIJ [7.1] pointed out that the catalytic qualities of transition elements arise from the fact that the potential of chemical forces $U(r)$ in the bonds formed by d electrons falls off with distance r much more slowly than in the bonds formed by s and p electrons, e.g., the value of dU/dr is smaller. Analysis of the potential diagrams of the catalytic reactions indicates that the extent and diffuseness of d orbitals result in the considerable lowering of activation energy for the catalysis on transition metals as compared with the catalysis on nontransition metals [7.1-3].

To illustrate this point we reproduce the potential energy diagram for the dissociative chemisorption of H_2 on the transition and the nontransition elements (Fig.7.1) [7.4]. The H-H bond, having the energy of 4.48 eV, is ruptured, and two M-H bonds are formed. The potential curves for 2H + M and H_2 + M intersect with one antoher, thus allowing the activated chemisorption of H_2. The bond energy M-H is about the same for the metals Na, Cu, and Ni; nevertheless, the chemisorption of H_2 goes quickly on nickel, and is almost zero on Na and Cu. As seen

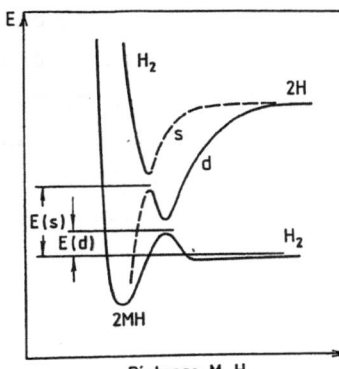

Fig.7.1. The potential energy curves for the dissociative chemisorption of the H_2 molecule. The activation energy is E(d) for the transition metals and E(s) for the nontransition metals

in the diagram, owing to the slow falloff of the potential of the d orbitals, the activation energy is lower for the transition metals even with the same depth of the potential well: E(s) > E(d).

Similar reasons – the relationship between the vibrational motions of nuclei and the configuration of d orbitals – were cited by COSSEE [7.2] as responsible for the specificity of the catalytic action of compounds of transition elements in the reactions of olefins, e.g., catalyzed polymerization (Sect.4.1).

The analysis of the particular features of complexing of transition metals with ligands given in Chaps.4,6 allows one to name several reasons for the high catalytic activity of the compounds of transition elements [7.1-10].

1) <u>The ability of transition metals to stabilize a large number of ligands by coordination.</u> When dealing with the complexes of transition metals ML_n one should speak not of the properties of the individual bond M-L (M being the transition metal atom, L being the ligand), but rather of the properties of the collection of bonds $M-L_1$, $M-L_2$, $M-L_3$, etc., or of the properties of the bond $M-L_1$ in the presence of $M-L_2$, $M-L_3$, etc. [7.3,11]. In contrast to the valence bonds, the coordination bonds are not manifested characteristically in the IR spectra. All this results in the high diversity of the properties of ML_n complexes and in the dependence of their catalytic properties

(e.g., specificity) on the configuration of the complex. For one thing, this property allows the existence of complexes of one and the same metal with varying numbers of ligands. The change in coordination complies with the rules described in Sects.4.1,2, and is very important for catalysis. The change of coordination accompanies complexing, adsorption, desorption, and sometimes the catalytic act itself.

2) <u>The multiorbital nature of each M–L bond</u>. We have more than once indicated that the participation of many orbitals in a single M–L bond results in some cases in chemical changes in the ligand L, and in the catalytic activity of metal M. The high electron acceptor qualitites of the transition metals are responsible for their high activity in nucleophilic reactions. In such cases the transition metal acts as a Lewis acid or "super-acid." However, the transition metals are capable of accelerating electrophilic reactions thanks to the back donation of electrons. In consecutive reactions the catalyst M can simultaneously form the donor-acceptor and the dative bonds.

3) <u>The existence of different oxidation levels, differing little from one another in energy</u>. This feature gives rise to the high catalytic activity of transition elements in redox reactions. The electron transport from the reactant to the catalyst and back again must take less time than the transport – in the absence of catalyst – from the reductant to the oxidant. If oxidation requires more than one electron, the role of the catalyst is especially indispensable: it helps to replace the one-electron (radial) process with the two-electron (or many-electron) process. In this connection let us observe also that in the compounds of transition metals not only the electron transfer between M atoms at different levels of oxidation (7.1a) is facilitated, but also the charge transfer (7.1b) occurs more readily than in the compounds of nontransition elements:

$$M^{(n+1)+} \underset{\longleftarrow}{\overset{e}{\longrightarrow}} M^{n+} \quad (a) \qquad L \overset{e}{\longrightarrow} M \quad (b). \qquad (7.1)$$

The width of the forbidden band in the oxides of nontransition elements is usually 5 – 10 eV, whereas in the transition oxides this width is smaller: 2p levels of oxygen lie 2 – 4 eV below the d level of the transition metal.

Some recently obtained data indicate that it is this property – the existence of several oxidation levels – that is of foremost importance in catalysis by transition metals and their compounds. In Sect.4.9 we have quoted from our results, which indicate the possibility of catalysis via the simple electron transfer from the solid (not necessarily a conductor or a semiconductor) to the reacting particles. The latter may be attached to the surface by a weak (not necessarily coordinative) bond.

4) The capability of atoms or ions of transition metals to bond with various ligands, including low-reactivity ones, sometimes resulting in peculiar effects of steric selectivity. The blocking of one or more coordinative vacancies may switch the reaction to a different route.

5) The existence of relatively stable, and at the same time highly reactive, complexes of transition metals. For example, the five coordinated complexes d^7 are similar in their properties to free radicals, the four-coordinated d^8 to carbenes, the five-coordinated d^8 to carbanions, etc. [7.10]. This is especially important for activating certain saturated molecules, such as H_2 and paraffins.

All the above-mentioned reasons ultimately stem from the peculiar symmetry and energy structure of d orbitals, and the specific angular and spatial distribution of their wave functions.

Progress in the study of the structure and properties of the surface and the configuration of surface complexes is due to the extensive employment of modern physical investigative techniques: IR and UV spectroscopy, EPR, LEED, UPS, XPS, EXAFS, etc. The development of these techniques, much advanced in the past decade, led to the identification of the structure of the active sites on the surface of transition metals and their oxides, and the structure of the chemisorptive complexes of simple molecules (H_2, O_2, CO), as well as of complexes or more complicated molecules. Improved quantum-chemical calculation procedures have interpreted the observed structures, computed the energy of the molecular orbitals in the complex, and determined the participation of given orbitals in the chemisorptive bonding.

Paying due attention to the local interaction between the adsorbed molecules and the adsorptive sites, one should not neglect the electronic processes on the solid surface which accompany adsorptive and catalytic phenomena. As we have more than once emphasized, these investigations were hampered by the use of high-dispersion catalyst powders. The barrier effects between particles greatly limit the usefulness of the measurements of conductivity , work function, or thermal emf. Our understanding of the role of collective effects in catalysis at the present stage can be greatly promoted by the use of model systems. As we have pointed out [7.12], such model systems can be represented by MDS, MDM, and MSM structures (M standing for metal, S for semiconductor, D for dielectric). The adsorptive phenomena taking place on the completely or partly oxidized metallic surfaces can be successfully simulated by the Schottky barriers (M1 structures). It is possible to make the thickness of layers in sandwich structures equal to the thickness of phases in the real catalysts. Much knowledge of the electron processes taking place in sandwich structures has been accumulated by semiconductor surface science and microelectronic research, together with the adequate investigative techniques. Unfortunately, many of these advances have been overlooked by chemists concerned with chemisorption and catalysis. As we have demonstrated [7.12], many of these techniques (the measurement of currents through MDS and MDM structures, light-induced charging of surface traps) can supply valuable information about the parameters of defects of the dielectric film and their variations caused by adsorption. The study of electron traps in dielectric catalysts so far did not employ the absorption and the photoelectron emission spectra obtained with highly coherent synchrotron radiation.

The study of surface structures and surface complexes in the course of a catalytic reaction is a major experimental problem; and reliable experimental results are few. That is why our main concern was with the mechanism of chemisorption, while the mechanism of catalysis was discussed on a more general basis. Recently, however, the methods of IR spectroscopy and EPR started being employed under actual catalytic conditions. Used originally by TAMARU [7.13], who studied the decomposition of HCOOH and CH_3OH on oxide catalysts, IR spectroscopy was later employed successfully for identifying surface compounds in the course of catalytic reactions. EPR spectroscopy was also employed for studying the active sites on the surface of oxides of transition metals in the course of catalysis [7.14].

The techniques of LEED, XPS, UPS, Auger spectroscopy, FEM, and FIM, which all require conditions of high vacuum, were less lucky. However, certain advances have been made in these fields as well. By letting the reaction proceed in the adsorbed layer, one can study its mechanism in high vacuum. For example, it was possible to study the mechanism of the decomposition of formic acid in the adsorbed layer on Ni using LEED and Auger spectroscopy [7.15], and the same on Cu, Ni, and Au using UPS and XPS [7.16]. In this connection let us observe that these techniques require maintaining "high", but not "ultrahigh" vacuum, the only latter being necessary for precleaning the surface. Therefore the mechanism of a catalytic reaction at low pressure (10^{-2}-10^{-4} Pa, or 10^{-4}-10^{-6} torr) can be examined while simultaneously studying the structure of the adsorbed layer by other physical techniques. Auger spectroscopy under catalytic conditions was used for studying the oxidation of ammonia [7.17] and CO [7.18] on platinum.

The detection of some structure or other in the course of a catalytic reaction does not yet prove its relevance to catalysis. The proof can be obtained by correlating the rates of the catalytic reactions with the rates of creation and relaxation of surface structures. So far, however, the kinetics of catalytic reactions has been studied in the steady-state regime, and investigations of this kind are not capable of elucidating the peculiar features of the mechanism of the catalytic reactions; in particular, the participation of intermediate active forms: complexes, ions, and radicals.

The principal advances in the study of chemical kinetics of homogeneous reactions were made in the sixties through the use of improved methods of investigation of nonstationary processes, based mainly on the so-called relaxation and pulse techniques. A rapid change in some external parameter, which affects equilibrium or steady-state process (e.g., temperature, proportion of components, pressure) compels the system to go to a new equilibrium (stationary) state. The relaxation time depends on the characteristic times of individual stages of the reaction, which can thus be made visible.

The nonstationary techniques have been used, if limitedly, in catalytic investigations [7.19]. Now the use of nonstationary techniques has become more extensive [7.20,21]. Today kinetic investigations of heterogeneous catalytic reactions are facing the necessity of employing the nonstationary techniques on a full-scale basis. These techniques must be supplemented by measurements of the changing composition of the products in the liquid or gaseous phase above the catalyst and measurements of the composition and structure of the adsorbed phase (and the catalyst's bulk); such measurements must rely on modern spectral and structural techniques.

The complete description of the mechanism of the catalytic reaction must single out all the individual steps of the net reaction, stating all the intermediate forms and the respective reaction rates. Some stages of catalysis can be very fast.

Under active development now are the methods of investigation of very rapid elementary processes, down to the picosecond (10^{-12} s). Such techniques will be of great help for the study of heterogeneous catalysis. Very promising is the molecular-beam technique, by which a collimated beam of molecules in high vacuum is directed onto the surface and either is reflected or reacts with the solid or with the adsorbed molecules.

The catalytic reaction must be treated as a complex dynamic system that includes the reacting medium and the catalyst. Such a system may exhibit several alternative steady-state regimes for the same external conditions, and sometimes will produce self-maintained oscillations of the reaction rate. A new theoretial approach, which studies the behavior of catalytic systems on the basis of equations of chemical kinetics and mathematical physics, was called by SLIN'KO [7.22] the "dynamics of heterogeneous catalytic systems" and includes "macrodynamics" and "microdynamics". Today the dynamics of heterogeneous catalysis is studied on the atomic and molecular level.

The investigations of various interactions between the reactants and the catalyst now include studying the processes of energy exchange. The energy of the exothermal stage of catalysis (e.g., the stage of chemisorption) can be used for the excitation of molecules or for the reconstruction of the surface. Both these effects can have great influence on the rate of the catalytic reaction.

For a long time the choice of catalysts for a given reaction was based on the method of trial and error. This method was succeeded by semiempirical techniques, using all kinds of chemical analogies and correlations (number of d electrons, width of the forbidden band, the metal-oxygen bonding energy, etc.). The real industrial catalysts are usually complicated multi-phase systems, and can hardly be conjured through the use of simple correlations. The large scale integration of modern physical investigative techniques, together with advances in theory, will provide the basis for a truly scientific choice of catalysts.

References

Chapter 2

2.1 V.F. Kiselev, O.V. Krylov: *Adsorption Processes on Semiconductor and Dielectric Surfaces I*, Springer Ser. Chem. Phys. Vol.32 (Springer, Berlin, Heidelberg 1985)

2.2 V.F. Kiselev, O.V. Krylov: *Electronic Phenomena in Adsorption and Catalysis on Semiconductors and Dielectrics*, Springer Ser. Surf. Sci., Vol.7 (Springer, Berlin, Heidelberg 1987)

2.3 H.A. Bethe: Ann. Phys. 3, 1929 (1929)

2.4 J.A. Van Vleck: J. Chem. Phys. B3B, 803, 807 (1935)

2.5 C.J. Ballhausen: *Molecular Electronic Structure of Transition-Metal Chemistry* (Mc Graw-Hill, New York 1979)

2.6 L.F. Orgel: An Introduction to Transition Metal Chemistry, (Wiley, New York 1965)

2.7 C.K. Jorgensen: Absorption Spectra and Chemical Bonding in Complexes (Pergamon, Oxford 1962)

2.8 C.K. Jorgensen: *Modern Aspects of Ligand Field Theory* (North-Holland, Amsterdam 1971)

2.9 I.B. Bersuker: *Elektronnoe Stroenie i Svojstva koordinatsionnykh soedinenij. Vvedenie v teoriju* (Khimija, Leningrad 1986) (in Russian)

2.10 J.D. Dunitz, L.E. Orgel: in *Advances in Inorganic Chemistry and Radiochemistry,* Vol.2 (Academic, New York 1960) p.1

2.11 R.G. Pearson: Proc. Nat. Acad. Sci. USA, 72, 2104 (1975)

2.12 L.E. Orgel: Disc. Farad. Soc. 26, 138 (1956)

2.13 O.V. Krylov: *Catalysis by nonmetals* (Academic, New York 1970)

2.14 K.B. Yatsimirsky: Zhur. neorg. khim. (J. Inorg. Chem.) 11, 2429 (1966) (in Russian)

2.15 J.K. Burdett: In *Advances in Inorganic Chemistry and Radiochemistry* 21, 113 (Academic, New York 1978)

2.16 M.E. Djatkina, E.L. Rosenberg: *Kvantovo-khimicheskie raschety soedinenij perekhodnykh elementov* (VINITI, Moscow 1974) (in Russian)

2.17 C.J. Ballhausen: Int. J. Quant. Chem. 5, 373 (1971)

2.18 J.C. Slater: In *Advances in Quantum Chemistry* 6, 1 (Academic, New York 1972)

2.19 K.J. Johnson: In *Advances in Quantum Chemistry* 7, 143 (Academic, New York 1973)

2.20 K.J. Johnson, R.P. Messmer: Int'l. J. Quant. Chem., Symp.**10**, 647 (1976)

2.21 H.P. Lefter, M.C. Hobson: In *Advances in Catalysis* **14**, 1 (Academic, New York 1964)

2.22 F.S. Stone: In *Surface Properties and Catalysis by Non-Metals*, ed. by J.P. Bonnelle, B. Delmon (Reidel, Dordrecht 1983) p.83

2.23 H. Luth: J. Appl. Phys. **8**, 1 (1975)

2.24 R.S. Mulliken: J. Am. Chem. Soc. **74**, 211 (1952)

2.25 J.C. Slater: *Insulators, Semiconductors and Metals*, (McGraw-Hill, New York 1967)

2.26 K. Siegbahn: Electron spectroscopy of atoms, molecules and condensed mater – an overview, in Proc. Int'l. Conf. on X-Ray and Inner-Shell Processes in Atoms, Molecules and Solids, ed. by A. Meisel, J. Finster (Leipzig 1984) pp.1-13

2.27 T.N. Rhodin, J.W. Gadzuk: In The Nature of the Surface Chemical Bond, ed. by T.N. Rhodin, G. Ertl (North Holland, Amsterdam, New York 1979)

2.28 D.P Woodruff, T.A. Delchar: Modern Techniques of Surface Science (Cambridge Univ. Press, Cambridge 1986) p.435

2.29 C. Defosse: In Characterization of Heterogeneous Catalysts, ed. by I. Delanney (Dekker, New York 1984) Vol.15, pp.225-298

2.30 T.H. Barr: Application of electron spectroscopy to herogeneneous catalysts, in Practical Surface Analysis by Auger and X-Ray Photoelectron Spectroscopy, ed. by D. Briggs, M.P. Seah (Wiley, New York 1983) pp.283-358

2.31 J. Stöhr: In Emission and Scattering Techniques, Proc. NATO Adv. Study Inst. Alghero, 1980 (Reidel, Dordrecht 1981) p.213

2.32 W.E. Spicer: In Electron and Ion Spectroscopy of Solids, ed. by L. Fiermans, J. Vennik, W. Dekeyser (Plenum, New York 1978) pp.54-92

2.33 R. Haight, J. Baker, R.R. Freeman, P.H. Bucksbaum: J. Vac. Sci. Techn. A**4**, 1481 (1986)

2.34 R.P. Messmer, H.J. Freund: Surf. Sci. **158**, 58 (1985)

2.35 H.D. Hagstrum, J.E. Rowe, J.C. Tracy: in Experimental Methods in Catalytic Research, Vol.3 ed. by R.B. Anderson (Academic, New York 1976) p.42

2.36 M. Thompson, A.D. Baker, A. Christie, J.F. Tyson: Auger Electron Spectroscopy, Chem. Phys., Vol.74 (Wiley, New York 1985)

2.37 C.R. Brundle, A.D. Baker (eds.): Electron Spectroscopy Theory, Techniques and Applications (Academic, New York 1984) Vol.5

2.38 B.E. Koel, G.A. Somorjai: In Catalysis, Science and Technology 7, 159 (Springer, Berlin, Heidelberg 1985)

2.39 G. Ertl: Low Energy Electrons and Surface Chemistry (VCH, Weinheim 1985)

2.40 K. Heinz: Appl. Phys. A **41**, 3 (1986)London 1974)

2.41 G.A. Somorjai: Chemistry in Two Dimensions: Surfaces (Cornell Univ. Press, Ithaca, NY 1981)
2.42 G.A. Somorjai, M.A. van Hove: Adsorbed Monolayers on Solid Surfaces (Springer, Berlin, Heidelberg 1979)
2.43 J.A. Strozier, D.W. Jepsen, F. Jona: In Surface Physics of Materials (Academic, London 1975) Vol.1, p.2
2.44 J.E. Demuth: Surf. Sci. **58**, 184 (1977)
2.45 A.A. Maradudin: Theory of Lattice Dynamics in the Harmonic Approximation (Academic, New York 1963)
2.46 R.F. Wallis: In Structure and Chemistry of Solid Surface, ed. by G. Somorjai (Wiley, New York 1969) p.17
2.47 R.D. Young: Phys. Rev. **113**, 110 (1959)
2.48 J.W. Gadzuk: Phys. Rev. **131**, 2110 (1970)
2.49 T.V. Vorburger, D. Penn, E.W. Plummer: Surf. Sci. **48**, 417 (1975)
2.50 E.W. Plummer: In Interactions on Metal Surfaces, ed. by R. Gomer, Topics Appl. Phys., Vol.4 (Springer, Berlin, Heidelberg 1975) p.143
2.51 L.W. Swanson, A.E. Bell: Adv. Electron Phys. **32**, 193 (197?)
2.52 J.W. Gadzuk, E.W. Plummer: Rev. Mod. Phys. **45**, 753 (1973)
2.53 N.J. Dionne: Field Emission Electron Spectroscopy of the Platinum Group Metals. Ph.D.Thesis. Research Group of Prof. T. Rhodin (Cornell University, Ithaca, NY 1975)
2.54 A. Modinas: Surf. Sci. **70**, 52 (1978)
2.55 E.W. Plummer, J.W. Gadzuk, D.K. Penn: Physics Today **28**, 343 (April 1975)
2.56 H.D. Hagstrum, G.E. Becker: Proc. Roy. Soc. **A331**, 395 (1972)
2.57 J.C. Tully: Phys. Rev. B **16**, 4324 (1977)
2.58 H.D. Hagstrum: In *Electron and Ion Spectroscopy of Solids*, ed. by L. Fiermans, J. Vennik, W. Dekeyser (Plenum, New York 1978)
2.59 P.D. Johnson, T.A. Dellchar: Surf. Sci. **77**, 400 (1978)
2.60 H. Conrad, G. Ertl, J. Kuppers, W. Sasselmann, H. Haber-baad: In *Inelastic Particle-Surface Collisions*, ed. by E. Taglauer, W. Heiland, Springer Ser. Chem. Phys., Vol.17 (Springer, Berlin, Heidelberg 1981) p.83
C. Binnig, H. Rohrer: Surf. Sci. **152/153**, 17 (1985)
2.61 L.J.E. Hofer: In *Experimental Methods in Catalytic Research*, ed. by R.B. Anderson (Academic, New York 1968) Vol.1, pp.432-435
2.62 P.W. Selwood: *Chemisorption and Magnetization* (Academic, New York 1975)
2.63 G.S. Krinchik: *Fizika magnitnykh javlenij* (MGU, Moscow 1977) (in Russian)
2.64 S. Methfessel, P.C. Mattis: *Magnetic Semiconductors* (Springer, Berlin, Heidelberg 1968)
2.65 P.W. Selwood: *Magnetochemistry*, 2nd ed. (Interscience, New York 1956)
2.66 P.W. Selwood: Advances in Catalysis **3**, 28 (1951)
2.67 W.L. Roth: Phys. Rev. **111**, 772 (1958)
2.68 A.G. Gurevich: *Magnitnyj rezonans v ferritakh i antiferro-magnetikakh* (Nauka, Moscow 1970)

2.69 G.S. Krinchik, V.E. Zubov, L.V. Nikitin: Poverkhnost' 1, 22 (1982) (in Russian)

2.70 A.Abraham, B. Bleaney: *Electron Paramagnetic Resoance of Transition Ions* (Clarendon, Oxford 1970)

2.71 H.A. Kuska, M.T. Rogers: *Electron Spin Resonance of First-Row Transition Metal Complex Ions* (Interscience, New York 1968)

2.72 R.J.Kokes: In *Experimental Methods in Catalytic Research* 1, 436-476 (Academic, New York 1968)

2.73 J.H. Lunsford: In *Spectroscopy in Heterogeneous Catalysis* 6, 183-235 (Academic, New York 1979)

2.74 J. Jafet: Solid State Phys. 14, 1 (1963)

2.75 K.N. Spiridonov, G.B. Parijskij, O.V. Krylov: Kinetika i Kataliz 12, 1448 (1971)

2.76 I. Tsuneki, J. Hiroshi, K. Larry: J. Chem. Phys. 75, 2485 (1981)

2.77 C.P. Slichter: *Principles of Magnetic Resonance*, 3rd. ed., Springer Ser. Solid-State Sci., Vol.1 (Springer, Berlin, Heidelberg 1988)

2.78 V.B. Kazansky, V.Ju. Borovkov: Kinetika i Kataliz 14, 1093 (1973) (in Russian)

2.79 D.R. Eton, V.D. Philips: J. Str. Chem. 9, 1353 (1968)

2.80 A.D.H. Clague: High resolution solid state NMR, in *Catalysis*, ed. by G.C. Bond, C. Webb (Burlinton House, London 1987)

2.81 W.H. Dawsch, S.W. Kaiser, P.D. Ellis, R.R. Inners: J. Am. Chem. Soc. 103, 6780 (1981)

2.82 K.I. Zamaraev: *Sbornik lektsij na 3 vsesojuznoj konferentsii po mekhanizmu kataliza* (SO AN SSSR, Novosibirsk 1983)

2.83 R.W. Vaughan: Ann. Rev. Mater. Sci. 4, 21 (1974)
      U. Haebergen: *High-Resolution NMR in Solids: Selective Averaging,* (Academic, New York 1976)

2.84 W.H. Delgass: Mössbauer Spectroscopy, in *Spectroscopy in Heterogeneous Catalysis*, ed. by W.H. Delgass (Academic, New York 1979) Chap.5, pp.132-182

2.85 V.I. Goldanskii, Yu.V. Maksimov, I.P. Suzdalev: Proc. Int. Conf. Mössbauer Spectroscopy, Crakow, 2, 163 (1973)

2.86 B.J. Tatarchuk, J.A. Dumesik: In *Chemistry and Physics of Solid Surfaces V*, ed. by R. Vanselow, R. Howe, Springer Ser. Chem. Phys., Vol.35 (Springer, Berlin, Heidelberg 1984) p.65

2.87 O.V. Krylov, L.Ja. Margolis: Kinetika i Kataliz 11, 432 (1970)

2.88 A.A. Davydov: *IK-spectroscopija v khimii poverkhnosti okislov* (IR spectroscopy in chemistry of oxide surfaces) (Nauka, Novosibirsk 1984)

2.89 A.T. Bell, M.L. Hair (eds.): Vibrational spectroscopies for adsorbed species, Am. Chem. Soc. Symp. Ser. (1980)

2.90 G.L. Haller: Infrared Spectroscopy, in *Spectroscopy in Heterogeneous Catalysis*, ed. by W.H. Delgass (Academic, New York 1979) Chap.2, pp.19-57

2.91 G.L. Haller: Catalysis Rev. 23, 477 (1981)

2.92 R.G. Greenler: J. Chem. Phys. **50**, 1963 (1969)
2.93 M.U. Kutyrev, V.A. Matyshak: Khimicheskaja Fizika **2**, 373 (1983) (in Russian)
2.94 R.P. Cooney, G. Curthoys, N.T. Tam: Adv. Catalysis **24**, 293 (1975)
2.95 T.A. Egerton, A.H. Hardin: Catalysis Rev. **11**, 71 (1975)
2.96 G.L. Haller: Raman Spectroscopy, in *Spectroscopy in Heterogeneous Catalysis*, ed. by W.H. Delgass (Academic, New York 1979) Chap.3, pp.58–95
2.97 S. Efrima: J. Chem. Phys. **83**, 1356 (1985)
2.98 C.R. Brundle, H. Morawitz (eds.): *Vibrations at Surfaces*, Proc. 3rd. Int'l. Conf., Asilomar 1982 (Elsevier, Amsterdam 1983)
R.F. Willis (ed.): *Vibrational Spectroscopy of Adsorbates*, Springer Ser. Chem. Phys., Vol.15 (Springer, Berlin, Heidelberg 1980)
2.99 J.L. Erskine: J. Vac. Sci. Techn. A **4**, 1982 (1986)
2.100 M.A. van Hove, W.H. Weinberg, C.-M. Chan: *Low-Energy Electron Diffraction*, Springer Ser. Surf. Sci., Vol.6 (Springer, Berlin, Heidelberg 1986)
2.101 G.P. Zhizhin, M.A. Moskaleva, P.A. Shafranovsky, B.R. Shub: Poverkhnost: Fisika, Khimija, Mekhanika (Surface: Physics, Chemistry, Mechanics) no.7, 141 (1987)

Chapter 3

3.1 P.W. Selwood: Chemisorption and Magnetization (Academic, New York 1975)
3.2 P.W. Selwood: Magnetochemistry, 2nd ed. (Interscience, New York 1956)
3.3 P.W. Selwood: Advances in Catalysis **3**, 28 (1956)
3.4 K.N. Spiridonov, G.B. Parijskij, O.V. Krylov: Kinetika i Kataliz **12**, 1448 (1971)
3.5 O.V. Krylov, L.Ja. Margolis: Kinetika i Kataliz **11**, 432 (1970)
3.6 A. Cimino, M. Schiavello, F.S. Stone: Disc. Faraday Soc. **41**, 390 (1966)
3.7 A. Cimino: Chimica e Industria **56**, 27 (1974)
3.8 R.I. Bickley, F.S. Stone: In Elektronnye javlenija v adsorbtsii i katalize na poluprovodnikakh, ed. by F.F. Volkenstein, Chap.11 (MIR, Moscow 1969) pp.211–226 (in Russian)
3.9 S.Z. Roginskij, V.A. Seleznev: Kinetika i Kataliz **8**, 1342 (1967) (in Russian)
3.10 G.F. Gerasimova, I.S. Sazonova, A.V. Rosljakova, G.M. Alikina, G.V. Bunina: Kinetika i Kataliz **17**, 1009 (1976) (in Russian)
3.11 S. Angelov, G.F. Gerasimova, V.M. Mastikin, N.P. Kejer: Kinetika i Kataliz **12**, 1533 (1971) (in Russian)

3.12 G.K. Boreskov: In Proc. 6th Int. Congress on Catalysis, London 1976, Preprint A13
3.13 F. Gesmundo, P.F. Rossy: Solid State Commun. **8**, 287 (1973)
3.14 S.M. Arija, N.L. Lukinykh: Fiz. Tverd. Tela **8**, 260 (1966) (in Russian)
3.15 Z. Kluz, J. Stoch, T. Creppe: Z. Phys. Chem., Neue Folge I**34**, I25, (1983)
3.16 V.N. Vorob'ev, T.E. Pursenkova, A.E. Martirosov, G.Sh. Talipov: Zh. Fiz. Khim. **50**, 1465 (1976) (in Russian)
E.E. Platera, G. Spoto, A. Zecchina: J. Chem. Sci., Faraday Trans. I, **81**, 1283 (1985)
3.17 P. Pomoris, J.C. Vickerman: J. Catalysis **55**, 88 (1978)
3.18 A. Andreev, E. Proikov, N. Nesnev, D. Shopov: J. Catalysis **74**, 1 (1982)
3.19 S.Y. Lin, H.H. King: Surf. Sci. **110**, 504 (1981)
3.20 T.M. Jur'eva, L.I. Kuznetsova, G.K. Boreskov: Kinetika i Kataliz **23**, 264 (1982)
3.21 A. Cimino, V. Indovina, F. Pepe, M. Schiavello: In Proc. 4th Int. Congress on Catalysis, Vol.I, Moscow, June 1968 (Akademiai Kiado, Budapest 1971) p.187
3.22 A. Cimino, M. Lo Jacono, P. Porta, M. Valigi: Z. Phys. Chem. (BRD) **70**, 166 (1970)
3.23 F.S. Stone: Anales. Real Soc. Espan. Fis. y Quim. **B61**, 109 (1965); Chimia **23**, 490 (1969)
3.24 G.N. Asmolov, O.V. Krylov: Kinetika i Kataliz **11**, 1028 (1970); **13**, 188 (1972)
3.25 V.A. Matyshak, M.Ja. Kushnerev, A.A. Kadushin: Kinetika i Kataliz **17**, 188 (1976)
3.26 K.N. Spiridonov, G.N. Parijskij, O.V. Krylov: Izv. Akad. Nauk SSSR, Ser.Khimia **8**, 2161 (1971) (in Russian)
3.27 G.N. Asmolov, V.A. Matyshak, A.A. Kadushin, O.V. Krylov: Kinetika i Kataliz **18**, 1506 (1976) (in Russian)
3.28 A.F. Shestakov, V.A. Matyshak, A.A. Kadushin, O.V. Krylov: Kinetika i Kataliz **20**, 189 (1979) (in Russian)
3.29 K. Dyrek: Bull. Acad. Sci. Polon., Ser. Sci. Chim. **21**, 673 (1973)
3.30 G.C.M. van Leuwen: Rec. Trav. Chim. Pays-Bas **92**, 195 (1973)
3.31 A. Zecchina, G. Spoto, S. Coluccia, E. Garrone: J. Chem. Phys. **80**, 463 (1984)
3.32 A. Bielanski, Z. Kluz, M. Jagiello: Z. Phys. Chem. (BRD) **97**, 207 (1975)
3.33 F. Pepe, F.S. Stone: J. Catalysis **56**, 160 (1979)
3.34 V. Indovina, A. Cimino, M. Inversi, F. Pepe: J. Catalysis **58**, 396 (1979)
3.35 R.B. Akhverdiev, A.P. Mamedov, F.B. Aliev, G.N. Asmolov, O.V. Krylov, Kinetika i Kataliz **21**, 999 (1980) (in Russian)
3.36 V.A. Shvets, V.B. Kazansky: Problemy Kinetiki i Kataliza **13**, 217 (1968)
3.37 Ju.I. Pecherskaja, V.B. Kazansky: Problemy Kinetiki i Kataliza **13**, 236 (1968) (in Russian)
3.38 D.E.O'Reilly: Adv. Catalysis **12**, 31 (1960)

3.39 P. Cossee: J. Catalysis 3, 80 (1964)

3.40 P. Cossee, L.L. van Reijen: Disc. Faraday Soc. 4, 277 (1966)

3.41 L.L. van Reijen, W.M.H. Sachtler, P. Cossee, D.A. Brouwer: In Proc. 3rd Int. Congress Catalysis, Vol.2, Amsterdam 1964 (North-Holland, Amsterdam 1965) p.820

3.42 J. Masson, B. Delmon: In Proc. 5th Int. Congress Catalysis, Vol.1, Miami Beach 1972 (North Holland, Amsterdam 1973) p.183

3.43 Y. Okamoto, M. Fujii, T. Imanaka: Bull. Chem. Soc. Japan 49, 859 (1976)

3.44 T.A. Egerton, F.S. Stone, J.C. Vickerman: J. Catalysis 33, 299, 307 (1974)

3.45 F.S. Stone, J.C. Vickerman: Proc. R. Soc. A354, 331 (1977)

3.46 M.P.McDaniel: J. Catalysis 67, 81 (1981)

3.47 T.M. Sabine, E.R. Vance: Solid State Chem. 1, 554 (1970)

3.48 W. Wintruff, N. Enden: Z. Wiss. Friedrich-Schiller Univ. Jena 22, 749 (1974)

3.49 J.W. Allen: Phys. Rev. Lett. 36, 1249 (1976)

3.50 V.F. Kiselev, O.V. Krylov: Electronic Phenomena in Adsorption and Catalysis, Springer Surf. Sci., Vol.7 (Springer, Berlin, Heidelberg 1987)

3.51 A. Cimino, M. Schiavello: J. Catalysis 20, 202 (1971)

3.52 A. Cimino, B.A. de Angelis, A. Lughetti, G. Minelli: J. Catalysis 45, 316 (1976)

3.53 A. Cimino, V. Indovina, M. Valigi: In Proc. 6th Int. Congress Catalysis, London 1976, Preprint A13

3.54 A. Cimino, F. Pepe, M. Schiavello: In Proc. 5th Int. Congress Catalysis, Vol.1, Miami Beach 1972 (North Holland, Amsterdam 1973) p.125

3.55 G.N. Asmolov, O.V. Krylov, Kinetika i Kataliz 12, 463 (1971) (in Russian)

3.56 D.S. McClure: J. Phys. Chem. Solids 3, 311 (1957)

3.57 A. Navrotsky, O.J. Kleppa: J. Inorg. Nucl. Chem. 29, 2701 (1967)

3.58 V.N.J. DeBeer, M.J.M. Van der Aalst, G.C.L. Schuit, C.J. Machiels: J. Catalysis 43, 78 (1976)

3.59 J. Grimblot, J.P. Bonnelle, J.P. Beaufils: J. Electron Spectroscopy 8, 437 (1976); 9, 449 (1976)

3.60 R.M. Friedman, J.J. Freeman, F.W. Lyttle: J. Catalysis 55, 10 (1978)

3.61 A.A. Awe, G. Miliades, J.C. Vickerman: J. Catalysis 62, 202 (1980)

3.62 K. Tarama, S. Yoshida, Y. Doi: In Proc. 4th Int. Congress Catalysis, Vol.1, Moscow, June 1968 (Academiai Kiado, Budapest 1971) p.197

3.63 B. Fubini, G. Ghiotti, L. Stradella, E. Garrone, C. Monterro: J. Catalysis 66, 200 (1980)

3.64 M.P. McDaniel: J. Catalysis 67, 71 (1981); 76,17, 29, 37 (1982)

3.65 K.N. Spiridonov, O.V. Krylov: Problemy Kinetika i Kataliza 16, 7 (1976) (in Russian)

3.66 E.L. Aptekar, M.G. Chudinov, A.M. Alekseev, O.V. Krylov: Reaction Kinetics and Catalysis Letters **1**, 493 (1979)
3.67 J.H. Ashley, P.C.H. Mitchell: J. Chem. Soc. **A**, 2730 (1969)
3.68 J.M. Lipsch, G.C.A. Schuit: J. Catalysis **15**, 174 (1969)
3.69 N. Giordano, J.C.J. Bart, A. Vaghi, A. Castellan, C. Martinotti: J. Catalysis **36**, 81 (1975); **37**, 204 (1975); **38**, 11 (1976)
3.70 F.E. Massoth: Adv. Catalysis **27**, 266 (1968)
3.71 J. Mason, J. Nechtschein: Bull. Soc. Chim. France **10**, 3933 (1968)
3.72 W.K. Hall, M.Lo Jacono: In Proc. 6th Int. Congress Catalysis, London 1976, Preprint A16
3.73 T. Kohno, T. Yokono, Y. Sanada, K. Yamashita, H. Hattori, K. Makino: Catalysis **22**, 201 (1986)
3.74 P. Gajardo, P. Grange, B. Delmon: J. Catalysis **61**, 66 (1980)
3.75 K. Jaganattan, A. Srinivasan, C.N.R. Rao: J. Catalysis **69**, 418 (1981)
3.76 T.S. Ismailov, L.N. Vorob'ev, G.Sh. Talipov: Kinetika i Kataliz **21**, 1028 (1980) (in Russian)
3.77 S.B. Nikishenko, A.A. Slinkin, E.S. Shpiro, G.V. Antoshin, Kh.M. Minachev: Kinetika i Kataliz **20**, 524 (1979)
3.78 V.A. Khalif, E.L. Aptekar', K.N. Spiridonov, O.V. Krylov: Kinetika i Kataliz **19**, 1231 and 1238 (1978) (in Russian)
3.79 H. Praliaud, M.-V. Mathieu: J. Chem. Phys. **73**, 689 (1976)
3.80 R. Bienert, W. Hanke, U. Illgen, G.H. Jerschkewitz, G. Lischke, G. Ohlman, I.W. Schulz: In Trudy vsesojuznoj konferentsii po mekhamizmu kataliza, Moscow 1974, Preprint 73
3.81 V.M. Villalba, K.N. Spiridonov, O.V. Krylov: Kinetika i Kataliz **20**, 1305 (1979) (in Russian)
3.82 Y. Murakami, M. Inomata, A. Migamoto, K. Mori: in Proc. 7th Int. Congress Catalysis, Tokyo, July 1980, Preprint B-49
3.83 V.B. Aleskovskij: Stekhiometrija i sintez tverdykh soedinenii (Stoichiometry and Synthesis of Solid-State Compounds) (Nauka, Leningrad 1976) (in Russian)
3.84 V.F. Kiselev, O.V. Krylov: Processes of Adsorption on the Surfaces of Semiconductors and Dielectrics, Springer Ser. Chem. Phys., Vol.32 (Springer, Berlin, Heidelberg 1985)
3.85 M.A. Eremeeva, A.P. Nechiporenko, G.N. Kuznetsova, S.I. Kol'tsch, V.B. Aleskovskij: Zh. Prikl. Khim. **43**, 2332 (1974) (in Russian)
3.86 V.B. Tolstoj, G.N. Kuznetsova, S.I. Kol'tsov, V.B. Aleskovskij: Zh. Prikl. Khim. **52**, 299, 2353 (1980) (in Russian)
3.87 V.N. Koval'kov, E.P. Smirnov, S.I. Kol'tsov, V.B. Aleskovskij: Zh. Obshchej Khim **46**, 2151 (1976) (in Russian)
3.88 H. Topsoe, B.S. Clausen, N.Y. Topsoe, E. Peterson: Ind. Eng. Chem. Fundamentals **25**, 25 (1986)
3.89 P. Rathnasami, H. Knozinger: J. Catalysis **54**, 153 (1978)
3.90 M.A. Apecetche, B. Delmon: Reac. Kinet. Catal. Lett. **12**, 385 (1979)

3.91 J.M. Dale, J.D. Hulett, F.D. Fuller, H.L. Richards, R.L. Sherwood: J. Catalysis **61**, 66 (1980)

3.92 P.F. Chester: J. Appl. Phys. Suppl. **32**, 866, 2233 (1961)

3.93 J. Kerson, J. Volger: Physics **69**, 535 (1973)

3.94 V.N. Bogomolov, L.S. Sochava: Fiz. Tverd. Tela **9**, 3355 (1967)

3.95 R.D. Iyengar, M. Codell, S.Y. Carra: J. Am. Chem. Soc. **88**, 5055 (1966)

3.96 A.I. Mashchenko, V.B. Kazansky, G.B. Parijskij, V.M. Sharapov: Kinetika i Kataliz **8**, 353 (1967) (in Russian)

3.97 P. Meriadeau, M. Che, P. Gravelle: Bull. Soc. Chim. France **1**, 13 (1971)

3.98 Lu Tung Sin, V.L. Rapoport: Vestnik MGU, Ser. Fiz. Khim. **10**, 45 (1966) (in Russian)

3.99 V.E. Henrich, H.J. Zeiger, D. Dresselhaus: Nat. Bureau of Standards, Publ."55 (1976)

3.100 Y. Mizokawa, S. Nakamura: Japan J. Appl. Phys. **14**, 779 (1975)

3.101 L.I. Burbuljavichus, Ju.A. Zarif'jants, S.N. Karjagin, V.F. Kiselev: Kinetika i Kataliz **14**, 1526 (1973) (in Russian)

3.102 S.N. Karjagin, Ju.A. Zarif'jants, V.F. Kiselev: Vestnik MGU, Ser. Fiz. **2**, 236 (1975)

3.103 V.F. Kiselev, Ju.A. Zarif'jants, S.P. Kozlov: In Problemy fizicheskoj khimii poverkhnostej monokristallicheskikh poluprovodnikov (The Problem of Physical Chemistry of Monocrystalline Semiconductor Surfaces) (Nauka, Novosibirsk 1978) p.200–246 (in Russian)

3.104 Ja.S. Lebedev, V.I. Muromtsev: EPR i relaksatsija stabilizirovannykh radikalov (EPR and Relaxation of Stabilized Radicals) (Khimija, Moscow 1972) (in Russian)

3.105 V.S. Vavilov, A.E. Kiv, O.P. Niyazova: Mekhanizmy obrazovaniya i migratsii defektov v poluprovodnikakh (The mechanisms of formation and migration of defects in semiconductors) (Nauka, Moscow 1981)

3.106 A.R. Allnat, E.Loffus: J. Chem. Phys. **59**, 2541 (1973)

3.107 D.H. Olson, W.O. Haag, R.M. Lago: J. Catalysis **61**, 390 (1980)

3.108 J.A. Rabo (ed.): Zeolite Chemistry and Catalysis, Monograph 171, Am. Chem. Soc. (Washington 1976)

3.109 V.D. Atanasova, V.A. Shvets, V.B. Kazansky: Uspekhi Khimii **50**, 385 (1981) (in Russian)

3.110 N.N. Tikhomirova, L.V. Nikolaeva, V.V. Demkin, E.N. Rosolovskaj, K.V. Topchieva: J. Catalysis **29**, 500 (1973)

3.111 N.N. Tikhomirova, I.V. Nikolaeva: Zh. Fiz. Khim. **55**, 2441 (1981) (in Russian)

3.112 H. Bremer, R. Schodel, F. Vogt: Z. Chem. **13**, 350 (1973)

3.113 T.A. Egerton, A. Hagan, F.S. Stone: J. Chem. Soc. Faraday Trans. **43**, 292 (1972)

3.114 P.J. Hutta, J.H. Lunsford: J. Chem. Phys. **66**, 4716 (1977)

3.115 P. Gallezot, B. Imelik: J. Phys. Chem. **77**, 652 (1973)

3.116 I.D. Mikheikin, G.N. Zhidomirov, V.B. Kazansky: Uspekhi khimii **41**, 909 (1982) (in Russian)

3.117 C. Naccache, Y. Ben Taarit: Chem. Phys. Lett. **11**, 11 (1971)

3.118 Y. Turkevich, Y. Ono, J. Soria: J. Catalysis **25**, 44 (1972)

3.119 J.W. Smith, J.M. Bennett, E.M. Flanigen: Nature **215**, 241 (1967)

3.120 W.H. Delgass, R.L. Garten, M. Boudart: J. Phys. Chem. **73**, 2970 (1969); J. Catalysis **18**, 90 (1970)

3.121 J.W. Ward: J. Catalysis **22**, 230 (1971)

3.122 G.K. Boreskov: Kinetika i kataliz **19**, 7 (1973) (in Russian)

3.123 S. Methfessel, P.C. Mattis: Magnetic Semiconductors, in Handbuch der Physik, Vol.18/1 (Springer, Berlin, Heidelberg 1968)

3.124 G.A. Smolenskij, V.V. Lemanov, G.M. Nedlin, M.P. Petrov, V.V. Pisarev: Fizika Magnitnykh Dielektrikov (Physics of Magnetic Insulators) (Nauka, Leningrad 1974) (in Russian)

3.125 D.J. Sith, L.A. Bursill, D.A. Jefferson: Surf. Sci. **175**, 673 and 684 (1986)

3.126 G.S. Krinchik, A.P. Khrebtov, A.A. Askochenskij, V.E. Zubov: Pis'ma v Zh. Eksp. Teor. Fiz. **17**, 446 (1973) (in Russian)

3.127 G.S. Krinchik, V.E. Zubov: Zh. Eksp. Teor. Fiz. **69**, 707 (1972) (in Russian)

3.128 M.I. Kaganov: Zh. Eksp. Teor. Fiz. **62**, 1196 (1972) (in Russian)

3.129 M.A. Van Hove, W.H. Weinberg, C.-M. Chan: Low-Energy Electron Diffraction, Springer Ser. Surf. Sci. Vol.6 (Springer, Berlin, Heidelberg 1986)

3.130 M.A. van Hove, P.M. Echenique: Surf. Sci. **82**, 2238 (1979)

3.131 M. Prutton, J.A. Walker, M.R. Walton.Cook, R.C. Felton, J.A. Ramsey: Surface Sci. **89**, 101 (1979)
R.C. Felton, M. Prutton, S.P. Tear, M.R. Welton-Cook: Surf. Sci. **88**, 474 (1979)

3.132 V.L. Vinetskij, G.A. Kholodar': *Statisticheskoe vzaimodejstvie elektronov i defektov v poluprovodnikakh* (Statistical Interaction of Electrons and Defects in Semiconductors) (Naukova Dumka, Kiev 1969) (in Russian)

3.133 K. Hauffe: Reaktionen in und an festen Stoffen, 2. Aufl. (Springer, Berlin, Göttingen, Heidelberg 1955)

3.134 O.V. Krylov: Catalysis by Nonmetals (Academic, New York 1970)

3.135 E.J.W. Verwey, J.H. De Boer: Rec. Trav. Chim. Pays-Bas **55**, 531 (1936)

3.136 G. Shobaky, P.C. Gravelle, S.J. Teichner: Bull. Soc. Chim. France **9**, 3244 (1977)

3.137 A. Bielanski, R. Dziembaj, J. Sloczynski: Bull. Acad. Polon. Sci. Ser. Sci. Chim. **14**, 569 (1966)

3.138 A.E. Cherkashin, A.N. Goldobin, V.I. Savchenko, N.P. Keier, G.L. Semin: React. Kinet. Catal. Lett. **1**, 411 (1974)

3.139 J.S. Anderson: In Problems of Nonstoichiometry, Chap.1, ed. by A. Rabenau (North Holland, Amsterdam 1970) p.1-85

3.140 R.J.D. Tiley: In Surface Properties and Catalysis by Non-Metals, ed. by J.P. Bonnelle, B. Delmon, E. Deronane (Reidel, Dordrecht 1982) pp.83-121

3.141 O.V. Krylov: In Khimija poverkhnoti okisnykh katalizatorov (Chemistry of Oxide Catalysts' Surface), ed. by A.Ja. Rozovskij (Nauka, Moscow 1979) p.33–52 (in Russian)

3.142 Y.W. Chung, W.J. Lo, G.A. Somorjai: Surface Sci. **64**, 588 (1977)

3.143 L.E. Firment: Surf. Sci. **116**, 205 (1981)

3.144 I.S. Kotousova, S.M. Poljakov: Kristallografija **17**, 661 (197?)

3.145 R.L. Burwell, G.L. Haller, K.C. Taylor, J.F. Read: Adv. Catalysis **20**, 1 (1969)

3.146 A.A. Davydov, Ju.M. Shchekochiknin, N.P. Kejer: Kinetika i Kataliz **13**, 1088 (1972) (in Russian)

3.147 E. Borello, S. Coluccia, C. Morterra: Ind. Chim. Belge **38**, 508 (1973)

3.148 J. Deren, J. Haber: Studies on the Physico-Chemical and Surface Properties of Chromium Oxides (Panstwowe Wyd. Krakow 1969)

3.149 P. Cossee, L.L. van Reijen: In Actes 2^e Congres Int. de Catalyse, Vol.2, Paris 1960 (Technipress, Paris 1961) p.1679

3.150 I. Aso, M. Nakao, T. Seiyama: J. Catalysis **57**, 285 (1979)

3.151 M.A. Poraj-Koshits, L.O. Atovmjan: *Kristallokhimija i stereokhimija koordinatsionnykh soedinenij molibdena* (Crystallochemistry and Stereochemistry of Molybdenum Coordination Compounds) (Moscow, Nauka 1974) (in Russian)

3.152 F. Trifiro, P. Centola, I. Pasquon, P. Jiru: In Proc. 4th Int. Congress Catalysis, Vol.1, Moscow, June 1968 (Akademiai Kiado, Budapest 1971) p.252

3.153 O.V. Krylov: Problemy Kinetiki i Kataliza **16**, 129 (1975) (in Russian)

3.154 L.C. Dufour, O. Bertrand, N. Floquet: Surf. Sci. **147**, 396 (1984); and **164**, 305 (1985)

3.155 G. Martino: J. Magn. Res. **15**, 262 (1974)

3.156 K.N. Spiridonov, O.V. Krylov, D. Gati: Problemy kinetiki i kataliza **17**, 149 (1978) (in Russian)

3.157 J. Grimblot, J.-P. Bonnelle: Compt. Rend. **286**, 399 (1976)

3.158 J. Haber, W. Marczewski, J. Stoch: Ber. Bunsengesellsch. **79**, 970 (1975)

3.159 A. Cimino, B.A. de Angelis: J. Catalysis **36**, 11 (1975)

3.160 L.Ja, Erman, E.L. Gal'perin, I.K. Kolchin, G.F. Dobrzhanskij, K.S. Chernyshev: Zh. Neorg. Khim. **9**, 1174 (1964) (in Russian)

3.161 G.C.A. Schuit: J. Less-Common Met. **36**, 329 (1974)

3.162 B. Gates, J.R. Ketzer, G.C.A. Schuit: Chemistry of Catalytic Processes, Chap.4, (McGraw-Hill, New York 1979)

3.163 J.E. Brazdil, D.D. Suresh, R.K. Grasselli: J. Catalysis **66**, 347 (1980)

3.164 R.K. Grasselli, J.D. Burrington: Adv. Catalysis **30**, 133 (1981)

3.165 I. Matsuura: Proc. 7th Int. Congress Catalysis, Tokyo, July 1980, Preprint B31

3.166 L.K. Yong, R.F. Howe, G.W. Keulks, W.K. Hall: J. Catalysis **52**, 544 (1978)

3.167 H. Miura, T. Otsuro, T. Shirasaki, Y. Morikawa: J. Catalysis **56**, 81 (1978)

3.168 I. Matsuura, R. Schuit, K. Hirakawa: J. Catalysis **63**, 152 (198?)

3.169 O.V. Krylov: Vestnik AN SSSR **1**, 26 (1983)

3.170 T. Noterman, G. Keulks, A.V. Skljarov, L.Ja. Margolis, O.V. Krylov: Kinetika i kataliz **17**, 758 (1976) (in Russian)

3.171 J. Haber: J. Less-Common Met. **54**, 243 (1977)

3.172 L.Ja. Margolis: Okislenie uglevodorodov na geterogennykh katalizatorakh (Oxidation of Hydrocarbons on Heterogeneous Catalysts) (Khimija, Moscow 1976) (in Russian)

3.173 K.Ju. Adzhamov, V.Ja. Dolgov, K.N. Spiridonov, Ju.N. Kafarov, T.G. Alkhazov, L.Ja. Margolis, O.V. Krylov: Zh. Fiz. Khim. **47**, 1112 (1973); Doklady AN SSSR **209**, 1127 (1973) (in Russian)

3.174 F. Trifiro, P. Centola, J. Pasquon: J. Catalysis **10**, 86 (1968)

3.175 M. Akimoto, E. Echigoya: J. Catalysis **29**, 191 (1973)

3.176 F. Trifiro, G. Capito, P.L. Villa: J. Less-Common Met. **36**, 305 (1974)

3.177 O.V. Krylov, Yu.V. Maksimov, L.Ya. Margolis: J. Catalysis **96**, 205 (1985)

3.178 G.A. Vorob'eva, B.V. Rozentuller, Yu.V. Maksimov, M.Yu. Kutyrev, L.Ya. Margolis: J. Catalysis **71**, 405 (1981)

3.179 G.S. Krinchik: Fizika Magnitnykh Javlenij (MGU, Moscow 1977) (in Russian)

3.180 E.L. Nagajev: Magnitnye poluprovodniki (Magnetic Semiconductors) (Nauka, Moscow 1980) (in Russian)

3.181 L. Neel: J. Phys. Rad. **15**, 225 (1954)

3.182 V.M. Fridkin: Fotosegnetoelektriki (Photoferroelectrics) (Nauka, Moscow 1979) (in Russian)

3.183 P. Mark: Catalysis Rev. **1**, 166 (1965)

3.184 T. Wolfram, F.J. Morin: J. Appl. Phys. **8**, 125 (1975)

3.185 C.J. Ballhausen: Int.J. Quant. Chem. **5**, 373 (1971)

3.186 J.C. Slater: In Advances in Quantum Chemistry **6**, 1 (1972)

3.187 F. Morin: Bell Syst. Techn. J. **37**, 1047 (1958)

3.188 N.F. Mott: Metal-Insulator Transitions (Taylor & Francis, London 1974)

3.189 S. Hüfner, T. Riesterer: Phys. Rev. B **33**, 7267 (1986)

3.190 R.R. Heikes, W.D. Johnston: J. Chem. Phys. **26**, 582 (1957)

3.191 J.B. Goodenough: Magnetism and the Chemical Bond (Interscience, New York 1966)

3.192 S.I. Pekar: Issledovanija po elektronnoj teorri kristallov (Studies in the Electron Theory of Crystals) (AN SSSR, Moscow, Leningrad 1951) (in Russian)

3.193 Ju.A. Firsov: In Poljarony, ed. by Ju.A. Firsov (Polarons)

3.194 J.G. Austin, N.F. Mott: Adv. Phys. **18**, 41 (1969)

3.195 P.W. Anderson: Phys. Rev. **109**, 1492 (1958)

3.196 S.R. Morrison: The Chemical Physics of Surfaces (Plenum, New York 1977)

3.197 O.V. Krylov: Problemy Kinetiki i Kataliza **13**, 141 (1968) (in Russian)
3.198 C.K. Jorgensen: Inorganic Complexes (Academic, New York 1963)
3.199 C.K. Jorgensen: Progr. Inorg. Chem. **12**, 101 (1970)
3.200 T. Ziegler, P. Rauk: J. Chem. Phys. **16**, 209 (1976)
3.201 E. Verdonck, L.G. Vanquickenborne: Inorg. Chim. Acta **23**, 67 (1977)
3.202 D.S. McClure: Optical Properties of Ions in Solids (Academic, New York 1971) p.401–418
3.203 F.S. Stone: In Surface Properties and Catalysis by Nonmetals, eds. J.P. Bonnelle, B. Delmon (Reidel, Dordrecht 1983) p.83
3.204 V.L. Bonch-Bruevich: In Statisticheskaja fizika i kvantovaja teorija polja (Statistical Physics and Quantum Field Theory) ed. by N.N. Bogoljubov (Nauka, Moscow 1971) p.337–391 (in Russian)
3.205 R. Newman, R.M. Chrenko: Phys. Rev. **114**, 1507 (1959)
3.206 A.E. Cherkashin, F.I. Vilesov: Fiz. Tverd. Tela **11**, 319 (1969) (in Russian)
3.207 V.Ja. Dolgov, K.N. Spiridonov, L.Ja. Margolis, O.V. Krylov: Zh. Fiz. Khim. **46**, 2418 (1972) (in Russian)
3.208 C.P. Poole, D.S. MacIver: J. Chem. Phys. **41**, 1500 (1964)
3.209 A. Bartecki, P. Dembicka: J. Inorg. Nucl. Chem. **29**, 2907 (1967)
3.210 M.A. Boll: J. Phys. Chem. Solids **5**, 13 (1972)
3.211 S.J. Cochran, F.P. Larkins: J. Chem. Soc., Faraday Trans., **81**, 2179 (1985); and **82**, 1721 (1986)
3.212 R. Holm, S. Storp: J. Appl. Phys. **9**, 217 (1976)
3.213 K.S. Kim: Phys. Rev. **B11**, 2177 (1975)
3.214 C.R. Brundle: In Electronic States of Inorganic Compounds, ed. by P. Day (Reidel, New York 1975) p.361–392
3.215 T.J. Chuang, C.R. Brundle, D.W. Rice: Surf. Sci. **59**, 413 (1976)
3.216 B. Grzhibovska, J. Haber, W. Marczewski: J. Catalysis **25**, 25 (1972); **46**, 327 (1972)
3.217 W.E. Spicer: In Electron and Ion Spectroscopy of Solids, ed. by L. Fiermans, J. Vernik, W. Dekeyser (Plenum, New York 1978)
3.218 V.F. Henrich, R.L. Kurtz: Phys. Rev. B **25**, 6280 (1981)
3.219 T. Wolfram, R. Hurst, F.J. Morin: Phys. Rev. **B15**, 1151 (1977)

Chapter 4

4.1 D.A. Dowden, D. Wells: In Actes 2^e Congres Int. de Catalyse, Vol.2, Paris 1960 (Technipress, Paris 1961) p.1489
4.2 D.A. Dowden, N. Mackenzie, B.M.W. Trapnell: Proc. Roy. Soc. **A237**, 247 (1956)
4.3 F. Basolo, R.G. Pearson: *Mechanisms of Inorganic Reactions*, 2nd ed. (Wiley, New York 1967)

4.4   D.A. Dowden: Anal. Real. Soc. Espanol Fis. y Quim. **B61**, 177 (1965)

4.5   K.S. De, M.J. Rossiter, F.S. Stone: In Proc. 3rd. Int'l Congress Catalysis, Vol.1, Amsterdam 1964 (North-Holland, Amsterdam 1965) p.520

4.6   G.M. Dixon, D. Nicholls, H. Steiner: In Proc. 3rd Int'l Congress Catalysis, Vol.2, Amsterdam 1964 (North-Holland, Amsterdam 1965) p.815

4.7   O.V. Krylov: *Catalysis by Nonmetals* (Academic, New York 1970)

4.8   O.V. Krylov: Problemy kinetiki i kataliza **13**, 141 (1968) (in Russian)

4.9   D.A. Dowden: In Proc. 4th Int. Congress Catalysis, Vol.1, Moscow, June 1968 (Akademiai Kiado, Budapest 1971) p.163

4.10   K.B. Jatsimirskij: Zh. Neorg. Khim. **11**, 2429 (1966) (in Russian)

4.11   D. Shopov, A. Andreev: *Khimicheskaja svjaz' pri adsorbtsii i katalize* (Chemical Bond at Adsorption and Catalysis), Vol.2, *Okisly* (Oxides) (BAN, Sofia 1979) (in Russian)

4.12   P. Cossee: J. Catalysis **3**, 80 (1964)

4.13   J. Chatt, L.A. Duncanson: J. Chem. Soc. **??**, 2939 (1953)

4.14   P. Cossee: Rec. Trav. Chim. Pays-Bas, **85** 1151 (1966)

4.15   B. Gorewit, M. Tsutsui: Adv. Catalysis **27**, 227 (1978)

4.16   P. Cossee, P. Ros, J.H. Schachtschneider: In Proc. 4th Int. Congress Catalysis, Vol.1, Moscow, June 1968 (Akademiai Kiado, Budapest 1971) p.207

4.17   P. Cassoux, P. Grasnier, J.F. Labarre: J. Organometal. Chem. **165**, 303 (1979)

4.18   O. Novaro, B. Blaisten-Barojas, E. Clementi, G. Giunchi, M.E. Ruiz-Vizcaya: J. Chem. Phys. **68**, 2337 (1978)

4.19   V.E.Lvovskij, E.L. Fushman, F.S. Dijachkovskij: Zh. Fiz. Khim. **56**, 1864 (1982) (in Russian)

4.20   R.P. Messmer, K.H. Johnson: In *Electrocatalysis on Non-Metallic Surfaces*, Publ.#45, Nat. Bureau of Standards (Washington 1976) p.67-86

4.21   N.D. Chuvylkin, G.M. Zhidomirov: In *Kinetika i kataliz* (Kinetics and Catalysis), ed. by V.B. Kazanskij (VINITI, Moscow 1980) p.3-98

4.22   R.F.W. Bader: Canad. J. Chem. **40**, 1164 (1962)

4.23   R.G. Pearson: J. Am. Chem. Soc. **91**, 1252 (1969); J. Appl. Chem. **27**, 145 (1971); Proc. Nat. Acad. Sci. USA **72**, 2104 (1975)

4.24   R.B. Woodward, R. Hoffmann: J. Am. Chem. Soc. **87**, 395 (1965)

4.25   R.B. Woodward, R. Hoffmann: *The Conservation of Orbital Symmetry* (Verlag Chemie, Weinheim 1970)

4.26   F.D. Mango, J.H. Schachtschneider: J. Am. Chem. Soc. **89**, 2484 (1967)

4.27   W.Th.A.M. Van der Lugt: Tetrahedron Letters **26**, 2281 (1970)

4.28   G.L. Caldow, M. MacGregor: J. Chem. Soc. **A**, 1654 (1971)

4.29   D.P. Eaton: J. Am. Chem. Soc. **90**, 4275 (1968)

4.30 F.D. Mango: Chem. Technol. **758** (1971); Tetrahedron Letters **17**, 1509 (1973)

4.31 J. Manassen: J. Catalysis **18**, 38 (1970)

4.32 R. Messmer, A.J. Bennett: Phys. Rev. **B6**, 633 (1972)

4.33 I.B. Bersuker: Kinetika i Kataliz **18**, 1268 (1977) (in Russian)

4.34 O.V. Krylov: In *Mechanism of Catalysis*, ed. by G.K. Boreskov (Nauka, Novosibirsk 1984) (in Russian) pp.51–71; Kinetika i Kataliz **26**, 263 (1985)

4.35 J.C. Mol, J.A. Moulijn: Adv. Catalysis **75**, 131 (1975)

4.36 B.A. Dolgoplosk: Kinetika i Kataliz **22**, 807 (1981) (in Russian)

4.37 A.F. Shestakov, V.A. Matyshak, A.A. Kadushin, O.V. Krylov: Kinetika i Kataliz **20**, 189 (1979) (in Russian)

4.38 V.F. Kiselev, O.V. Krylov: *Electronic Phenomena in Adsorption and Catalysis*, Springer Ser. Surf. Sci., Vol.7 (Springer, Berlin, Heidelberg 1987)

4.39 V.L. Bonch-Bruevich: Zh. Fiz. Khim **25**, 1033 (1951) (in Russian)

4.40 P. Mark: Catalysis Rev. **12**, 74 (1975)

4.41 S.R. Morrison: Surface Sci. **13**, 85 (1969); **15**, 363 (1969); **27**, 586 (1971)

4.42 C.K. Jorgensen: Progr. Inorg. Chem. **12**, 101 (1970)

4.43 I.B. Bersuker: Problemy Kinetiki i Kataliza **13**,7 (1968) (in Russian)

4.44 I.B. Bersuker, S.S. Budnikov: In Proc. 4th Int. Congress Catalysis, Vol.1, Moscow, June 1968 (Akademiai Kiado, Budapest 1971) p.58

4.45 T. Wolfram, F.J. Morin: J. Appl. Phys. **8**, 125 (1975)

4.46 V.F. Kiselev: Problemy Kinetiki i Kataliza **13**, 249 (1968)

4.47 H. Watanabe, M. Wada: Japan J. Appl. Phys. **4**, 945 (1965)

4.48 K.M. Sancier, A. Aoshima, H. Wise: J. Catalysis **34**, 257 (1974)

4.49 D.P. Frankl: In Surface Science. Recent Progress and Perspectives, ed. by P.S. Jayadevaish, P. Vanselow (CRC Press, Cleveland 1974)

4.50 E.K. Putsejko, A.N. Terenin: Doklady AN SSSR **70**, 401 (1950) (in Russian)

4.51 O.V. Krylov, L.Ja. Margolis: Kinetika i Kataliz **11**, 432 (1970)

4.52 F. Trifiro, P. Forzatti, I. Pasquon: In *Catalysis Heterogeneous and Homogeneous* (Elsevier, Amsterdam 1975) p.509–520

4.53 A.M. Gritskov, V.A. Shvets, V.B. Kazansky: Kinetika i Kataliz **14**, 1062 (1978) (in Russian)

4.54 F.I. Vilesov, D.A. Sukhov, Uspekhi Fotoniki **6**, 38 (1977) (in Russian)

4.55 R.L. Kuntz, V.E. Henrich: J. Vac. Sci. Technol. **A2**, 842 (1984)

4.56 W.J. Lo, Y.W. Chung, G.A. Somorjai: Surface Sci. **71**, 199 (1979)

4.57 G.W. Rubloff, H. Luth, F. Grobman: J. Vac. Sci. Techn. **13**, 333 (1976)

4.58 D. Ross, W. Ranke, K. Jacobi: Surface Sci. **105**, 77 (1981)

4.59 P.D. Bringans, H. Hochst, H.R. Shanks: Surface Sci. **111**, 80 (1981)

4.60 M.V. Roberts, R.S. C. Smart: Surface Sci. **108**, 271 (1981)

4.61 V.N. Kondrat'ev (ed.): *Energija razryva khimicheskikh svjazej* (Energy of Rupture of Chemical Bonds) (Nauka, Moscow 1974) (in Russian)

4.62 W. Gopel: Surface Sci. **62**, 165 (1977)

4.63 K. Tanaka, G. Blyholder: Chem. Comm. **21**, 1343 (1971)

4.64 K.N. Spiridonov, O.V. Krylov: Problemy Kinetiki i Kataliza **16**, 7 (1976) (in Russian)

4.65 M. Kobayashi, H. Kobayashi: Bull. Chem. Soc. Japan **49**, 3009 (1976)

4.66 Y. Ishii, J. Matsuura: Nippon Kagaku Zasshi **89**, 553 (1968); **92**, 302 (1971)

4.67 K. Hauffe: *Reaktionen in und an Festen Stoffen*, 2.Aufl. (Springer, Berlin, Göttingen 1955)

4.68 J. Novotny: J. Material Sci. **12**, 1147 (1977)

4.69 J.E. Bennet, D.J.E. Ingram, M.C.R. Symons: Phil. Mag. **46**, 443 (1955)

4.70 M.L. Meistrich: J. Phys. Chem. Solids **29**, 1111 (1968)

4.71 W. Kanzig, M.H. Cohen: Phys. Rev. Lett. **3**, 509 (1959)

4.72 P.H. Kasai: J. Chem. Phys. **43**, 3322 (1965)

4.73 J.H. Lunsford, J.P. Jayne: J. Chem. Phys. **44**, 1487 (1966)

4.74 I.D. Mikhejkin, A.I. Mashchenko, V.B. Kazansky: Kinetika i Kataliz **8**, 1363 (1967) (in Russian)

4.75 V.A. Shvets, M.E. Sarichev, V.B. Kazansky: J. Catalysis **11**, 378 (1968)

4.76 C. Naccache, P. Meriaudeau, A.J. Tench: Trans. Faraday Soc. **67**, 506 (1971); J. Catalysis **21**, 208 (1971)

4.77 P. Meriaudeau, J.C. Vedrine: J. Chem. Soc. Faraday Trans. Part 2, **2**, 472 (1976)

4.78 A.Ju. Loginov, K.V. Topchieva, S.V. Kostikov: Doklady AN SSSR **232**, 1331 (1977) (in Russian)

4.79 A.J. Tench, P. Kolroyd: Chem. Comm. **8**, 471 (1968)

4.80 E.G. Derouane, V. Indovina: Chem. Phys. Lett. **14**, 455 (1972)

4.81 A.A. Gezalov, G.M. Zhabrova, V.V. Nikisha, G.B. Parijskij, K.N. Spiridonov: Kinetika i Kataliz **9**, 462 (1968) (in Russian)

4.82 J.H.C. van Hooff, J.F. van Helden: J. Catalysis **8**, 199 (1967); **11**, 277 (1968)

4.83 V.A. Shvets, V.B. Kazansky: J. Catalysis **25**, 123 (1972)

4.84 M. Dufux: Ph.D. Thesis, Inst. Catalyse, Lyon, CNRS (1968)

4.85 N.I. Lipatkina, V.A. Shvets, V.V. Kazansky: Kinetika i Kataliz **19**, 979 (1978) (in Russian)

4.86 F.P. Billingsley, C. Trindle: J. Phys. Chem. **76**, 2995 (1972)

4.87 P. Svejda, W. Hartmann, R. Haul: Ber. Bunsengesellsch. **80**, 1327 (1976)

4.88 M. Codell, J. Weisberg, H. Gisser, R.D. Iyengar: J. Am. Chem. Soc. **91**, 7762 (1969)
4.89 M.C.R. Symons: J. Phys. Chem. **76**, 3095 (1972)
4.90 M. Setaka, T. Kwan: Bull. Chem. Soc. Japan **43**, 2727 (1970)
4.91 M. Breusse, B. Claudel, P. Meriaudeau: J. Chem. Soc. Faraday Trans. 2(1), 1 (1976)
4.92 V.E. Shubin, V.A. Shvets, V.B. Kazansky: Kinetika i Kataliz **19**, 1270 (1978) (in Russian)
4.93 J.G. Brailsford, J.R. Morton, L.E. Vannotti: J. Chem. Phys. **49**, 2237 (1968)
4.94 F.T. Gamble, R.H. Bartram, P.W. Levy: Phys. Rev. **134A**, 589 (1964)
4.95 O.F. Schirmer: J. Phys. Chem. Solids **32**, 499 (1972)
4.96 A.L. Taylor, G.F. Filipovich, G.K. Lindberg: Solid State Comm. **8**, 1359 (1970)
4.97 V.A. Shvets, V.B. Kazansky: J. Catalysis **15**, 288 (1969)
4.98 M. Che, K. Dyrek, C. Louis: J. Phys. Chem. **83**, 4526, 4531 (1985)
4.99 N.I. Lipatkina, V.A. Shvets, V.B. Kazansky: Kinetika i Kataliz **19**, 979 (1978) (in Russian)
4.100 L.K. Przheval'skaja, V.A. Shvets, V.B. Kazansky: Kinetika i Kataliz **15**, 180 (1974) (in Russian)
4.101 A.J. Tench, T. Lowson: Chem. Phys. Lett. **7**, 459 (1970)
4.102 T. Ito, M. Kato, K. Toi, T. Shirakawa, I. Ikemoto, T. Tokuda: J. Chem. Soc., Faraday Trans. I. **82**, 2835 (1985)
4.103 A.D. McLachlan, M.C.R. Symons, M.G. Townsend: J. Chem. Soc. **?**, 952 (1959)
4.104 N.B. Wang, J.H. Lunsford: J. Chem. Phys. **56**, 2664 (1972)
4.105 N.I. Lipatkina, V.A. Shvets, N.D. Chuvylkina, V.B. Kazansky: Kinetika i Kataliz **21**, 747 (1980) (in Russian)
4.106 B.N. Shelimov, C. Naccache, M. Che: J. Catalysis **37**, 279 (1975)
4.107 W.B. Williamson, J.H. Lunsford, C. Naccache: Chem. Phys. Lett. **9**, 33 (1971)
4.108 R. Baumann, H. Zeller: Helv. Phys. Acta **40**, 363 (1967)
4.109 A.V. Griva, V.V. Nikisha, B.N. Shelimov, G.M. Zhidomirov, V.B. Kazansky: Kinetika i Kataliz **14**, 1246 (1973) (in Russian)
4.110 P. Mark: J. Phys. Chem. Solids **26**, 959 (1965)
4.111 M. Iwamoto, Y. Yoda, M. Egashiro, T. Seiyama: J. Phys. Chem. **80**, 1989 (1976)
4.112 H. Horiguchi, M. Setaka, K.M. Sancier, T. Kwan: In Proc. 4th Int. Congress Catalysis, Vol.1, Moscow, June 1968 (Akademiai Kiado, Budapest 1971) p.81
4.113 A.A. Davydov: Kinetika i Kataliz **20**, 1506 (1979) (in Russian)
4.114 N.I. Lipatkina, L.K. Przheval'skaja, V.A. Shvets, V.B. Kazansky: Doklady AN SSSR **242**, 1114 (1978) (in Russian)
4.115 V.I. Vladimirova, Ju.N. Rufov, O.V. Krylov: Kinetika i Kataliz **18**, 809 (1977) (in Russian)
4.116 J.P. Guillory, C.P. Shiblom: J. Catalysis **54**, 24 (1978)

4.117 K.N. Spiridonov, G.B. Parijskij, O.V. Krylov: Kinetika i Kataliz **12**, 1448 (1971)

4.118 K.N. Spiridonov, G.N. Parijskij, O.V. Krylov: Izv. Akad. Nauk SSSR, Ser. Khimia **8**, 2161 (1971) (in Russian)

4.119 G.N. Asmolov, V.A. Matyshak, A.A. Kadushin, O.V. Krylov: Kinetika i Kataliz **18**, 1506 (1976) (in Russian)

4.120 E.L. Aptekar, M.G. Chudinov, A.M. Alekseev, O.V. Krylov: Reaction Kinetics and Catalysis Letters **1**, 493 (1979)

4.121 V.A. Khalif, E.L. Aptekar', K.N. Spiridonov, O.V. Krylov: Kinetika i Kataliz **19**, 1238 (1978) (in Russian)

4.122 O.V. Krylov, G.B. Pariiskii, K.N. Spiridonov: J. Catalysis **23**, 301 (1971)

4.123 K.N. Spiridonov, O.V. Krylov, E.A. Fokina: Zh. Fiz. Khim. **47**, 2052 (1975) (in Russian)

4.124 M.Ja. Bykhovskij, A.A. Kadushin, O.V. Krylov, O.S. Morozova: Kinetika i Kataliz **15**, 1246 (1974) (in Russian)

4.125 V.A. Matyshak, A.A. Khalif, A.A. Kadushin, E.A. Aptekar', O.V. Krylov: Kinetika i Kataliz **18**, 715 (1977) (in Russian)

4.126 E.L. Aptekar', M.Ja. Bykhovskij, O.V. Krylov, Ju.A. Lebedev: Kinetika i Kataliz **15**, 1568 (1974) (in Russian)

4.127 V.A. Khalif, B.V. Rozentuller, A.M. Frolov, E.L. Aptekar', K.N. Spiridonov, O.V. Krylov: Kinetika i Kataliz **19**, 1231 (1968) (in Russian)

4.128 T.Z. Tabasaranskaja, A.A. Kadushin, K.N. Spiridonov, O.V. Krylov: Kinetika i Kataliz **15**, 1540 (1974) (in Russian)

4.129 E.L. Aptekar', O.V. Krylov: Kinetika i Kataliz **14**, 78 (1973) (in Russian)

4.130 R.B. Akhverdiev, A.P. Mamedov, F.B. Aliev, G.N. Asmolov, O.V. Krylov: Kinetika i Kataliz **21**, 999 (1980) (in Russian)

4.131 J.H. Lunsford, J.P. Jayne: J. Phys. Chem. **70**, 3484 (1966)

4.132 V.M. Vorotyntsev, V.A. Shvets, V.B. Kazansky: Kinetika i Kataliz **10**, 365 (1969); **15**, 1246 (1974) (in Russian)

4.133 S. Yoshida, K. Tarama, T. Suzuki, Y. Ishida: In Proc. 5th Soviet-Japan Seminar on Catalysis (FAN, Tashkent 1979) p.103

4.134 M. Che, A.J. Tench: Advances in Catalysis **31**, 272 (1982); **32**, 2 (1983)

4.135 N. Kotsev, D. Shopov: J: Catalysis **22**, 297 (1971)

4.136 K.I. Slovetskaja, A.M. Rubinstein: Kinetika i Kataliz **10**, 158 (1969) (in Russian)

4.137 R.J. Kokes, A.L. Dent: Adv. Catalysis **22**, 1 (1972)

4.138 J.L. Carter, D.J.C. Yates, P.J. Lucchesi, J.J. Elliott, V. Kevorkian: J. Phys. Chem. **70**, 1126 (1966)

4.139 A.L. Dent, R.J. Kokes: J. Phys. Chem. **73**, 3773 (1969)

4.140 G.C. Bond: Disc. Faraday Soc. **41**, 200 (1966)

4.141 A.A. Efremov, Y.A. Lokhov, A.A. Davydov: React. Kinet. Catal. Lett. **14**, 21 (1980)

4.142 V.A. Seleznev, A.A. Kadushin: Problemy Kinetiki i Kataliza **16**, 177 (1975) (in Russian)

4.143 R.J. Kokes: In Proc. 5th Int'l Congress Catalysis, Vol.1, Miami Beach 1972 (North-Holland, Amsterdam 1973) p.A1

4.144 A.L. Dent, R.J. Kokes: J. Am. Chem. Soc. 92, 1092, 6709 (1970)

4.145 A.L. Dent, R.J. Kokes: J. Phys. Chem. 75, 487 (1971)

4.146 T.Z. Tabasaranskaja, A.A. Kadushin: Kinetika i Kataliz 12, 1507 (1971) (in Russian)

4.147 C. Busca, T. Zerlia, V. Lorenzelli, A. Giralli: J. Catalysis 88, 125, 131 (1984)

4.148 F. Trifiro: Chimica e Industria 56, 541 (1974)

4.149 B.A. Dolgoplosk, K.L. Makovetskij, E.I. Tinjakova, O.K. Sharlev: *Polimerizatsija dienov pod vlijaniem π-allil'nykh kompleksov* (Dien Polymerization Influenced by π-Allyl Complexes) (Nauka, Moscow 1968) (in Russian)

4.150 K. Vrieze, C. Maclean, P. Cossee, C.W. Hilbers: Rec. Trav. Chim. 85, 1077 (1966)

4.151 M. Herberhold: *Metal π-Complexes* (Elsevier, New York 1972) Vol.2

4.152 K.C. Ramey, B.C. Lini, W.B. Wise: J. Am. Chem. Soc. 90, 4275 (1968)

4.153 H.C. Volger, K. Vrieze: J. Organometal. Chem. 6, 297 (1960)

4.154 M.L. Clarke: J. Organometal. Chem. 80, 369 (1974)

4.155 M.E. Djatkina, E.L. Rosenberg: *Kvantovo-khimicheskie raschety soedinenij perekhodnykh elementov* (VINITI, Moscow 1974) (in Russian)

4.156 C.K. Adams, T.J. Jennings: J. Catalysis 2, 63 (1963)

4.157 W.M.H. Sachtler: Rec. Trav. Chim. 82, 243 (1965)

4.158 H.H. Voge, C.D. Wagner, D.P. Stevenson: J. Catalysis 2, 58 (1963)

4.159 T.Z. Tabasaranskaja, A.A. Kadushin: Kinetika i Kataliz 13, 167 (1972)

4.160 S.J. Ashcroft, C.T. Mortimer: J. Chem. Soc. ??, 781 (1971)

4.161 M. Krivanek, P. Jiru: J. Catalysis 27, 461 (1972)

4.162 N.K. Kotsev, A.A. Kadushin, D.M. Shopov, O.V. Krylov: React. Kinet. Catal. Lett. 2, 211 (1975)

4.163 S.V. Gerey, E.V. Rozhkova, Ya.B. Gorokhovatsky: J. Catalysis 28, 341 (1973)

4.164 V.G. Mikhal'chenko, V.D. Sokolovskij, A.A. Filippova, A.A. Davydov: Kinetika i Kataliz 14, 1253 (1973) (in Russian)

4.165 A.A. Rudneva, A.A. Davydov, V.G. Mikhal'chenko: Kinetika i Kataliz 16, 480, 486 (1975) (in Russian)

4.166 A.A. Davydov, A.A. Efremov: Kinetika i Kataliz 20, 1242 (1979) (in Russian)

4.167 V.G. Mikhalchenko, A.A. Davydov, A.A. Budneva, B.N. Kuznetsov, V.D. Sokolovskii: React. Kinet. Catal. Lett. 2, 163 (1975)

4.168 T. Kondo, S. Saito, K. Tamaru: In Proc. 2nd Japan–Soviet Catalysis Seminar, Tokyo 1973, p.26–35

4.169 T. Seyama, M. Egashiro, T. Sakamoto, I. Asa: J. Catalysis 24, 76 (1972)

4.170 O.V. Krylov, L.Ya. Margolis: Intern. Revs. Phys. Chem. 3, 305 (1983)

4.171 D. Cordischi, V. Indovina: In *Adsorption and Catalysis on Oxide Surfaces*, ed. by M. Che, G.C. Bond (Elsevier, Amsterdam 1985) p.193

4.172 T.A. Gordymova, A.A. Davydov: Kinetika i Kataliz **20**, 733 (1978) (in Russian)

4.173 M.P. Leftin: In *Carbonium Ions* (Interscience, New York 1968) Vol.1, pp.353-412

4.174 G.N. Asmolov, O.V. Krylov: Problemy Kinetiki i Kataliza **16**, 78 (1975) (in Russian)

4.175 A.G. Whitney, I.D. Gay: J. Catalysis **28**, 176 (1972)

4.176 V.B. Kazansky, V.Ju. Borovkov: Kinetika i Kataliz **15**, 1283 (1974) (in Russian)

4.177 Y. Moro-oka: Trans. Faraday Soc. **67**, 3381 (1971)

4.178 I.C. Higatsune: J. Catalysis **74**, 18 (1982)

4.179 C.C. Chang, W.C. Connor, R.J. Kokes: J. Phys. Chem. **77**, 1957 (1973)

4.180 Ja.B. Gorokhovatskij: Kinetika i Kataliz **14**, 83 (1973) (in Russian)

4.181 A.A. Davydov: *IK-spectroscopija v knimii poverkhnosti okislov* (IR - Spectroscopy in Chemistry of Oxide Surfaces) (Nauka, Novosibirsk 1984)

4.182 R.P. Eischens, W.A. Pliskin: Adv. Catalysis **9**, 662 (1957); **10**, 1 (1958)

4.183 G. Blyholder: J. Phys. Chem. **68**, 2772 (1964)

4.184 S.Z. Roginskij, V.A. Seleznev, A.A. Kadushin: Zh. Fiz. Khim. **43**, 1075 (1969); Doklady AN SSSR **194**, 864 (1970) (in Russian)

4.185 R.I. Soltanova, E.A. Paukshtis, E.N. Jurchenko: Kinetika i kataliz **23**, 164 (1982) (in Russian)

4.186 N.M. Neshev, A.A. Andreev, D.M. Shopov: Dokl. Bulg. Akad. Nauk **27**, 519 (1974) (in Russian)

4.187 L.M. Roev, I.G. Voroshilov: Doklady AN Ukr. SSR **9**, 830 (1973) (in Russian)

4.188 P. Politzer, S.D. Kaster: Surface Sci. **36**, 186 (1973)

4.189 R.A. Gardner: In Proc. 4th Int. Congress Catalysis, Vol.2, Moscow, June 1968 (Akademiai Kiado, Budapest 1971) p.466

4.190 J.A. Rabo, C.L. Angeli, P.H. Kasai, V.Shomaker: Disc. Faraday Soc. **41**, 328 (1966)

4.191 J.W. London, A.T. Bell: J. Catalysis **31**, 32 (1973)

4.192 W. Hertl: J. Catalysis **31**, 232 (1973)

4.193 H.G. Tompkins, R.G. Greenler: Surface Sci. **28**, 194 (1971)

4.194 D.V. Pozdnjakov, V.N. Filippov: Zh. Fiz. Khim. **46**, 1011 (1972) (in Russian)

4.195 J.B. Peri: J. Phys. Chem. **78**, 588 (1974)

4.196 I.G. Voroshilov, L.M. Roev, G.M. Kozub, M.T. Rusov, N.K. Lunev: Kinetika i kataliz **16**, 1267 (1975) (in Russian)

4.197 Ju.A. Lokhov, A.A. Davydov: Kinetika i kataliz **21**, 1523 (1980) (in Russian)

4.198 Y.Y. Huang: J. Catalysis **30**, 187 (1973)

4.199 T.A. Bregadze, V.A. Seleznev, A.A. Kadushin, O.V. Krylov: Izv. AN SSSR, Ser. Khim. **12**, 2701 (1973) (in Russian)

4.200 N.N. Bobrov, A.A. Davydov, N.G. Maksimov, K.G. Ione: Izvestija AN SSSR, Ser. Khim. 4, 845 (1975); Kinetika i kataliz 16, 1272 (1975) (in Russian)

4.201 V.A. Matyshak, A.A. Kadushin, O.V. Krylov: Kinetika i kataliz 22, 461 (1981) (in Russian)

4.203 A.I. Mashchenko, M.Ja. Kon', V.A. Shvets, V.B. Kazansky: Zh. Teor. Eksp. Khim. 8, 801 (1972) (in Russian)

4.204 P. Meriaudeau, M. Breusse, R. Claudel: J. Catalysis 15, 484 (1974)

4.205 P. Pichat, G. Bray: J. Chem. Phys. 66, 724 (1969)

4.206 V.I. Jakerson, L.I. Lafer, A.M. Rubinstein: Problemy kinetiki i kataliza 16, 49 (1973) (in Russian)

4.207 V.A. Matyshak, A.A. Kadushin, O.V. Krylov, D.R. Mekhandzhiev: Izvestija AN SSSR, Ser. Khim. 10, 2677 (1977) (in Russian)

4.208 P.G. Gravelle, S.J. Teichner: Adv. Catalysis 20, 167 (1969)

4.209 T.A. Bregadze, M.Ja. Bykhovskij, V.A. Seleznev: Izvestija AN SSSR, Ser. Khim. 2, 442 (1973) (in Russian)

4.210 M. Breysse, M. Guenin, B. Claudel, J. Veron: J. Catalysis 28, 54 (1973)

4.211 R.I. Bickley, F.S. Stone: In *Elektronnye javlenija v adsorbtsi i katalize na poluprovodnikakh*, ed. by F.F. Volkenstein (MIR, Moscow 1969) Chap.11, pp.211–226 (in Russian)

4.212 F. Pepe, F.S. Stone: J. Catalysis 56, 160 (1979)

4.213 V. Indovina, A. Cimino, M. Inversi, F. Pepe: J. Catalysis 58, 396 (1979)

4.214 F.S. Stone, J.C. Vickerman: Proc. Roy. Soc. A354, 331 (1977)

4.215 F.S. Stone: J. Solid State Chem. 12, 271 (1975)

4.216 A. Cimino, M. Schiavello, F.S.Stone: Disc. Faraday Soc. 41, 390 (1966)

4.217 A. Cimino: Chimica e Industria 56, 27 (1974)

4.218 A. Cimino, V. Indovina: J. Catalysis 17, 54 (1970)

4.219 M. Schiavello, A. Cimino, J.M. Criado: Gazz. Chim. Ital. 101, 47 (1971)

4.220 A. Cimino, D. Cordischi, G. Guarino, A. Micheli: Trans. Faraday Soc. 67, 1776 (1971)

4.221 T.A. Egerton, J.C. Vickerman: J. Catalysis 19, 74 (1970)

4.222 P. Pomonis, J.C. Vickerman: J. Catalysis 55, 88 (1978)

4.223 A. Andreev, E. Proinov, N. Neshev, D. Shopov: J. Catalysis 74, 1 (1982)

4.224 A. Cimino, V. Indovina, F. Pepe, M. Schiavello: In Proc. 4th Int. Congress on Catalysis, Vol.1, Moscow, June 1968 (Akademiai Kiado, Budapest 1971) p.187

4.225 E.G. Derouane, J.P. Pirard, G.A. L'Homme: In *Catalysis Heterogeneous and Homogeneous*, ed. by S. Delmon (Elsevier, Amsterdam 1975) p.275–288

4.226 V. Indovina, A. Cimino, M. Valigi: In Proc. 6th Int. Congress Catalysis, London 1976, Preprint A13

4.227 V. Indovina, A. Cimino, M. Inversi, F. Pepe: J. Catalysis 58, 396 (1979)

4.228 O.V. Krylov: Problemy kinetiki i kataliza **16**, 129 (1975) (in Russian)

4.229 G.K. Boreskov: In Proc. 6th Int. Congress on Catalysis, London 1976, Preprint A13

4.230 K.N. Spiridonov, A.A. Kadushin, O.V. Krylov: Kinetika i kataliz **13**, 542 (1972) (in Russian)

4.231 T.Z. Tabasaranskaja, A.A. Kadushin, K.N. Spiridonov, O.V. Krylov: Kinetika i kataliz **13**, 542 (1972) (in Russian)

4.232 G. Davies, A.G. Sykes: J. Chem. Soc.(A), 2841 (1968)

4.233 Ju.N. Rufov: Problemy Kinetiki i Kataliza **14**, 184 (1970) (in Russian)

4.234 A.M. Gasymov, V.A. Shvets, V.B. Kazansky: Kinetika i kataliz **22**, 153 (1981) (in Russian)

4.235 V.B. Kazansky, V.A. Shvets, M.Ja. Kon, V.V. Nikisha, B.N. Shelimov: In Proc. 5th Int. Congress Catalysis, Vol.2, Miami Beach, July 1972 (Elsevier, Amsterdam 1973) p.1423

4.236 A.A. Kadushin, N.A. Kutyreva, V.A. Matyshak, O.S. Morozova: Kinetika i kataliz **22**, 233 (1981) (in Russian)

4.237 R. Bienert, W. Hanke, U. Illgen, H.G. Jerschkewitz, G. Lischke, G. Ohlman, I.W. Schulz: In Trudy vsesojuznoj konferentsii po mekhanizmu kataliza, Moscow 1974, Preprint 73

4.238 V.B. Aleskovskij: *Stekhiometrija i sintez tverdykh soedinenij* (Stoichiometry and Synthesis of Solid-State Compounds) (Nauka, Leningrad 1976) (in Russian)

4.239 V.F. Kiselev, O.V. Krylov: *Adsorption Processes on Semiconductor and Dielectric Surfaces I*, Springer Ser. Chem. Phys., Vol.32 (Springer Berlin, Heidelberg 1985)

4.240 G.F. Golovanova, A.S. Petrova, E.A. Silaev: Kinetika i kataliz **23**, 1275 (1982) (in Russian)

4.241 O.V. Krylov: Problemy Kinetiki i Kataliza **18**, 5 (1981) (in Russian)

4.242 S.Z. Roginskij, O.V. Al'tshuler, O.M. Vinogradova, I.L. Tsitovskaja: Doklady AN SSSR **126**, 872 (1972) (in Russian)

4.243 O.V. Al'tshuler, O.M. Vinogradova, V.A. Seleznev, I.L. Tsitovskaja: Problemy kinetiki i kataliza **15**, 56 (1973) (in Russian)

4.244 O.V. Al'tshuler, I.L. Tsitovskaja, V.A. Seleznev, O.M. Vinogradova: Izvestija AN SSSR, Ser. Khim. 2145 (1971) (in Russian)

4.245 T.A. Egerton, F.S. Stone: Trans. Faraday Soc. **69**, 22 (1973)

4.246 R.L. Garten, W.H. Delgass, M. Boudart: J. Catalysis **18**, 90 (1970)

4.247 G.K. Boreskov: Kinetika i kataliz **19**, 7 (1973) (in Russian)

4.248 K.G. Ione: *Polifunktsional'nyj kataliz na tseolitakh* (Multifunctional Catalysis on Zeolites) (Nauka, Novosibirsk 1982) (in Russian)

4.249 S. Tsuruya, Y. Okamoto, T.Kuwada: J. Catalysis **56**, 52 (1979)

4.250 P.W. Selwood: *Chemisorption and Magnetization* (Academic, London 1975)

4.251 P.W. Selwood: *Magnetochemistry*, 2nd ed. (Interscience, New York, London 1956)
4.252 P.W. Selwood: Advances in Catalysis **3**, 28 (1951)
4.253 S.M. Arija, N.L. Lukinykh: Fiz. Tverd. Tela **8**, 260 (1966) (in Russian)
4.254 E.L. Nagaev, G.L. Lazarev: Surface Sci. **54**, 101 (1976)
4.255 R.J.H. Voorhoeve, D.W. Johnson, J.P. Nemeira, J. Gallagher: Science **195**, 827 (1977)
4.256 V. Srinivasan, C.S. Swaney, G. Muralidhar, S.L. Raj, R. Pitcher, K.H. Vijayakumar: In Proc. 7th Int. Congress Catalysis, Tokyo, June 1980, Preprint D17
4.257 B.V. Rozentuller, K.N. Spiridonov, O.V. Krylov: Kinetika i kataliz **22**, 797 (1981); Doklady AN SSSR **259**, 895 (1981) (in Russian)

Chapter 5

5.1 N.W. Ashcroft, N.D. Mermin: *Solid State Physics* (Holt, Rinehart & Winston, New York, Chicago, San Francisco 1976)
5.2 D. Shopov, A.Andreev: *Khimicheskaja svjaz'pri adsorbtsii i katalize*, Vol.1 "Metally" (Chemical Bond at aDsorption and Catalysis Metals) (BAN, Sofia 1975) (in Russian)
5.3 I.M. Lifshits, M.Ja. Azbel, M.I. Kaganov: *Elektronnaja teorija metallov* (The Electron Theory of Metals) (Nauka, Moscow 1971) (in Russian)
5.4 S.V. Vonsovskij: *Magnetism* (Nauka, Moscow 1967) (in Russian)
5.5 J.B. Goodenough: *Magnetism and the Chemical Bond* (Interscience, New York, London 1966)
5.6 G.C. Bond: Surface Sci. **18**, 11 (1969)
5.7 W. Trost: Canad. J. Chem. **37**, 460 (1959)
5.8 O.K. Anderson: Phys. Rev. **B2**, 883 (1970)
5.9 T.N. Rhodin, J.W. Gadzuk: In *The Nature of the Surface Chemical Bond*, ed. by T.N. Rhodin, G. Ertl (North Holland, Amsterdam 1979)
5.10 G.K. Wertheim: In *Electron and Ion Spectroscopy of Solids*, ed. by L. Fiermans, J. Vennik, W. Dekeyser (Plenum, New York 1978) p.190–229
5.11 G.C. Bond: Disc. Faraday Soc. **41**, 200 (1966)
5.12 W.H. Weinberg: J. Vac. Sci. TEchn. **10**, 89 (1973)
5.13 C.F. Melius: J. Chem. Phys. Lett. **39**, 287 (1976)
5.14 Z. Knor: Surface Sci. **70**, 286 (1978)
5.15 J.C. Riviere: In *Solid State Surface Science*, ed. by M. Green (Dekker, New York 1969) Vol.1
5.16 G.A. Somorjai: *Chemistry in two Dimensions: Surfaces* (Cornell Univ. Press, Ithaca 1981)
5.17 J. Holzl, F.K. Schulte, H. Wagner: In *Solid Surface Physics* (Springer, Berlin, Heidelberg 1979) p.5
5.18 T.B. Grimley: J. Physique **31**, C1–85, C3–934 (1970)

5.19 T.B. Grimley: In *The Nature of the Surface Chemical Bond,* ed. by T.N. Rhodin, G. Ertl (North Holland, Amsterdam 1979) p.1–50
5.20 A. Van der Avoird: Surface Sci. **18**, 159 (1969)
5.21 J.R. Schrieffer, R. Gomer: Surface Sci. **25**, 315 (1971); J. Vac. Sci. Techn. **9**, 561 (1972)
5.22 C. Allen, M. Lanno, G. Leman: La Vide **28**, 89 (1971)
5.23 I. Terakura, K. Terakura, N. Hamada: Sruface Sci. **103**, 103 (1981)
5.24 G.A. Somorjai, M.A. van Hove: *Adsorbed Monolayers on Solid Surfaces* (Springer, Berlin, Heidelberg 1979)
5.25 K.A.R. Mitchell: Contemp. Phys. **14**, 251 (1973)
5.26 M.A. van Hove: Surface Sci. **80**, 1 (1979)
5.27 C.M. Chen, S.L. Cunningham, K.L. Luke, W.H. Weinberg, S.P. Withrow: Surface Sci. **78**, 15 (1978)
5.28 M. Alft, W. Moritz: Surface Sci. **80**, 24 (1979)
5.29 H.C. Davies, J.R. Noonan, L.H. Jenkins: Surface Sci. **83**, 59 (1979)
5.30 D.L. Adams, H.B. Nielsen, M.A. van Hove, A. Ignatiev: Surface Sci. **104**, 39 (1981)
5.31 J. Kirschner, B. Feder: Surface Sci. **79**, 179 (1979)
5.32 J.A. Davies, T.E. Jackman, D.P. Jackson, P.R. Norton: Surface Sci **109**, 20 (1981)
5.33 M.N. Read, C.J. Russell: Surface Sci. **88**, 95 (1979)
5.34 P. Heilmann, K. Heinz, R. Müller: Surface Sci. **83**, 487 (1979)
5.35 M.A. van Hove, S.Y. Tong: Surface Sci. **4**, 91 (1976)
5.36 F. Jona: Disc. Faraday Soc. **60**, 213 (1975)
5.37 H.D. Shih, F. Jona: Surface Sci. **104**, 39 (1981)
5.38 J.A. Davies, D.P.Jackson, N. Matsunami, P.R. Norton, J.H. Anderson: Surface Sci. **78**, 274 (1978)
5.39 C.M. Chen, P.A. Thiel, W.H. Weinberg: Surface Sci. **76**, 296 (1978)
5.40 J.F. van der Veen, R.G. Smeenk, R.M. Thorp, F.W. Saris: Surface Sci. **79**, 219 (1979)
5.41 J.W. May: Adv. Catalysis **21**, 191 (1970)
5.42 G. Ertl: *Low Energy Electrons and Surface Chemistry* (VCH, Weinheim 1985)
5.43 P.J. Estrup: In *Chemistry and Physics of Solid Surfaces V,* ed. by R. Vanselow, R. Howe, Springer Ser. Chem. Phys., Vol.44 (Springer, Berlin, Heidelberg 1984) p.205
5.44 D.W. Blakely, G. Somorjai: Surf. Sci. **65**, 419 (1977)
5.45 T.N. Rhodin, G. Broden: Surf. Sci. **60**, 466 (1976)
5.46 H.B. Lyon, G.A. Somorjai: J. Chem. Phys. **46**, 2539 (1967)
5.47 H. Niehus: Surface Sci. **I45**, 407 (1984)
5.48 P.R. Norton, J.A. Davies, D.P. Jackson, N. Matsunami: Surface Sci. **85**, 269 (1979)
5.49 M.A. van Hove, R.J. Koestner, P.C. Stair, J.P. Biberian, L.L. Kesmodel, S.J. Bartu, G.A. Somorjai: Surface Sci. **103**, 189, 218 (1981)
5.50 R.J. Behm, W. Hösler, E.B. Ritter, G. Binning: J. Vac. Sci. Technol. **A4**, 1330 (1986)

M. Salmeron, G.A. Somorjai: Surface Sci. **91**, 373 (1980)

5.51 J.F. Wendelken, D.M. Zehrer: Surface Sci. **71**, 178 (1978)

5.52 M.A. van Hove: In *The Nature of the Chemical Bond*, ed. by T.N. Rhodin, G. Ertl (North Holland, Amsterdam 1979) p.275-312

5.53 S.L. Cunningham, W. Ho, W.H. Weinberg, J. Dobrzinsky: Appl. Surface Sci. **1**, 37 (1979)

5.54 R.D. Young: Phys. Rev. **113**, 110 (1959)
G.A. Somorjai: Treat. Solid State Chem. **6A**, 1 (1976)

5.55 V.I. Savchenko: Kinetika i kataliz **13**, 1583 (1972) (in Russian)

5.56 J.C. Shelton, H.R. Patil, J.M. Blakely: Surface Sci. **43**, 493 (1974)

5.57 A.Ja. Tontegode, E.Ja. Zandberg, F.K. Jusifov: Zh. Tekhn. Fiz. **45**, 1320 (1975) (in Russian)

5.58 G. Ehrlich: In *Surface Science. Recent Progress and Perspectives*, ed. by T.S. Jayadeviaiah, P. Vanselow (CRC, Cleveland 1974)

5.59 W.R. Graham, G. Ehrlich: Phys. Rev. Lett. **31**, 1407 (1973)

5.60 C.L. Kellogg, T.T. Tsong, P. Cowan: Surface Sci. **70**, 485 (1978)

5.61 H. Jaeger, J.V. Sanders: J. Res. Inst. CAtal. Hokkaido Univ. **16**, 287 (1969)

5.62 D.E. Eastman, J.K. Cashion: Phys. Rev. Lett. **27**, 1520 (1971)

5.63 C. Defosse: In *Characterization of Heterogeneous Catalysts*, Vol.15, ed. I. Delanny (Dekker, New York 1984) p.225-298

5.64 R. Haight, J. Baker, R.R. Freeman, P.H. Bucksbaum: J. Vac. Sci. Technol. **A4**, 1481 (1986)

5.65 E.W. Plummer: In *Interactions on Metal Surfaces*, ed. R. Gomer, Toics Appl. Phys., Vol.4 (Springer, Berlin, Heidelberg 1975) p.143

5.66 K.N. Johnson, R.P. Messmer: J. Vac. Sci. TEchn. **11**, 236 (1974)

5.67 B. Feuerbacher, B. Futton: Phys. Rev. Lett. **29** 186 (1972)

5.68 A.M. Bradshaw, D. Menzel, M. Steinkilberg: Jpn. J. Appl. Phys. Suppl.2, 841 (1974)

5.69 P.C. Stephenson, D.W. Bullett: Surface Sci. **139**, I (1984)

5.70 G.W. Rubloff, J. Anderson, M.A. Passler: Phys. Rev. Lett. **32**, 67 (1974)

5.71 T.A. Clarke, I.D. Gray, R. Mason: Chem. Phys. Lett. **27**, 172 (1974)

5.72 P.R. Norton: Surface Sci. **47**, 98 (1974)

5.73 N.J. Dionne: Field Emission Electron Spectroscopy of the Platinum Group Metals, Ph.D. Thesis, Research group of Prof. T. Rhodin, Cornell Univ. Ithaca (1975)

5.74 O.K. Anderson: Phys. Rev. **82**, 883 (1970)

5.75 L. Pauling: Proc. Roy. Soc. **A196**, 343 (1949)

5.76 D.A. Dowden, P.W. Reynolds: Disc. Faraday Soc. **8**, 184 (1950)

5.77 G.A. Somorjai, S.H. Overbury: Faraday Disc. **6**, 279 (1975)

5.78 W.M.H. Sachtler: Appl. Surf. Sci. **19**, 167 (1984)
5.79 M. Kelley: J. Catalysis **57**, 125 (1975)
5.80 D.A. Dowden: In Proc. 5th Int. Congress Catalysis, Vol.1, Miami Beach 1972 (North Holland, Amsterdam 1973) p.621
5.81 W.M.H. Sachtler, P. van der Planck: Surf. Sci. **12**, 35 (1968); **18**, 62 (1969)
5.82 W.M.H. Sachtler, C.J.H. Dorgelo: J. Catalysis **4**, 654 (1965)
5.83 A.A. Slinkin: *Struktura i kataliticheskie svojstva getero-gennykh katalizatorov* (Structure and Catalytic Properties of Heterogeneous CAtalysts) (VINITI, Moscow 1971) (in Russian)
5.84 P.R. Weller, C.E. Rojas, P.J. Dobson, P. Chadwick: Surface Sci. **105**, 20 (1981)
5.85 A. Jablonski, S.H. Overbury, G. Somorjai: Surf. Sci. **65**, 578 (1977)
5.86 D.T. Ling, J.N. Miller, J. Lindon, W.E. Spicer, P.M. Stefan: Surface Sci. **74**, 612 (1978)
5.87 G. Maire, L. Hilaire, P. Legare, F.G. Gault, A. O'Cinneide: J. Catalysis **44**, 293 (1976)
5.88 H.H. Brongersma, M.J. Spaarnay, T.M. Buck: Surf. Sci. **71**, 657 (1978)
5.89 R. Bouwmann, W.M.H. Sachtler: J. Catalysis **10**, 127 (1970); **28**, 63 (1972)
5.90 J.R. Chalikowsky: Surface Sci. **119**, L197 (1984)
5.91 G.P. Schwarz: Surface Sci. **76**, 113 (1978)
5.92 F.J. Kujiers, V. Ponec: J. Catalysis **60**, 100 (1979)
5.93 J.K. Anderson, K. Foger, R.J. Breakspere: J. Catalysis **57**, 485 (1979)
      L.L. Kesmodel, G. Somorjai: Acc. Chem. Res. **9**, 392 (1976)
5.94 D. Gupta, D.R. Campbell, P.S. Ho: In *Thin Films. Interdiffusion and Reactions*, ed. by J.M. Poat (Wiley, New York 1978) p.161
5.95 G.S. Krinchik, L.V. Nikitin: Fiz. Tverd. Tela **20**, 2545 (1978) (in Russian)
5.96 G.S. Krinchik, L.V. Nikitin, V.V. Lukin, P.A. Chernavskij: Fiz. Tverd. Tela **21**, 599 (1979) (in Russian)
5.97 J.W.E. Coenen, R.Z.C. van Meerten, H.T. Rijnten: In Proc. 5th Int. Congress Catalysis, Vol.1, Miami Beach 1972 (North Holland, Amsterdam 1973) p.671-678
5.98 S. Ladas, H. Poppa, M. Boudart: Surface Sci. **102**, 151 (1981)
5.99 J.B. Andersen, J. Shimoyama: In Proc. 5th Int. Congress Catalysis, Vol.1, Miami Beach 1972 (North Holland, Amsterdam 1973) p.695-701
5.100 R. van Hardeveld, F. Hartog: Surface Sci. **15**, 189 (1969); Adv. Catalysis **22**, 75 (1972)
5.101 O.M. Poltorak, V.S. Boronin, A.N. Mitrofanova: In Proc. 4th Int. Congress Catalysis, Vol.2, Moscow, June 1968 (Akademiai Kiado, Budapest 1971) p.276
5.102 G.C. Bond: Ibid. p.266
5.103 E.G. Schlosser: Ibid. p.312
5.104 G.C. Bond: Platinum Met. Rev. **19**, 126 (1975)

5.105 M.B. Gordon, F. Gyrot-Lackmann, M.C. Desjouqueres: Surface Sci. **80**, 159 (1979)
5.106 A. Renou, M. Gillet: Surface Sci. **106**, 27 (1981)
5.107 M.J. Yacaman, J.M. Dominguez: J. Catalysis **64**, 213 (1980)
5.108 M.J. Yacaman: In *Chemistry and Physics of Solid Surfaces V*, ed. by R. Vanselow, R. Howe, Springer Ser. Chem. Phys., Vol.44 (Springer, Berlin, Heidelberg 1984) p.183
5.109 N.M. Zajdman: Kinetika i kataliz **13**, 1012 (1972) (in Russian)
5.110 D. Briggs: J. Electron Spectr. **9**, 487 (1976)
5.111 P.W. Selwood: *Chemisorption and Magnetization* (Academic, New York 1975)
5.112 A.A. Slinkin, E.A. Fedorovskaja: Uspekhi Khim. **40**, 1057 (1971) (in Russian)
5.113 T.E. Whyte: Catal. Rev. **8**, 117 (1973)
5.114 H. Spindler: Z. Chemie **13**, 1 (1973)
5.115 J.L. Carter, J.H. Sinfelt: J. Catalysis **10**, 134 (1968)
5.116 M.C. Hobson: Surface Membrane Sci. **5**, 1 (1972)
5.117 B.S. Clausen, S. Morup, H. Topsoe: Surface Sci. **106**, 35 (1981)
5.118 B.S.S. Clausen, S. Morup, H. Topsoe: Surface Sci. **106**, 438 (1981)
     J.A. Dumesik, H. Topsoe, M. Boudart: J. Catalysis **37**, 513 (1975)
5.119 Ju.F. Krupjanskij, I.P. Suzdalev: Zh. Eksp. Teor. Fiz. **65**, 1715 (1973): **67**, 736 (1974) (in Russian)
5.120 L. Figueras, B. Mensier, L. de Mourgues, C. Naccache, Y. Trambouze: J. Catalysis **19**, 315 (1970)
5.121 S.V. Markevich, L.S. Kravchuk, A.A. Kupcha: Doklady AN Bel. SSR **23**, 833 (1977) (in Russian)
5.122 A.V. Skljarov, O.V. Krylov, G. Keulks: Kinetika i Kataliz **18**, 1487 (1977) (in Russian)
5.123 J.L. Ogilvie, A. Wolberg: J. Appl. Spectr. **26**, 401 (1972)
5.124 T. Edmonds: In *Characterization of Industrial Catalysts*, ed. by M. Thomas (Wiley, New York 1980) p.50
5.125 G.J. den Ofter, F.M. Dautzenberg: J. Catalysis **53**, 116 (1978)
5.126 P. Meriadeau, O.H. Ellestad, M. Dufaux, C. Naccache: J. Catalysis **75**, 243 (1982)
5.127 S.J. Tauster, S.C. Fung, R.L. Garter: J. Am. Chem. Soc. **100**, 170 (1978)
5.128 M.G. Mason, R.C. Baetzold: J. Chem. Phys. **64**, 271 (1976)
5.129 J.K. Anderson: In Proc. 5th Int. Congress Catalysis, Vol.1, Miami Beach 1972 (North Holland Publ. Amsterdam 1973) p.766
5.130 A.A. Slinkin, A.M. Rubinshtein: Problemy Kinetiki i Kataliza **16**, 166 (1975) (in Russian)
5.131 R.A. Dalla Betta, M. Boudart: In Proc. 5th Int. Congress Catalysis, Vol.2, Miami Beach 1972 (North Holland, Amsterdam 1973) p.1329
5.132 C. Naccache, M. Prime, M.V. Mathieu: Adv. Chem. Sci **121**, 66 (1973)

5.133 J.H. Lunsford, D.S. Treybig: J. Catalysis **68**, 192 (1981)
5.134 P.H. Lewis: J. Catalysis **69**, 511 (1981)
5.135 M.D. Baker, G.A. Ozin, J. Godber: J. Phys. Chem. **89**, 305, 2299 (1985)
5.136 P. Gallezot: Surface Sci. **106**, 459 (1981)
5.137 R. Moraweck, G. Clugnet, A.J. Renouprez: Surface Sci. **81**, L631 (1979)
5.138 J.R. Anderson, R.W. Howe: Nature **268**, 129 (1977)
5.139 P. Galin, Y. Ben Taarit, C. Naccache: J. Catalysis **59**, 357 (1979)
5.140 D.J.C. Yates, L.L. Murrell, E.B. Partridge: J. Catalysis **57**, 47 (1979)
5.141 E.B. Partridge, D.J.C. Yates: Nature **234**, 345 (1971)
5.142 J.H. Sinfelt: In *Surface Science. Recent Progress and Perspective*, ed. by T.S. Jaiadevaiah, R. Vanselow (CRC, Cleveland 1974)
5.143 P.D. Gonzales: Appl. Surface Sci. **19**, 181 (1984)
5.144 R.C. Baetzold: J. Catalysis **29**, 129 (1973)
5.145 P.N. Ross, K. Kinoshita, P. Stonehart: J. Catalysis **32**, 163 (1974)
5.146 R.P. Messmer, K.H. Johnson: In *Electrocatalysis on Non-Metallic Surfaces*, Publ.#55, Nat. Bureau of Standards (Washington 1976) p.67–86
5.147 F. Freund: Japan. J. Appl. Phys., Suppl.2, 847 (1974)
5.148 R.P. Messmer: Surface Sci. **106**, 225 (1981)
5.149 Many reviews in Surface Sci. **156**, I (1985)
5.150 Yu.I. Petrov: *Clusters and Small Particles* (Nauka, Moscow 1986) p.360 (in Russian)

Chapter 6

6.1   G.C. Bond: Disc. Faraday Soc. **41**, 200 (1966)
6.2   S.Z. Roginskij: Zh. Fiz. Khim **6**, 334 (1935) (in Russian)
6.3   O. Beeck: Disc. Faraday Soc. **8**, 118 (1950)
6.4   D.A. Dowden, P.W. Reynolds: Disc. Faraday Soc. **8**, 184 (1950)
6.5   J.J. Rooney: J. Catalysis **3**, 486 (1964)
6.6   D. Shopov, A. Andreev: *Khimicheskaja svjaz'pri adsorbtsii i katalize*, Vol.1 "Metally" (Chemical Bond at Adsorption and Catalysis, Metals) (BAN, Sofia 1975) (in Russian)
6.7   M.McD. Baker, G.I. Jenkins: Adv. Catalysis **7**, 1 (1955)
6.8   G.C. Bond: Surface Sci. **18**, 11 (1969)
6.9   G.C. Bond: Platinum Met. Rev. **19**, 126 (1975)
6.10 H.D. Hagstrum, J.E. Rowe, J.C. Tracy: In *Experimental Methods in Catalytic Research*, Vol.3, ed. by R.B. Anderson (Academic, New York 1976) p.42
6.11 J.W. May: Adv. Catalysis **21**, 191 (1970)
6.12 G.A. Somorjai: *Chemistry in Two Dimensions: Surfaces* (Cornell University, Ithaca 1981)
6.13 G.A. Somorjai, M.A. van Hove: *Adsorbed Monolayers on Solid Surfaces* (Springer, Berlin, Heidelberg 1979)

6.14 V.F. Kiselev, O.V. Krylov: *Adsorption Processes on Semiconductor and Dielectric Surfaces I*, Springer Ser. Chem. Phys., Vol.32 (Springer, Berlin, Heidelberg 1985)

6.15 A.D. Berman, O.V. Krylov: Problemy Kinetiki i Kataliza 17, 102 (1978) (in Russian)

6.16 T.N. Rhodin, P.W. Palmberg, E.W. Plummer: In *Structure and Chemistry of Solid Surfaces*, ed. by G. Somorjai (Wiley, New York 1969) p.22-1-22-28

6.17 T. Toya: J. Vac. Sci. Techn. 9, 890 (1972)

6.18 G. Ertl: In *Electron and Ion Spectroscopy of Solids*, ed. by L. Fiermans, J. Vennik, W. Dekkeyser (Plenum, New York 1978) pp.144-189

6.19 T.B. Grimley: In *The Nature of the Surface Chemical Bond*, ed. by T.N. Rhodin, G. Ertl (North-Holland, Amsterdam 1979) pp.1-50

6.20 D.A. King: Surface Sci. 47, 384 (1975)

6.21 S.Z. Roginskij: *Adsorbtsija i kataliz na neodnorodnykh poverkhnostjakh* (Adsorption and Catalysis on Nonuniform Surfaces) (AN SSSR, Moscow, Leningrad 1948) (in Russian)

6.22 Ju.K. Ustinov: *Fazovye perekhody v submonoslojnykh plenkakh na poverkhnosti perekhodnykh metallov* (Phase Transitions in Submonolayer Films on Transition Metal Surfaces) (LIJaF, Leningrad 1979) (in Russian)

6.23 D. Menzel: Surface Sci. 47, 370 (1975)

6.24 T.E. Madey: J. Vac. Sci. Technol. A4, 257 (1986)

6.25 J.J. Czyzewski: Acta Univ. Wratislav. 29, 7 (1977)

6.26 T.E. Madey: In *Inelastic Particle-Surface Collisions*, ed. by E. Taglauer, W. Heiland, Springer Ser. Chem. Phys., Vol.17 (Springer, Berlin, Heidelberg 1981) pp.80-103

6.27 T.E. Madey, J.T. Yates: Surface Sci. 76, 397 (1978)

6.28 J.J. Czyzewski, T.E. Madey, J.T. Yates: Phys. Rev. Lett. 32, 777 (1974)

6.29 R. Haight, J. Baker, R.R. Freeman, P.H. Bucksbaum: J. Vac. Sci. Technol. A4, 1481 (1986)

6.30 R.P. Messmer, H.J. Freund: Surface Sci. 158, 58 (1985)

6.31 R. Haight, J. Baker, R.R. Freeman, P.H. Bucksbaum: J. Vac. Sci. Technol. A$, 1481 (1986)

6.32 J.C. Fuggle, M. Steinkelberg, D. Menzel: Phys. Rev. Lett. 51, 163 (1975); J. Chem. Phys. 11, 307 (1975)

6.33 R. Suhrmann, Y. Mizushima, A. Hermann: Z. Phys. Chem. (BRD) 20, 332 (1959)

6.34 J.C. P. Mignolet: *Chemisorption* (Butterworth, London, Washington 1957)

6.35 I. Muller: Izvestija AN SSSR, Ser. Fiz. 38, 345 (1974) (in Russian)

6.36 I.I. Tretjakov: Problemy kinetiki i kataliza 10, 164 (1960) (in Russian)

6.37 W.J. Power, G.A. Ozin: Adv. Inorg. Chem. Radiochem. 23, 80 (1980)

6.38 J.E. Halse, M. Moscovits: Surface Sci. 57, 125 (1976)

6.39 R. Gomer: Solid State Phys. 90, 93 (1975)

6.40 R.P. Messmer: In *The Nature of the Surface Chemical Bond,* ed. by T.N. Rhodin, G. Ertl (North Holland, Amsterdam 1979) p.51–111

6.41 J.R. Schrieffer, R. Gomer: Surface Sci. **25**, 315 (1971); J. Vac. Sci. Techn. **9**, 561 (1972)

6.42 K.F. Wojcechowski: Le Vide **167**, 197 (1974)

6.43 D.A. Dowden: J. Chem. Soc. **??**, 242 (1950)

6.44 D.A. Dowden: Anal. Real. Soc. Espanol Fis. y Quim. **B61**, 177 (1965)

6.45 D. Shopov, A. Andreev, D. Petkov: J. Catalysis **13**, 123 (1969)

6.46 D.J.M. Fassert, H. Veerbeck, A. van der Avoird: Surface Sci. **29**, 501 (1971)

6.47 C.R. Abeledo, P.W. Selwood: J. Catalysis **8**, 375 (1967)

6.48 I.I. Zakharov, V.D. Sutula: Kinetika i kataliz **10**, 631 (1969) (in Russian)

6.49 P.W. Anderson. Phys. Rev. **124**, 41 (1961)

6.50 C.F. Melius: J. Chem. Phys. Lett. **39**, 287 (1976)

6.51 T.B. Grimley: J. Am. Chem. Soc. **90**, 3016 (1968); J. Vac. Sci. Techn. **178**, 1123 (1969)

6.52 D.M. Newns: Phys. Rev. **178**, 1123 (1969)

6.53 P. Nordlender, S. Holloway, J.K. Norskov: Surface Sci. **136**, 59 (1984)

6.54 E. Anda, N. Majlis, D. Grempe: Solid State Phys. **10**, 2365 (1977)

6.55 A. van der Avoird: Surface Sci. **18**, 159 (1969)

6.56 J.W. Gadzuk: Phys. Rev. **131**, 2110 (1970)

6.57 J.W. Gadzuk, E.W. Plummer: Rev. Mod. Phys. **45**, 753 (1973)

6.58 J.W. Gadzuk, J.K. Hartman, T.N. Rhodin: Phys. Rev. **B4**, 241 (1971)

6.59 V.L. Bonch-Bruevich, V.B. Glasko: Problemy kinetiki i kataliza **10**, 141 (1960) (in Russian)

6.60 J.C. Slater, K.H. Johnson: Physics Today **27**, 1034 (1975)

6.61 C.F. Melius, J.W. Moskowitz, A.P. Mortola, M.B. Baillie, M.A. Ratner: Surface Sci. **59**, 279 (1976)

6.62 Z. Knor: Surface Sci. **70**, 286 (1978)

6.63 D.E. Eastman, J.K. Cashion: Phys. Rev. Lett. **27**, 1520 (1971)

6.64 T.N. Rhodin, J.W. Gadzuk: In *The Nature of the Surface Chemical Bond,* ed. by T.N. Rhodin, G. Ertl (North Holland, Amsterdam 1979)

6.65 D.P. Woodruff, T.A. Delchar: *Modern Techniques of Surface Science* (niversity Press, Cambridge 1986) p.453

6.66 C. Defosse: In *Characterization of Heterogeneous Catalysts,* Chem. Industr., Vol.15, ed. by I. Delanney (Dekker, New York 1984) p.225–298

6.67 T.H. Barr: In *Practical Surface Analysis by Auger and X-ray Protoelectron Spectroscopy,* ed. by D. Briggs, M.P. Seah (Wiley, New York 1983) p.283–358

6.68 J. Stohr: In *Emission and Scattering Techniques,* Proc. NATO Adv. Study Inst. Alghero, Sept.14–25, 1980 (Dordrecht 1981) p.213

6.69 W.E. Spicer: In *Electron and Ion Spectroscopy of Solids*, ed. by L. Fiermans, J. Vernik, W. Dekeyser (Plenum, New York 1978) pp.54-92

6.70 R.W. Joyner: Surface Sci. **63**, 291 (1977)

6.71 J. Küppers, H. Konrad, G. Ertl: Japan J. Appl. Phys. Suppl. 2/2, 225 (1974)

6.72 G. Ertl, J. Küppers: *Low Energy Electrons and Surface Chemistry* (VCH, Weinheim 1985)

6.73 K.E. Lu, R.R. Rye: Surface Sci. **48**, 677 (1974)

6.74 C.R. Helms, P. Bonzel, S. Kelemen: J. Chem. Phys. **65**, 1773 (1976)

6.75 E.G. Seebaner, A.C.F. Kong, L.D. Schmidt: Surf. Sci. **176**, 134 (1986)
      T.N. Rhodin, G. Broden: Surf. Sci. **60**, 466 (1976)

6.76 B.E. Niewenhaus: Surf. Sci. **59**, 430 (1976)

6.77 K. Christman, G. Ertl, T. Pignet: Surface Sci. **54**, 365 (1976); **60**, 365 (1976)

6.78 J. Marien: Bull. Soc. Roy. Sci. Liege **45**, 103 (1976)

6.79 H. Conrad, G. Ertl, J. Kuppers, E.E. Latta: Surface Sci. **58**, 578 (1976)

6.80 K.Y. Yu, W.E. Spicer, J. Lindau, P. Pianetta, S.F. Lin: Surface Sci. **57**, 157 (1976)

6.81 J.E. Demuth: Surface Sci. **65**, 369 (1977)

6.82 D.C. Castner, B.A. Sexton, G.A. Somorjai: Surface Sci. **71**, 519 (1978)

6.83 A.M. Baro, H. Ibach, H.D. Bruchmann: Surface Sci. **88**, 384 (1979)

6.84 B. Poelsma, G. Mechterscheimer, G. Comsa: Surface Sci. **114**, 519 (1981)

6.85 A.M. Baro, H. Ibach: Surface Sci. **92**, 237 (1980)

6.86 V.I. Savchenko, G.K. Boreskov: Kinetika i kataliz **9**, 142 (1968) (in Russian)

6.87 V. Penka, K. Christman, G. Ertl: Surface Sci. **136**, 307 (1984)

6.88 R. Wortman, R. Gomer, R. Lundy: J. Chem. Phys. **27**, 1099 (1957)

6.89 R.F. Willis, W. Ho: Surface Sci. **80**, 593 (1979)

6.90 A.H. Smith, R.A. Barker, P.J. Estrup: Surface Sci. **136**, 327 (1984)

6.91 G.C. Wang, J. Unguris, P.J. Pierre, R.J. Calotta: Surface Sci. **114**, L35 (1982)

6.92 M.W. Holmes, D.A. King: Surface Sci. **110**, 120 (1981)

6.93 N.J. Dionne: Field Emission Electron Spectroscopy of the Platinum Group Metals, Ph. D. Thesis, Research group of Prof. T. Rhodin (Cornell Univ., Ithaka 1975)

6.94 D.J.M. Fassaert, A. van der Avoird: Surface Sci. **55**, 291 (1976)

6.95 C.F. Melius, J.W. Moskowitz, A.P. Mortola, M.B. Bailie: Surface Sci. **59**, 279 (1976)

6.96 J.E. Demuth, D.E. Eastman: J. Vac. Sci. Techn. **13**, 283 (1976)

6.97 T.A. Delchar, F.C. Tomkins. Proc. Roy. Soc. **A300**, 141 (1967)

6.98 P.R. Norton: Surface Sci. **47**, 98 (1974)

6.99 C.N. Derry, P.N. Ross: J. Chem. Phys. **82**, 2772 (1985)

6.100 G.K. Hall, C.H.B. Mee: Surface Sci. **28**, 598 (1971)

6.101 D.A. Gorodetskij, Ju.P. Mel'nik: Izvestija AN SSSR, Ser. Fiz. **35**, 1064 (1971) (in Russian)

6.102 G.D.W. Smith, J.S. Anderson: Disc. Faraday Soc. **60**, 1231 (1975)

6.103 N.G. Krishnan, W.N. Delgass, W.D. Robertson: Surface Sci. **57**, 1 (1976)

6.104 T.A. Delchar: Surface Sci. **27**, 11 (1971)

6.105 J. Völter, M. Prokop, H. Berndt: Surface Sci. **39**, 453 (1973)

6.106 W.J. Rootsaert, L.L. van Reijen, W.M.H. Sachtler: J. Catalysis **1**, 416 (1962)

6.107 R. Lewis, R. Gomer: Surface Sci. **12**, 157 (1968)

6.108 R.F. Scarr: J. Electrochem. Soc. **116**, 1528 (1969)

6.109 M.J. Braithwaite, R.W. Joyner, M.W. Roberts: Disc. Faraday Soc. **60**, 89 (1975)

6.110 M. Wilf, P.T. Dawson: Surface Sci. **65**, 399 (1977)

6.111 R. Lang, R.W. Joyner, G. Somorjai: Surface Sci. **30**, 454 (1972)

6.112 H. Michel, R. Opila, R. Gomer: Surface Sci. **105**, 48 (1981)

6.113 O.V. Krylov: Kinetika i Kataliz **22**, 15 (1981) (in Russian)

6.114 P. Vishnu Kamath, C.N.R. Rao: J. Phys. Chem. **88**, 464 (1984)

6.115 E. Fromm, O. Mayer: Surface Sci. **74**, 259 (1978)

6.116 J.L. Gland, V.N. Korchak: Surface Sci. **75**, 733 (1978)

6.117 M.R. McClellan, J.L. Gland, F.R. McFeely: J. Electron Spectroscopy **29**, 213 (1985)

6.118 H.P. Bonzel, G. Comsa: Le Vide **189**, 130 (1977)

6.119 S.M. Davies, G.A. Somorjai: Surface Sci. **91**, 73 (1980)

6.120 V.L. Tataurov, V.I. Savchenko, A.N. Salanov: React. Kinet. Catal. Lett. **17**, 305 (1981)

6.121 F.P. Netzer, R.A. Wiele: Surface Sci. **74**, 547 (1978)

6.122 C.E. Smith, J.P. Biberian, G.A. Somorjai: J. Catalysis **57**, 426 (1979)

6.123 G.K. Boreskov, V.I. Savchenko, K.I. Dadajan, V.P. Ivanov, V.P. Bulgakov: Problemy Kinetiki i Kataliza **17**, 115 (1978) (in Russian)

6.124 J.G. Bouillaid, M. Sotto: Surface Sci. **152/153**, 392 (1985)

6.125 J.W.M. Frenken, N.G. Smeenk, J.F. van der Veen: Surface Sci. **135**, 147 (1984)

6.126 H.H. Brongersma, J.B. Theeten: Surface Sci. **54**, 519 (1976)

6.127 G. Dalmai-Imelik, J.C. Bertolini, J. Rousseau: Surface Sci. **63**, 67 (1977)

6.128 J.M. Gallagher, R. Haydock: Surface Sci. **83**, 117 (1979)

6.129 H.J. Grabke, H. Viefhaus: Surface Sci. **111**, L779 (1981)

6.130 M. Maglietta, E. Zanazzi, U. Bardi, F. Jona, D.W. Jepsen, P.M. Marcus: Surface Sci. **77**, 101 (1978)

6.131 A.L. Pignocco, G.E. Pelissa: Surface Sci. **43**, 141 (1974)

6.132 T. Horiguchi, S. Nakanishi, Japan J. Appl. Phys. Suppl.2, **89**, 501 (1974)

6.133 C.F. Brucker, T.N. Rhodin: Surface Sci. **57**, 523 (1976)

6.134 T.W. Orent, S.D. Bader: Surface Sci. **115**, 323 (1981)

6.135 E. Bauer, E. Poppa: Surface Sci. **88**, 31 (1979)

6.136 A.M. Bradshaw, D. Menzel, M. Steinkilberg: Japan J. Appl. Phys. Suppl. 2 **??**, 841 (1974)

6.137 K. Griffitts, T.E. Jackman, J.A. Davies, R.R. Norton, P.E. Binder: Surface Sci. **138**, 113, 125 (1984)

6.138 G.A. Somorjai, F.J. Szalkowski: J. Chem. Phys. **54**, 389 (1971)

6.139 M.L. Yu: Surface Sci. **71**, 121 (1978)

6.140 S. Prigge, H. Niehus, E. Bauer: Surface Sci. **75**, 635 (1978)

6.141 T.M. Christensen, C. Raoul, J.M. Blakely: Appl. Surf. Sci. **26**, 408 (1986)

6.142 T. Narysawa, W.M. Gibson, E. Tornqvist: Surface Sci. **114**, 331 (1982)

6.143 P.A. Thiel, J.T. Jates, W.H. Weinberg: Sruface Sci. **82**, 22 (1979)

6.144 N. Niehus: Surface Sci. **80**, 245 (1975)

6.145 R. Ducros, R.P. Merril: Surface Sci. **55**, 227 (1976)

6.146 M. Salmeron, G.A. Somorjai: Surface Sci. **91**, 373 (1980)

6.147 L.K. Verheij, J.A. van den Bergh, D.G. Armour: Surface Sci. **84**, 408 (1979)

6.148 H. Niehus, G. Comsa: Surface Sci. **151**, L171 (1985)

6.149 V.I. Savchenko, G.K. Boreskov, K.A. Dadajan: Kinetika i Kataliz **20**, 741 (1979) (in Russian)

6.150 G.R. Gonzalski, D.M. Zehner, J.F. Wendelken, R.S. Hathcock: Surface Sci. **151**, 430 (1985)

6.151 R.P.N. Bronckers, A.G.J. Dewit: Surface Sci. **112**, 133 (1981)

6.152 J. Jupille, B. Bigeard, J. Fusy, A. Cassuto: Surface Sci. **84**, 190 (1979)

6.153 W.Y. Ching, D.L. Huber, M.G. Lagally, G.C. Wang: Surface Sci. **77**, 550 (1978)

6.154 C. Theodorou: Surface Sci. **81**, 379 (1979)

6.155 G. Ertl, D. Schillinger: J. Chem. Phys. **66**, 2569 (1977)

6.156 J.C. Buchholz, G.A. Somorjai: Acc. Chem. Res. **9**, 333 (1976)

6.157 G. Ertl: Surface Sci. **7**, 309 (1967)

6.158 G. Ertl, P. Rau: Surface Sci. **15**, 443 (1969)

6.159 K. Moliero, R. Portele: In *Structure and Chemistry of Solid Surfaces*, ed. by G.A. Somorjai (Wiley, New York 1969) p.69-1 - 69-21

6.160 H. Conrad, G. Ertl, J. Kuppers: Surface Sci. **76**, 323 (1978)

6.161 R.Kh. Burstein, N.A. Shurmovskaya: Surface Sci. **7**, 261 (1967)

6.162 T.E. Madey, H. Engelhardt, D. Menzel: Surface Sci. **48**, 304 (1975)

6.163 P. Hofmann, R. Unwin, W. Wyrobisch, A.M. Bradshaw: Surface Sci. **72**, 635 (1978)

6.164 H.P.M. Habraken, G.A. Bootsma, P. Hofmann, S. Hachicha, A.M. Bradshaw: Surface Sci. **88**, 285 (1979)

6.165 B.M. Zykov, D.S. Ikonnikov, V.I. Tskhakaja: Fiz. Tverd. Tela **17**, 274 (1975) (in Russian)

6.166 R.Kh. Burshtein, N.A. Shurmovskaya: Uspekhi Khim. **34**, 1153 (1965) (in Russian)

6.167 R.Kh. Burshtein, T.V. Kalish: Doklady AN SSSR **81**, 1093 (1951) (in Russian)

6.168 M.A.H. Lanyon, B.M.W. Trapnell: Proc. Roy. Soc. **A255**, 387 (1955)

6.169 C.T. Campbell, D.C. Foyt, J.M. White: J. Phys. Chem. **81**, 491 (1977)

6.170 T. Fleisch, N. Winograd, W.N. Delgass: Surface Sci. **78**, 141 (1978)

6.171 K.H. Muller, P. Beckmann, M. Schemmer, A. Benninghoven: Surface Sci **80**, 325 (1979)

6.172 P.H. Dawson, W.C. Tam: Surface Sci. **81**, 164 (1979)

6.173 K.H. Rieder: Appl. Surface Sci. **2**, 74 (1970)

6.174 D.E. Eastman, J.E. Demuth: Japan J. Appl. Phys. Suppl. 2/2, 827 (1974)

6.175 J. Küppers, G. Ertl: Surface Sci. **77**, L647 (1978)

6.176 P.R. Norton, R.L. Tapping, J.W. Goodale: Surface Sci. **65**, 13 (1977)

6.177 S. Evans, J. Pielaszek, J.M. Thomas: Surface Sci. **56**, 644 (1976)

6.178 S. Evans, D.E. Parry, J.M. Thomas: Disc. Faraday Soc. **60**, 102 (1975)

6.179 V.I. Savchenko: Kinetika i Kataliz **13**, 1583 (1972) (in Russian)

6.180 G. Broden, T.N. Rhodin: Disc. Faraday Soc. **60**, 112 (1975)

6.181 J.N. Miller, D.T. Ling, I. Linda, D.M. Collins, W.E. Spicer: Surface Sci. **77**, L651 (1978)

6.182 D.L. Weissman, M.L. Shak, W.E. Spicer: Surface Sci. **92**, L59 (1980)

6.183 C. Boizian, C. Garot, R. Nuvolone, J. Roussel: Surface Sci. **91**, 313 (1980)

6.184 J.H. Block: Japan J. Appl. Phys. Suppl.2, 505 (1974)

6.185 P.A. Cilty, N.C. Rol, W.M.H. Sachtler: In Proc. 5th Int. Congress Catalysis, Vol.2, Miami Beach 1972 (North Holland, Amsterdam 1973) p.929

6.186 S.V. Gerej, K.M. Kholjavenko, M.Ja. Rubanik: Ukr. Khim. Zh. **31**, 449 (1965) (in Russian)

6.187 C.T. Campbell, M.T. Paffett: Surface Sci. **143**, 517 (1984)

6.188 R.A. Marbrow, R.M. Lambert: Surface Sci. **71**, 107 (1978)

6.189 N. Rosch, D. Menzel: J. Chem. Phys. **13**, 243 (1976)

6.190 R.B. Clarkson, A.C. Cirilbo: J. Catalysis **33**, 163 (1974)

6.191 S.P. Kajumov, N.V. Kul'kova, M.I. Temkin: Kinetika i Kataliz **15**, 157 (1974) (in Russian)

6.192 A.A. Davydov: *IK-spectroscopija v, khivmii poverkhnosti okislov* (IK-Spectroscopy in Chemistry of Oxide Surfaces) (Nauka, Novosibirsk 1984)

6.193 R.P. Eischens, W.A. Pliskin: Adv. Catalysis **9**, 662 (1957); **10**, 1 (1958)

6.194 G. Blyholder: J. Phys. Chem. **68**, 2772 (1964)

6.195 R.A. Gardner: In Proc. 4th Int'l Congress Catalysis, Vol.2, Moscow, June 1968 (Akademiai Kiado, Budapest 1971) p.466

6.196 M.I. Kottke, R.G. Greenler, H.G. Tompkins: Surface Sci. 32, 231 (1972)

6.197 A. Palazov, C.C. Chang, R.J. Kokes: J. Catalysis 36, 338 (1975)

6.198 B.E. Hayden, K. Kretzschmar, A.M. Bradshaw, R.G. Greenler: Surface Sci. 149, 394, 406 (1985)

6.199 M.W. Severson, W.H. Tornquist, J. Overend: J. Phys. Chem. 88, 469 (1984)

6.200 K. Horn, J. Pritchard: Surface Sci. 55, 701 (1976); 63, 244 (1973)

6.201 B.E. Hayden, K. Kretzschmar, A.M. Bradshaw, R.G. Greenler: Surface Sci I49, 394 (1985); 155, 553 (1985)

6.202 J.N. Allison, W.A. Goddard: Surface Sci. 115, 553 (1982)

6.203 W.F. Barholzer, R.I. Mazel: Surface Sci. 137, 339 (1984)

6.204 A.M. Bradshaw, F.M. Hoffmann: Surface Sci. 72, 513 (1978)

6.205 A.M. Bradshaw: Surface Sci. 80, 215 (1979)

6.206 M. Trenary, K.J. Uram, F. Bozco, J.T. Yates: Surface Sci. 146, 269 (1981)

6.207 J. Pritchard: Surface Sci. 79, 231 (1979)

6.208 J.C. Campuzano, R.C. Greenler: Surface Sci. 83, 301 (1979)

6.209 P. Hollins, J. Pritchard: Surface Sci. 89, 486 (1979)

6.210 R. Ryberg: Surface Sci. 114, 627 (1982)

6.211 C.F. Brucker, T.N. Rhodin: Surface Sci. 105, 305 (1981)

6.212 B.E. Niewenhuys: Surface Sci. 105, 305 (1981)

6.213 P.R. Norton, J.W. Goodale, E.B. Selkirk: Surface Sci. 83, 189 (1979)

6.214 P.L. Young, R. Gomer: J. Chem. Phys. 61, 4955 (1975)

6.215 J.C. Tracy: J. Chem. Phys. 56, 2748 (1972)

6.216 H. Froitzheim, H. Ibach, S. Lehwald: Surface Sci. 63, 56 (1977); J. Appl. Phys. 13, 147 (1977)

6.217 S. Anderson: Solid State Comm. 21, 75 (1977)

6.218 T.E. Madey: Surface Sci. 79, 575 (1979)

6.219 A. Arnot, J.D. Carette: Surface Sci. 75, 109 (1978)

6.220 J.C. Bertolini, B. Tardy: Surface Sci. 102, 131 (1981)

6.221 C. Berndorf, R. Goetz, K.H. Grossmann, J. Hassler, F. Thieme: Surface Sci. 76, 509 (1978)

6.222 A. Brown, J.C. Vickerman: Surface Sci. 151, 319 (1985)

6.223 A. Matolin, S. Charnakhove, F. Gillet: Surface Sci. 164, 209 (1985)

6.224 K.E. Foley, N. Winograd: Surface Sci. 116, 1 (1981)

6.225 R. Jaeger, D. Menzel: Surface Sci. 93, 71 (1980)

6.226 T.E. Madey, J.T. Yates, A.M. Bradshaw, M.A. Hoffman: Surface Sci 89, 370 (1979)

6.227 G. Doyen, G. Ertl: Surface Sci. 43, 197 (1974)

6.228 R.M. Lambert, C.M. Lomrie: Surface Sci. 38, 197 (1973); 49, 325 (1975)

6.229 K. Christmann, O. Schober, G. Ertl: J. Chem. Phys. 51, 4852 (1969)

6.230 P.A. Threl, R.J. Behm, P.R. Norton, G. Ertl: Surface Sci. 121, L553 (1982); 147, 143 (1984)

6.231 T.N. Taylor, R.J. Estrup: J. Vac. Sci. Techn. **10**, 26 (1973)
6.232 C.G. Goymour, D.A. King: Faraday Trans. Part 1, 736 (1973)
6.233 C.G. Goymour, D.A. King: Faraday Trans. Part 1, 736 (1973)
6.234 G. Ertl, J. Koch: In Proc. 5th Int. Congress Catalysis, Vol.2, Miami Beach 1972 (North Holland, Amsterdam 1973) p.964
6.235 H. Conrad, G. Ertl, J. Küppers, E.E. Latta: Surface Sci. **57**, 475 (1976); Solid State Comm. **17**, 613 (1975)
6.236 C.M. Lomrie, R.M. Lambert: Faraday Trans. **72**, 1659 (1975)
6.237 G. Ertl, M. Neumann, K.M. Streit: Surface Sci. **64**, 393 (1977)
6.238 C.M. Lomrie, W.H. Weinberg: J. Chem. Phys. **64**, 250 (1976)
6.239 C. Steinbrugel, R. Gomer: Surface Sci. **67**, 21 (1977)
6.240 T.J. Vink, O.L.J. Gijzeman, J.W. Geus: Surface Sci. **150**, 14 (1985)
6.241 T.E. Felter, P.J. Estrup: Surface Sci. **54**, 179 (1976)
6.242 S. Anderson, J.B. Pendry: Surface Sci. **71**, 75 (1978)
6.243 C. Broden, C. Pirug, H.P. Bonzel: Surface Sci. **72**, 45 (1978)
6.244 W. Erley, H. Wagner, H. Ibach: Surface Sci. **74**, 333 (1978); **80**, 612 (1979)
6.245 H. Hopster, H. Ibach: Surface Sci. **77**, 109 (1978)
6.246 E.D. Williams, W.H. Weinberg: Surface Sci. **82**, 93 (1979)
6.247 W. Englert, W. Heiland, E. Taglauer, D. Menzel: Surface Sci. **83**, 243 (1979)
6.248 R.J. Koestner, M.A. van Hove, G.A. Somorjai: Surface Sci. **107**, 439 (1981)
6.249 J.E. Demuth: Surface Sci. **58**, 184 (1977)
6.250 Z. Schay, A.V. Skljarov, M. Prokop, M. Völter: Kinetika i Kataliz **13**, 1234 (1972) (in Russian)
6.251 W.L. Winterbottom: Surface Sci. **37**, 195 (1973)
6.252 M.A. Burteau, E.I. Ko, R.J. Madix: Surface Sci. **102**, 99 (1981)
6.253 S.R. Keleman, T.E. Fischer, J.A. Schwarz: Surface Sci. **81**, 440 (1979)
6.254 G. Ehrlich: In *Surface Science. Recent Progress and Perspectives*, ed. by T.S. Jayadeviaiah, P. Vanselow (CRC Press, Cleveland 1974)
6.255 G. Ertl: Surface Sci. **47**, 86 (1973)
6.256 J.P. Batra, O. Robaux: J. Vac. Sci. Techn. **12**, 248 (1975)
6.257 C.L. Allyn, T. Gustafsson, E.W. Plummer: J. Chem. Phys. Lett. **47**, 127 (1977); Solid State Comm. **24**, 531 (1977)
6.258 P.M. Williams, P. Butcher, J. Wood: Phys. Rev. **B14**, 321 (1976)
6.259 T.N. Rhodin, M.H. Tsai, R.V. Kasawski: Appl. Surface Sci. **22/23**, 426 (1985)
6.260 P.R. Norton, R.L. Tapping: J. Chem. Phys. Lett. **38**, 207 (1976)
6.261 K. Horn, A.M. Bradshaw, K. Jakobi: Surface Sci. **72**, 719 (1978)
6.262 A. Rosen, E.J. Baerends, D.E. Ellis: Surface Sci. **82**, 139 (1979)

6.263 T.N. Rhodin, C.F. Bruker: Solid State Comm. **23**, 275 (1977)
6.264 M. Textor, J.D. Gay, R. Mason: Proc. Roy. Soc. **A356**, 37 (1977)
6.265 K.Y. Yu, W.E. Spicer, J. Lindau, P. Pianetta, S.F. Lin: J. Vac. Sci. Techn. **13**, 277 (1976)
6.266 J.E. Demuth, D.E. Eastman: Solid State Commun. **18**, 105 (1976)
6.267 S.A. Isa, R.W. Joyner, M.W. Roberts: Chem. Commun. **4**, 377 (1977)
6.268 C.N.R. Rao, M.K. Rajmon, K. Prabhakaran, M.S. Heide, P.V. Kamath: Chem. Phys. Letters **129**, 130 (1986)
6.269 G. Broden, T.N. Rhodin, C.F. Brucker, R. Benbow, Z. Hurych: Surface Sci. **59**, 593 (1976)
6.270 W.H. Egelhoff, J.W. Linnett, D.L. Perry: Disc. Faraday Soc. **58**, 35 (1974)
6.271 G. Broden, T.N. Rhodin: Solid State Comm. **18**, 105 (1976)
6.272 G. Broden, G. Gafner, H.P. Bonzel: J. Appl. Phys. **13**, 333 (1977); J. Chem. Phys. Lett. **51**, 250 (1977)
6.273 T.N. Rhodin, J. Kanski, C. Bruker: Solid State Comm. **23**, 723 (1977)
6.274 P.A. Zhdan, G.K. Boreskov, A.J. Boronin, A.P. Schepelin, W.F. Egelhoff, W.H. Weinberg: Surface Sci. **71**, 267 (1978)
6.275 J.N. Miller, J. Lindau, W.E. Spicer: Surface Sci. **111**, 595 (1981)
6.276 M. Trenany, S.L. Tang, R.J. Simonson, F.R. McFeely: J. Chem. Phys. **79**, 6349 (1983)
6.277 T.A. Clarke, J.D. Gay, B. Law: J. Chem. Phys. Lett. **31**, 29 (1975)
6.278 S. Ferres, K.H. Frank, B. Keihl: Surface Sci. **162**, 264 (1985)
6.279 H.P. Bonzel, T.E. Fischer: Surface Sci. **51**, 213 (1975)
6.280 A. Liebsch: In *Electron and Ion Spectroscopy of Solids*, ed. by L. Fiermans, J.Vernik, W. Dekeyser (Plenum, New York 1978) pp.93–143
6.281 A. Spitzer, H. Luth: Surface Sci. **102**, 29 (1981)
6.282 S.D. Bader, J.M. Blakely, M.B. Brodsky, B.J. Friddle, K.L. Panosh: Surface Sci. **74**, 405 (1978)
6.283 K. Akimoto, Y. Savisaka, N. Nishijima, M. Onchi: Surface Sci. **88**, 109 (1979)
6.284 L. Hilaire, P. Legau, Y. Holl, G. Maire: Surface Sci. **103**, 125 (1981)
6.285 F. Chehab, W. Kirstein, F. Thieme: Surface Sci. **108**, L419 (1981)
6.286 V. Ponec: Surface Sci. **80**, 352 (1979)
6.287 W.M.H. Sachtler, R.A. van Santen: Adv. Catalysis **20**, 69 (1977)
6.288 W.M.H. Sachtler: Appl.Surface Sci. **19**, 167 (1984)
6.289 J.H. Sinfelt: Acc. Chem. Res. **10**, 15 (1977)
6.290 J.J. Burton, E. Hyman, D.G. Fedak: J. Catalysis **37**, 106 (1975)
6.291 A. Couper, D.D. Eley: Disc. Faraday Soc. **8**, 172 (1950)
6.292 J.K.A. Clarke: Chem. Rev. **75**, 291 (1975)

6.293 D.A. Dowden: In Proc. 5th Int. Congress Catalysis, Vol.1, Miami Beach 1972 (North Holland, Amsterdam 1973) p.621

6.294 M. Boudart: Adv. Catalysis **20**, 153 (1969)

6.295 W.M.H. Sachtler, P. van der Planck: Sruface Sci. **12**, 35 (1968); **18**, 62 (1969)

6.296 R.L. Moss, L. Whally: Adv. Catalysis **22**, 115 (1972)

6.297 G.A. Martin, J.A. Dalmon: J. Catalysis **75**, 233 (1982)

6.298 J.J. Burton, E. Hyman: J. Catalysis **37**, 114 (1975)

6.299 J.K.A. Clarke, J.J. Byrn: Nature **214**, 1009 (1967)

6.300 T. Takeuchi, Y. Tezuka, O. Takeyasu: J. Catalysis **14**, 126 (1969)

6.301 C.M.A.M. Mestens, A. De Koster, O.L.J. Gijzeman, J.W. Geus: Appl. Surface Sci. **20**, 13 (1984)

6.302 J.C.M. Harberts, A.F. Bourgonie, J.J. Stephan, V. Ponec: J. Catalysis **47**, 92 (1977)

6.303 K.Y. Yu, D.T. Ling, W.E. Spicer: J. Catalysis **44**, 373 (1976)

6.304 L. Whalley, D.H. Thomas, R.L. Moss: J. Catalysis **22**, 309 (1971)

6.305 A.R. Bowman, G.J.M. Lippens, W.M.H. Sachtler: J. Catalysis **25**, 360 (1972)

6.306 J. Soma-Noto, W.M.H. Sachtler: Japan J. Appl. Phys. Suppl.2, 241 (1974)

6.307 N.P. Sokolova: Zh. Fiz. Khim. **50**, 536 (1976) (in Russian)

6.308 M. Prime, M.V. Mathieu, W.M.H. Sachtler: J. Catalysis **44**, 324 (1976)

6.309 H. Wise: J. Catalysis **43**,373 (1976)

6.310 W. Palczewska, A. Frackiewicz, Z. Karpinski: In Proc. 4th Inct. Congress Catalysis, Vol.2, Moscow, June 1968 (Akademiai Kiado, Budapest 1971) p.72

6.311 J.J. Stephan, V. Ponec, W.M.H. Sachtler: Surface Sci. **47**, 40 (1973)

6.312 F.J. Kuijers, R.P. Dessing, W.M.H. Sachtler: J. cAtalysis **33**, 316 (1974)

6.313 R. Bowman, W.M.H. Sachtler: J. Catalysis **10**, 127 (1970); **28**, 63 (1972)

6.314 R.C. Baetzold: J. Catalysis **29**, 129 (1973)

6.315 R. van Hardeveld, F. Hartog: Surface Sci. **15**, 189 (1969); Adv. Catalysis **22**, 75 (1972)

6.316 O.M. Poltorak, V.S. Boronin, A.N. Mitrofanova: In Proc. 4th Int. Congress Catalysis, Vol.2, Moscow, June 1968 (Akademiai Kiado, Budapest 1971) p.276

6.317 G.C. Bond: Ibid. p.266

6.318 E.G. Schlosser: Ibid. p.312

6.319 M. Boudart: J. Vac. Sci. Technol. **12**, 329 (1975); J. Molec. Catalysis **30**, 27 (1985)

6.320 G.R. Wilson, W.K.Hall: J. Catalysis **24**, 306 (1972)

6.321 J.E. Benson, M. Boudart: J. Catalysis **4**, 704 (1965)

6.322 G.B. McVicker, J.J. Ziemiak: J. Catalysis **95**, 473 (1985)

6.323 S. Sakellson, M.M. McMillan, G.L. Haller: J. Phys. Chem. **90**, 473 (1985)

6.324 S.W. Weller, A.A. Montagne: J. Catalysis **20**, 394 (1971)

6.325 R.A. Dalla Betta, M. Boudart: In Proc. 5th Int. Congress Catalysis, Vol.2, Miami Beach 1972 (North Holland, Amsterdam 1973) p.1329

6.326 Tran Mann Tri, J. Massardier, P. Gallezot, B. Imelik: In Proc. 7th Int. Congress Catalysis, Tokyo, July 1980, Preprint A16

6.327 E. Rockova: J. Catalysis 48, 137 (1977)

6.328 J. Frell: J. Catalysis 25, 149 (1972)

6.329 D.J.C. Yates, L.L. Murrell, E.B. Partridge: J. Catalysis 57, 47 (1979)

6.330 Chan-Hwa Yang, J.C. Goodwin: React. Kinet. Catal. Lett. 20, 13 (1982)

6.331 R. Moraweck, G. Clugnet, A.J. Renouprez: Surface Sci. 81, L631 (1979)

6.332 P. Gallezot: Surface Sci. 106, 459 (1981)

6.333 L.M. Roev, I.G. Voroshilov: Doklady AN Ukr. SSR 9, 830 (1973) (in Russian)

6.334 R.A. Dalla Betta: J. Phys. Chem. 79, 2519 (1975)

6.335 M. Primet, J.M. Basset, E. Garbowski, K. Mutin: J. Am. Chem. Soc. 97, 3655 (1975)

6.336 G.A. Somorjai, J. Carazza: Ind. Eng. Chem., Fundamentals 25, 63 (1986)

6.337 G.K. Boreskov: Kinetika i Kataliz 8, 1020 (1967) (in Russian)

6.338 J.M. Basset, G. Dalmai-Imelik, M Primet: J. Catalysis 37, 22 (1975)

6.339 J.R. Anderson, N.R. Avery: J. Catalysis 5, 46 (1966)

6.340 J.B. Andersen, J. Shimoyama: In Proc. 5th Int. Congress Catalysis, Vol.1, Miami Beach 1972 (North Holland, Amsterdam 1973) p.695-701

6.341 J.W.E. Coenen, R.Z.C. van Meerten, H.T. Rijnten: In Proc. 5th Inct. Congress Catalysis, Vol.1, Miami Beach 1972 (North Holland, Amsterdam 1973) p.671-678

6.342 K.G. Ione: *Polifunktsional'nyj Kataliz na tseolitakh* (Multifunctional Catalysis on Zeolites) (Nauka, Novosibirsk 1982) (in Russian)

6.343 L.M. Carballo, E.E. Wolf: J. Catalysis 53, 363 (1978)

6.344 H. Topsoe, N. Topsoe, H. Bohlbro, J. Dumesik: In Proc. 7th Int. Congress Catalysis, Vol.1, Tokyo, July 1980, Preprint A15

6.345 Ju.B. Vasil'ev, V.S. Baletskij: Problemy kinetiki i Kataliza 16, 260 (1975) (in Russian)

6.346 G. Ertl, J. Koch: In Proc. 5th Int. Congress Catalysis, Vol.1, Miami Beach 1972 (North Holland Publ. Amsterdam 1973) p.969

6.347 S.J. Teichner, G.M. Pajonik, M. Lacroix: In *Surface Properties and Catalysis by Non-Metals*, ed. J.B. Bonnelle, B. Delmon (Reidel, Dordrecht 1983) p.457

6.348 D.W. Blakely, G. Somorjai: Surface Sci. 65, 419 (1977)

6.349 W.C. Neikam, M.A. Vannice: J. Catalysis 20, 260 (1971); 27, 207 (1972)

6.350 L.E. Hunt: J. Catalysis 23, 93 (1971)

6.351 P.A. Sermon, G.C. Bond: Faraday Trans. **72**, 730, 755 (1976)
6.352 K.M. Sancier: J. Catalysis **20**, 106 (1971)
6.353 W.C. Conner, J.F. Cevallos-Candau, N.Shah, V. Haensel: In *Spillover of Adsorbed Species*, ed. G.M. Pajonik, S.J. Teichner (Elsevier, Amsterdam 1983) p.135

Chapter 7

7.1   S.Z. Roginskij: Problemy Kinetiki i Kataliza **10**, 5 (1960) (in Russian)
7.2   P. Cossee: Rec. Trav. Chim. Pays-Bas **85**, 1151 (1966)
7.3   I.B. Bersuker: Problemy Kinetiki i Kataliza **13**, 7 (1968) (in Russian)
7.4   R.J.U. Voorhoeve: In *Magnetism and Magnetic Materials (Proc. 19th Ann. Conf. ASP, Boston 1972)* **4**, 19 (1973)
7.5   D. Shopov, A. Andreev: *Khimicheskaja svjaz' pri adsorbtsii i katalize* (Chemical bonding in adsorption and catalysis), Vol.2 *Okisly* (Oxides) (BAN, Sofia 1979) (in Russian)
7.6   J. Chatt, L.A. Duncanson: J. Chem. Soc. **??**, 2939 (1953)
7.7   B. Gorewit, M. Tsutsui: Adv. Catalysis **27**, 227 (1978)
7.8   P. Cossee, P. Ros, J.H. Schachtschneider: In Proc. 4th Int. Congress Catalysis, Vol.1, Moscow, June 1968 (Akademiai Kiado, Budapest 1971) p.207
7.9   R.S. Nyholm: In Proc. 3rd Int. Congress Catalysis, Vol.1, Amsterdam 1964 (North Holland, Amsterdam 1965) p.25
7.10 J. Halpern: Adv. Chem. Ser. **70**, 1 (1968)
7.11 E. Shustororich: Surface Sci. **176**, L83 (1986)
7.12 V.F. Kiselev, O.V. Krylov: *Electronic Phenomena in Adsorption and Catalysis*, Springer Ser. Surf. Sci., Vol.7 (Springer, Berlin, Heidelberg 1987)
7.13 K. Tamaru, Y. Noto, T. Onishi, I. Fukudaki: Trans. Faraday Soc. **63**, 2300 (1967); Bull. Chem. Soc. Japan **40**, 2459, 2722 (1967)
7.14 B.V. Rozentuller, K.N. Spiridonov, O.V. Krylov: Kinetika i Kataliz **22**, 797 (1981); Doklady AN SSSR **259**, 895 (1981) (in Russian)
7.15 S.W. Johnson, R.J. Madix: Surface Sci. **66**, 189 (1977)
7.16 R.W. Joyner, M.W. Roberts: Proc. Roy. Soc. **A350**, 107 (1976)
7.17 J.L. Gland, V.N. Korchak: J. Catalysis **53**, 9 (1973)
7.18 T. Matsushima, D.B. Almy, J.M. White: Surface Sci. **67**, 89, 122 (1977)
7.19 S.Z. Roginskij, M.I.Janowskij, A.D. Berman: *Osnovy primenenija khromatografii v katalize* (Basic applications of chromatography in catalysis) (Nauka, Moscow 1972) (in Russian)
7.20 O.V. Krylov: Kinetika i Kataliz **23**, 1391 (1983) (in Russian)
7.21 E. Fiolitakis, H. Hofmann: Catalysis Revs. **24**, 113 (1982)
7.22 M. G. Slin'ko, G.S. Jablonskij: Problemy Kinetiki i Kataliza **17**, 154 (1978) (in Russian)

Subject Index